PHYSICS AND PHILOSOPHY: SELECTED ESSAYS

# EPISTEME

A SERIES IN THE FOUNDATIONAL,
METHODOLOGICAL, PHILOSOPHICAL, PSYCHOLOGICAL,
SOCIOLOGICAL AND POLITICAL ASPECTS
OF THE SCIENCES, PURE AND APPLIED

*Editor:* MARIO BUNGE
*Foundations and Philosophy of Science Unit, McGill University*

*Advisory Editorial Board:*

RUTHERFORD ARIS, Chemistry, *University of Minnesota*
DANIEL E. BERLYNE, Psychology, *University of Toronto*
HUBERT M. BLALOCK, Sociology, *University of Washington*
GEORGE BUGLIARELLO, Engineering, *Polytechnic Institute of New York*
NOAM CHOMSKY, Linguistics, *MIT*
KARL W. DEUTSCH, Political science, *Harvard University*
BRUNO FRITSCH, Economics, *E.T.H. Zürich*
ERWIN HIEBERT, History of science, *Harvard University*
ARISTID LINDENMAYER, Biology, *University of Utrecht*
JOHN MYHILL, Mathematics, *SUNY at Buffalo*
JOHN MAYNARD SMITH, Biology, *University of Sussex*
RAIMO TUOMELA, Philosophy, *University of Helsinki*

VOLUME 6

HENRY MARGENAU

# PHYSICS AND PHILOSOPHY: SELECTED ESSAYS

**D. REIDEL PUBLISHING COMPANY**
DORDRECHT : HOLLAND / BOSTON : U.S.A.
LONDON : ENGLAND

Library of Congress Cataloging in Publication Data

Margenau, Henry, 1901–
    Physics and philosophy.

    (Episteme ; v. 6)
    Bibliography: p.
    Includes index.
    1.  Physics—Philosophy.  2.  Margenau, 1901–
3.  Physicists—United States—Biography.  I.  Title.
QC6.M3515        530′.01        78–16695
ISBN 90–277–0901–7

---

Published by D. Reidel Publishing Company,
P.O. Box 17, Dordrecht, Holland

Sold and distributed in the U.S.A., Canada, and Mexico
by D. Reidel Publishing Company, Inc.
Lincoln Building, 160 Old Derby Street, Hingham,
Mass. 02043 U.S.A.

All Rights Reserved
Copuright © 1978 by D. Reidel Publishing Company, Dordrecht, Holland
and copyrightholders as specified on appropriate pages
No part of the material protected by this copyright notice may be reproduced or
utilized in any form or by any means, electronic or mechanical,
including photocopying, recording or by any informational storage and
retrieval system, without written permission from the copyright owner

Printed in The Netherlands

*This book is dedicated with the author's deep gratitude to Professor F.S.C. Northrop, who stimulated his interest in philosophy of science and honored him with his friendship.*

HENRY MARGENAU
EUGENE HIGGINS PROFESSOR OF PHYSICS
AND NATURAL PHILOSOPHY EMERITUS
YALE UNIVERSITY

# TABLE OF CONTENTS

EDITORIAL FOREWORD ix

PREFACE xi

INTRODUCTION xiii

ACKNOWLEDGMENTS xxxvii

### PART I  Metascience: Philosophical Analysis of Scientific Truth

1  The Problem of Physical Explanation  3
2  Probability and Causality in Quantum Physics  21
3  Meaning and Scientific Status of Causality  39
4  Methodology of Modern Physics  52
5  Metaphysical Elements in Physics  90
6  Is the Mathematical Explanation of Physical Data Unique?  114

### PART II  Fundamental Problems of 20th Century Physics

7  Probability, Many-Valued Logics and Physics  125
8  On the Frequency Theory of Probability  143
9  Can Time Flow Backwards?  158
10  Causality in Quantum Electrodynamics  175
11  Relativity: An Epistemological Appraisal  186
12  Philosophical Problems Concerning the Meaning of Measurement in Physics  199
13  Bacon and Modern Physics: a Confrontation  211

## PART III  Science and Human Affairs

| | | |
|---|---|---|
| 14 | Western Culture, Scientific Method and the Problem of Ethics | 225 |
| 15 | Physical versus Historical Reality | 241 |
| 16 | The New View of Man in His Physical Environment | 165 |
| 17 | Science and Human Affairs | 283 |
| 18 | The New Style of Science | 295 |

## PART IV  Issues Beyond the Boundaries of Present Science

| | | |
|---|---|---|
| 19 | Phenomenology and Physics | 317 |
| 20 | Physics and Ontology | 329 |
| 21 | Faith and Physics | 333 |
| 22 | Metaethics | 339 |
| 23 | The Pursuit of Significance | 351 |
| 24 | Note on Quantum Mechanics and Consciousness | 373 |
| 25 | Religious Doctrine and Natural Science | 375 |

LIST OF PUBLICATIONS                                391

INDEX                                               401

# EDITORIAL FOREWORD

This book is intended for people interested in physics and its philosophy, for those who regard physics as an essential component of modern culture rather than merely a tool for industry or war. Indeed this volume is addressed to those students, teachers and research workers who enjoy learning, teaching or doing physics, and are in the habit of pausing once in a while to ponder over key physical concepts and hypotheses and to wonder whether received theories are as perfect as textbooks would have us believe and, if not, how they might be improved.

Henry Margenau, recently retired from Yale University as Eugene Higgins Professor of Physics and Philosophy, is the most important philosopher of physics of his generation, and indeed one of the most eminent philosophers of science of our century. He introduced and elucidated the notion of the correspondence rule. He claimed and showed, in the heyday of positivism, that physics has metaphysical presuppositions. He was the first to realize that quantum mechanics can do without von Neumann's projection postulate — and that was as far back as 1936. He clarified the physics and the philosophy of Pauli's exclusion principle at a time when it seemed mysterious. He was the first physicist to publish a philosophical paper in a physics journal, which he did as early as 1941. He was also one of the rare scientists who proclaimed the need for a scientific approach to value theory and ethics. And, unlike other scientist-philosophers, Margenau did not wait for his arteries to harden before philosophizing: he has been studying philosophy since his high school years, and has acquired technical competence in it. Nevertheless he has always managed to convey his philosophical ideas with a minimum of technical jargon.

When I first met Margenau, over two decades ago, I asked him how he managed to gain the ear of physicists when talking philosophy. He replied: "By continuing to do physics". How true: physicists listen to other physicists — often only to colleagues — and seldom to philosophers, even in philosophical matters. This is why Margenau's articles and books on the philosophy of science, like those of Planck, Bridgman, Bohr, and Einstein, are so widely read by scientists.

This volume gathers together some of Margenau's best papers on the philosophy of science, previously scattered in a number of journals. Their author has updated them and added a long bibliography. Moreover he has added a charming autobiographical Introduction that is at once amusing and instructive.

M. BUNGE

# PREFACE

When Browning spoke hopefully of "the last of life, for which the first was made" he was still fairly young and highly optimistic. The end of an academic career, called retirement, does bring with it certain advantages, among them leisure, freedom from deadlines and a choice of things one can do. In most cases, particularly among people whose earlier work has committed them to analytic precision, as distinct from historical circumspection and artistic or imaginative productivity, that choice is limited, for age usually diminishes the kind of creativity and originality required in the field of exact science. One therefore develops a penchant for reflection and retrospection, a tendency to assess one's previous work. Indulgence in it can be either reward or punishment, and hope clings to the former.

A friendly publisher has suggested that I collect a number of articles from my earlier pen, papers that deal with problems in the philosophy of physical science, which in restrospect I deem worthy of republication. No request could be more welcome, more in consonance with the faculties of an older mind, or more flattering to an author. The papers selected portray, I hope, a progressive trend toward a unifying epistemology of science, diversified by applications to some specific problems with which numerous scientific papers, not included here, have dealt. My endeavor to fulfill this hope and the desire to appeal primarily to philosophers and to scientists interested in larger issues have led me to exclude material of a technical and mathematical nature. The papers have been grouped under 4 headings which are meant to characterize their contents. The autobiography was written at the request of the publisher and is offered with due humility.

I should like to add a word concerning the papers dealing with certain metaphysical, regulative principles of science. They contain a methodological requirement called simplicity, Occam's economy of thought. Were I to write these articles today I would place additional emphasis upon this maxim of simplicity, expanding it to include elegance and even beauty. For these have become guiding ideals which have inspired important recent scientific theories. The essence of the creative act in science, art and even ethics and religion appears to be very much the same.

*October 6, 1977*                                                              H. MARGENAU

# INTRODUCTION

This autobiographical introduction has been added, at the editor's request, to a collection of articles dealing with the philosophy of physics. He thought it worthwhile to indicate how, in the early days of this century, when the separation between university departments in the United States was rigorous, one got from physics into philosophy. This seems to require an account of my educational history.

I was born in 1901 in Germany to parents who had recently moved from farms into the city of Bielefeld. They belonged to the working class, and we lived frugal and humble lives. Class distinctions were severe, and I looked upon people of means and erudition with reverence and envy. One of them, my father's employer, took an interest in me, rewarded me when I brought home a good report card and gave us Christmas presents. This one link to the higher stratum of society encouraged, almost inspired me and fired me with ambition at an early age. Had we been on anonymous welfare no occasion for an upward drive would have arisen.

The German grade school system had three branches. There was the Volksschule, the public school which offered training in the first eight grades. It was free, offered courses in the usual subjects, including religion, dealt extensively with German history (emphasizing dates), but taught no foreign languages. Its graduates were destined to become laborers or apprentices in manual trades. Access to higher learning and eventual admission to a university was practically precluded. I started as a pupil in this type of school.

On a somewhat higher level was the so-called middle school, which covered the first nine grades and led to a degree called the 'Einjährige' upon a somewhat rigorous final examination. The term implies that a youngster possessing this degree served only one year, not two, in the Emperor's army. This school taught a modern foreign language – French or English – and it required the payment of tuition which, as I recall it, was sixty marks a year. At the age of ten my father managed to send me to a middle school from which I graduated, aged nearly fifteen, during the first World War.

The best education was offered by the 'higher schools', the Gymnasium and the Realschule, where choices of higher mathematics, sciences, as well as Latin and Greek, were offered. But the tuition was high; hence it was open only to the elite. It covered the grades from four to thirteen. Admission was usually gained after three years of training in a special 'Vorschule' and an entrance examination. The degree it conferred was called Arbitur, and it assured entrance into a university. A transition from one of the lower schools to a higher school was practically impossible because of the incompatibility of their curricula.

I smarted under the realization that attendance at a Gymnasium was precluded for me by my family's poverty and became extremely jealous of boys who had the opportunity to learn Latin and Greek. Upon graduating from the Middle School my parents decided, with my enthusiastic concurrence, that I should become a grade school teacher. Instructors in higher schools had to have a university education. The normal schedule of training for an elementary school teacher consisted of attendance for three years at a preparatory school followed by three years in a teachers' 'seminary'. Entrance to the preparatory school required nothing more than graduation from the eighth grade of a public school. The courses offered emphasized German history, literature, religion, and one foreign language, French. Admission to the teacher's seminary required either completion of three years of preparatory school or the Einjährigen degree. But a strange law demanded that the candidate had to be seventeen years of age at entrance.

This meant that I, having acquired the necessary degree at the age of about fifteen, could not be admitted but had to spend two useless years in the "Präparandenanstalt", repeating courses I had already taken, suffering through nauseating reviews of all the dates of German military history, relearning the Lutheran catechism and sitting through a course in beginning French, a language I was already able to speak with fair fluency. One inadvertent benefit did result from this boring ordeal: our religion teacher once mentioned Nietzsche as a most seditious philosopher whose books we should not read. As a result, some of us rushed to the public library to take out his works; I had the good fortune of getting there first. "Thus spoke Zarathustra" was my first introduction to philosophy.

To relieve the tedium of these two years I decided to spend two hours every day in self-study of that elite classical language, Latin. Having regretted for years that knowledge of it had been denied me, I now studied it with fervor and persistence, using the text books and reading the authors (Caesar, Ovid, Cicero) which, as I learned from more fortunate friends, formed the curriculum of the Gymnasium.

There was another reason why I drowned myself in these endeavors. The first World War was raging and life was difficult; my father was fighting in the trenches and my mother was in the hospital, terminally ill. She died in 1918, when I entered the teachers' seminary for the final three-year stretch.

There the work became more interesting. Teaching in the life sciences, then called botany and zoology, with a distinct emphasis on ecology (a name which did not exist at the time), was added to the usual subjects. Beautiful courses in pedagogy, in methods of teaching together with the historical development of teaching techniques (Socrates, Rousseau, Pestalozzi, Herbart) were given by competent instructors and philosophical issues formed focal centers of discussion. One of my teachers, Dr. Richtmeyer, was a disciple of Rudolf Steiner, the founder of anthroposophy, and through his influence I, too, came under the spell of that doctrine. Furthermore, my eyes were opened to Oriental philosophy, where Steiner's ideas seemed to have originated, and some alignment with Eastern metaphysics characterized my thinking until, several years later, I entered upon a scientific career. Anthroposophy then fell apart.

One of the most useful features of my Seminary training was the teaching practice which accompanied the courses in learning theory. This was done under closest supervision by experts. In fact the grade school connected with the teacher's college, where this practice took place, had the reputation of being the best school in the city.

When I graduated in 1921, inflation had destroyed the German economy. The state was bankrupt, unable to pay its teachers, who were state employees. Hence my certificate was offically worthless. Fortune, however, came to my rescue. A member of the wealthy family that had befriended me in my early youth, Baroness von Khaynach, needed a private tutor for her nine year old son, a person who could teach the first year courses of the Gymnasium, including Latin, and she hired me. I thus spent a year of travelling with the Baroness and her son, mainly in the Bavarian Alps, a time in which I continued to improve my acquaintance with Latin literature while teaching the rudiments of the language. The authors I recall most clearly were Horace, Livy and St. Augustine.

At the end of that unforgettable year I was forced to make a difficult choice. Inflation was rampant, our savings had become worthless, no teaching positions were open. Hence I had 3 options: to wait for conditions to become normal (with a prospect of starving), change my profession, or to emigrate.

The kindness of a Nebraska farmer, distantly related to my family, made it possible for me to come to the U.S. by providing an affidavit of support. Thus in May of 1922, I arrived at the railroad station of Winside, Nebraska, where my "Uncle Peperkorn," a farmer, waited for me with his real, privately owned Buick. Astonishment at the affluent life style of American farmers filled the next few wonderful days. Then I took a job as farm hand on a neighbor's ranch. I learned to milk cows (the first one I tackled was named "Moses"), to feed, curry, drive and ride horses. My first major job was to cultivate long rows of corn. The change from private teaching in poor external circumstances to hard manual labor with abundant food was a bit traumatic and evoked nostalgia for home and friends, but it was a healthy and stabilizing experience. After one summer of farming I accepted a position as clerk in a general mechandise store in the small town of Winside with the incredible salary of $65 a month, the equivalent of about 100,000 Marks at that time.

The leisure afforded by free evenings and Sundays caused me to reconsider the possibility of a college education. The desire for it was strengthened when I acquired a few educated friends who lived in the town, among them the pastor of the Lutheran church, which I often attended. Then, in the summer of 1923, a curious encounter took place: a gentleman in clerical garb entered the store and bought a cigar. During the transaction he started a conversation which turned toward my personal affairs. The minister, evidently aware of my background, inquired why I did not attend college. I told him I was saving money to finance my further education. His next question, "how much have you saved?" was answered by "$300." The gentleman then identified himself as president of Midland Lutheran College visiting Winside as a guest preacher at the Lutheran church; he convinced me that my funds were sufficient if I worked while studying and urged me to come to Midland in the fall.

I went. The job waiting for me was that of paper boy and required me to distribute newspapers very early in the morning, before classes. This, however, lasted only a few months, until the College asked me to teach a class in French, thus giving me an easier life. My standing in the college system, presumably somewhat advanced beyond that of freshman, was uncertain and awaited an evaluation of credits I might have acquired in Germany, along with the results of examinations. As a major I chose Latin, a field in which I had advanced standing, and I graduated after one year upon completing a thesis, written in Latin, on the philosophy of Seneca. Other Latin authors I read during that year were Livy and St. Augustine, along with portions of the Vulgate.

Hopes were high that I would find a teaching position in a high school. In fact I was offered one in Norfolk, Nebraska, at the fantastic salary of $1500, as teacher of Latin. But Fortune's smile changed to a grin when the school board learned that I was not an American citizen — the waiting period of five years after immigration imposed upon foreigners who desired citizenship was not yet up, and the law did not permit foreigners to teach in public schools.

During the summer of 1924 I did odd menial jobs in Lincoln, the seat of the University of Nebraska. By this time I had been in the U.S. for about two years, and my mastery of the English language was adequate for academic purposes, even though the colloquial or vulgar meanings of certain phrases escaped me. As an illustration I venture to record a shameful experience. Next to the house in which I rented a room a dozen young ladies, most of them school teachers, had their board and rooms. Their landlady, mindful of my solitary existence, kindly invited me to a dinner party with her charges. During the meal her dog found his way into the dining room, sneaked under the table and, it seemed, occasionally brushed against the girls' legs. I sensed their annoyance and decided, since I was the only male guest, that it was probably my duty to call the dog's presence to the landlady's attention. Hence, when she entered, carrying a platter of food, I turned to her, saying in perfect innocence: "Mrs. O'Connell, there is a little son of a bitch under the table which appears to be annoying the ladies." Everybody looked at me aghast, and none of the girls spoke to me thereafter.

One of my Midland professors, who had kindly taken an interest in my affairs, also spent the summer in Lincoln and invited me to a party at her house. Among the guests were Professor and Mrs. Moore. The lady was, like myself, a recent immigrant from Germany, and she engaged me in a conversation which revealed my status as unemployed college graduate in need of sympathetic counsel. Later, when the guests were leaving, she introduced me to her husband, who identified himself as a research physicist at the University. After a brief conversation he startled me by offering me a position as his assistant. Flabbergasted, I attempted to correct what was evidently a misapprehension on his part regarding my field of study, informing him that I had never had a course in college physics. "I know," he said, "but I am informed that you are a hard worker, willing to undertake a strenuous task." And then he explained that he was an experimental spectroscopist, studying at the time the Zeeman effect in Lanthanum (whatever that might be!). This involved photographic exposures of over thirty hours and required the continual presence of a person in a small room adjacent a larger room with a diffraction grating and a photographic film, a person who would keep a carbon arc focused upon the slit of a spectrograph and adjust the current in the electromagnet producing the field. He offered to teach me the experiments to be performed in the freshman laboratory in exchange for help in connection with his research and suggested that I come to the physics laboratory two weeks before term time, a period in which he instructed me in the performance of the first two laboratory exercises to be carried out by beginning physics students. This sounded two good to be true, and I accepted eagerly.

Thus, during my first year in the physics department of the University of Nebraska, I spent many nights monitoring the light source, took courses in physics and mathematics and, through the kindness of Professor Moore, kept two jumps ahead of the students in the freshman lab, performing my assistant's duties. Intensive work, partly self-study, probably gave me the equivalent of a major in physics so that in my second year at the University, my assistantship being renewed, I could take graduate courses and do some independent research. The problem I chose, for fairly obvious reasons, was a study of the Zeeman effect in the spectrum of Cerium.

During the second year my beloved benefactor died. At the end of it I took the M.Sc. in physics, presenting as my master's thesis a paper on the theme just mentioned. I was also offered an instructorship for the following academic year. Evidently the department wanted Professor Moore's work continued, for this represented at the time its largest commitment of funds. And I was stuck with it, since I was the only person who had intimate experience with Moore's work. I gratefully accepted, and in the following summer celebrated my good fortune, tinged with sadness at the loss of a revered older friend, by visiting my family in Germany.

On returning in the fall of 1926 as a full-fledged instructor I spent my free time learning mathematics and scanning the American philosophical literature, for philosophy was still an obsession with me and introduced an unsatisfied craving into my thinking. Augustine's discussion of determinism and freedom in "De libero arbitrio" had impressed itself deeply on my mind. Further professional advancement, the attainment of a Ph.D. degree, however, was my most important objective; hence I applied for admission to several Eastern universities. Meanwhile my master's thesis had been published in the Physical Review, and I was offered and accepted a fellowship in the Yale physics department.

Then followed two more years of study and the completion of a Ph.D. thesis. The latter was an experimental investigation of the reflecting power of a silver surface for different wavelengths in its dependence on the amount of stress in the surface (from crystalline to amorphous to highly polished). The work was done in elaborate and cumbersome fashion with the use of a spectrophotometer and home-made photocells. After months of labor it was found that the minimum in the reflecting power, which occurs in the near ultraviolet, was shifted to shorter wavelengths by several tens of Ångstrom units when the surface was highly polished — that the shift was, in fact, a measure of the amount of 'cold work' in the surface. Having found this effect and observed its magnitude, I proceeded to photograph the reflected spectrum in one fell swoop, and within two days I obtained its photographic verification. In dismay I went to my advisor, Professor McKeehan, and showed him the plate, expecting him to rebuke me for my initial clumsy approach and to deprecate the effect as too trivial for the substance of an acceptable thesis. But he was kind and assured me that the work was acceptable.

During my first year at Yale I became acquainted with Ernest Lawrence, who had already conceived the idea of the cyclotron. He was attracted by an offer to Berkeley in 1928 and provided me an opportunity to follow him, assuring me the rank of instructor. Yale, however, countered with a similar promotion, and it seemed expedient for me to stay. In retrospect I observe that this decision, which seemed so trivial at the time, may have been decisive for all my later years.

The summer of 1928 afforded a little time for reflection, for a collection of thoughts on current issues in philosophy of science. It resulted in a paper entitled "The Problem of Physical Explanation" which was published in The Monist (**39**, p. 321, 1929). On re-reading it I find it contains in embryonic form some of the features of my later philosophy (The Nature of Physical Reality, McGraw Hill, 1950). It introduces in particular the germinal concept of 'rules of correspondence', the connection between direct, observational experience and the concepts that correspond to it. The direct experience of time, the sensed flow of consciousness, it maintains, is not the same as the symbol $t$ in the equations of motion, nor is the feeling of hotness in my finger tips as I touch an object identical with the reading of a thermometer. In making these distinctions I was strongly influenced by Bridgman's theory of operational definitions, which were later recognized as one important class among the rules of correspondence.

In the spring of 1929, upon completing my thesis, I became anxious to acquaint myself more thoroughly with the issues in theoretical physics of the day. The quantum theory had been established, and since it was not taught intensively at Yale my desire was to continue my work in Europe. Yale offered attractive post-doctoral fellowships, named after their donor, Sterling. They normally paid a stipend of $1000 per year, but the usual understanding was that their incumbents were to work at Yale. Only rarely was a 'travelling Sterling fellowship' awarded which permitted a recipient to work abroad. I applied for one of these, requesting only $800 in the strategic hope of improving my chance of getting one.

Wilbur Cross, later governor of Connecticut, was then dean of the graduate school. His custom was to interview the candidates for Sterling fellowships. In due time he called me into his office, and something like the following conversation took place. Dean Cross: "I see you are applying for a Sterling to go abroad." Answer: "Yes, sir." Then, "Where do you want to go?" "To Munich, Paris and possibly Berlin." The plan of my studies was, of course, incorporated in the application papers, which the Dean had before him. He then went on: "And you are asking for $800." "Yes, if that seems appropriate." After a moment of silence, enhanced by the glimmer of a faint smile on the Dean's face he asked: "Would you mind very much if we made it $1000?"

The extra $200 were spent travelling in Europe prior to settling in Munich for work in the institute of the famed Professor Sommerfeld whose latest book, known as the 'Ergänzungsband', a name characterizing it as addition to his previous volume on atomic physics, had served as my introduction to the quantum theory. Joining his institute required possession of a Ph.D. degree and mine, obtained in the U.S., had to be acknowledged and certified by the Bavarian department of education. This involved several visits to the town hall of Munich, and I was asked to produce documents to prove that Yale was a reputable institution, qualified to give a Ph.D. degree. The authorities wanted to see at least a catalog describing the university's requirements for the degree. At that point I grew a little indignant and suggested that, if the Bavarian ministry of education assumed the role of judging foreign institutions, it should be equipped with materials needed for that judgement. Somewhat depressed by this turn of events I spoke to Sommerfeld, who expressed sympathy with my sentiments. Two days later I received in the mail a document affirming the validity of my Yale degree and granting me the privilege of entering Sommerfeld's institute.

Its members included men like Bethe and Unsöld, and much of the work done at the time was focused on Sommerfeld's theory of metals; a formalism, now superannuated, which treated the conduction electrons in a metal in accordance with Fermi-Dirac statistics. On this I had recently published a small paper having to do with the change of the number of free electrons, their progressive release from metal atoms, with increasing temperature. Sommerfeld had seen it and asked me to be the first speaker in his weekly seminar, using the paper as topic for discussion. Needless to say I felt honored, embarrassed and alarmed at this request and prepared myself as one does for a presentation to a sympathetic audience. The result was devastating. Questions were asked throughout my talk, and after about one hour – the seminars lasted two hours – they bore down on my use of a certain formula not derived in my article, a formula I was asked to prove in order to justify its use. For this I was unprepared, and my performance ended prematurely with a disconsolate promise on my part that I would, after more mature consideration, return with an answer.

The blow to my ego was severe. I had never endured such an ordeal of questioning and in my perplexity I did not show up in the institute for several days after the disaster. When I did reappear Sommerfeld called me in and attempted to heal my wounds, suggesting that I should take another look at the problem with special attention to the questioned formula. More complete relief came only in the following week, when another member of the institute presented a paper which folded up under questioning after forty-five minutes. I learned that this was a common occurrence, and a healthy one, in this illustrious place.

Before the end of the semester I did succeed in obtaining my former result in another way, avoiding the hazard of the unproved formula. Bethe was then Sommerfeld's assistant, and members of the group were expected to clear their work with him. I remember placing my new notes on his desk, sitting next to him while he looked through them cursorily and then inspected the final result. He proceeded to scribble on a piece of paper. The details were beyond my range of vision. After awhile he turned to me with the apodictic decision: your result is correct. I inquired how he had been able to verify my answer so quickly. He did not satisfy my curiosity but repeated his affirmation. I thanked him and left with the comment that I had obviously wasted my time. To that he said: "Oh no, you must publish this new version of your work." I still feel that this was an act of kindness on Bethe's part; but I did publish a brief account of the new approach in the Zeitschrift für Physik. Bethe's quick insight, however, remains miraculous to me.

The institute group held a weekly meeting over coffee in a restaurant of the English Garden. The discussions there were extremely fruitful and often led to publications of articles soon after the topics discussed had been pondered. Indeed the two lasting impressions I carried away from my semester in Munich were the extreme severity of the seminars and the productive fertility of the informal coffee hours.

During the latter, Sommerfeld became aware of my interest in philosophy. This he felt to be a useless diversion, a hindrance to scientific success which I should try to overcome. I must admit that this was a serious disappointment to me. However, I can not leave this observation without recording a sequel. In the summer of 1936, during the ascendency of Nazi power in Germany, as I passed through Munich with my wife and baby son, I called at Sommerfeld's

Institute and was told by his secretary that he was spending the summer in Garmisch, in the Bavarian Alps. A few days later I received from him a note asking, if possible, to visit him there. Circumstances made it difficult to accept his invitation. Hence I respectfully declined but marvelled at the fact that he remembered me, a one-semester participant in his seminar of whose philosophic interests he disapproved. And then, years later, just before his death, I received from him a friendly letter indicating that he, too, had acquired an interest in philosophy in his later days and that he had read some of my publications. I now believe that this was more than flattery, and I regret that I was unable to thank him before he died.

After one semester in Munich I joined the Institute for Theoretical Physics at the University of Berlin, where I attended some of Schrödinger's lectures. While their substance was fascinating and profound his appearance was unusually informal for the Germany of that day: he lectured coatless, in an open-collared tennis shirt. But aside from an incident in which, when signing my admission card for his seminar, he splashed ink on his face and I offered my handkerchief to wipe it off, he took no notice of me. Later, however, in correspondence, he recalled that incident. This took place in 1960 when William Scott, who had come to Yale under a research grant, wrote a book on Schrödinger's philosophy under my administrative guidance. Among the illustrious people in Berlin at the time were Planck, Einstein, and Haber.

My work in Berlin was dominated almost entirely by the genius of Fritz London, who had just published a first paper on the quantum theory of Van der Waals forces. Realizing that the best way to learn quantum mechanics was to apply it, I extended London's work, computing higher terms in the series representing the interaction energy between two neutral atoms, terms in the reciprocal distance between them which become important at close approach. This was followed by application of the results to such problems as the equation of state of real gases and later the shape of pressure—broadened spectral lines. Interest in philosophy was dormant at the time — maintained, to be sure, by listening to lectures by Reichenbach — but subordinated to the attempt to learn as much as possible of the new physics at its source. Trips to Goettingen brought me into occasional contact with Born, whose interest in philosophy left his mark on my thinking.

Much later, after Born's retirement from the University of Edinburgh, where he had fled from Nazi persecution, he transferred his residence to the small town of Bad Oeynhausen in Westfalia, only a few miles from my birthplace. There I visited him on several occasions when travels brought me to Bielefeld. Fond memories connected Born with this final abode, a lovely spa, for it was here, he told me, that he met his wife. During one of my visits we walked through the garden and stopped in front of a fountain. He asked me to read the inscription on the plaque attached to its face. It said, in Westfalian dialect: De hillige Born (The holy fountain). Then, reminding me of the treatment he had received by an earlier generation of Germans he joked: "And now they have made me a saint!" In these final days his interest in the philosophy of physics was very intense.

In June of 1930 the Theoretical Institute in Berlin closed. I was called back to Yale as an instructor in physics (with a promise of promotion to an assistant professorship after one year) and spent the summer at the home of my parents in Bielefeld, writing down some recent thoughts on the problem of causality. They were published in the Monist vol. 41, p. 1, 1931. The fall found me back in New Haven, teaching elementary physics.

An amusing incident took place at about this time when I taught a course in elementary physics. Upon dealing with the principles of mechanics I sprang a ten-minute quiz upon my unsuspecting class, asking for a definition of 'mechanical advantage' and suggesting the example of a long pumphandle as a simple device to illustrate the answer.

A physics major gave the customary answer, defining mechanical advantage as the ratio of force output to input and proving it to be the ratio of the lengths of the two lever arms.

A philosophy major wrote: "The mechanical advantage of a machine provides a systematic amelioration of an unfortunate situation involving too much work (short pumphandle)" – and the sentence goes on, as most philosophical sentences do – "achieved by a categorical application of general mechanical principles: *a triumph of mind over matter*." The paper was impressive and scholarly; it contained footnotes to the literature including the quotation from Archimedes: "Dos moi pou stesomai, kai to pan kineso."

An engineering student wrote: "The mechanical advantage of a long pumphandle lies in the fact that you can get two people to pump."

In retrospect, perhaps this episode added strength to my intention to enlighten philosophers and engineers with respect to the basic ideas, the epistemology of physical science.

At Yale I came under the influence of two remarkable personalities, who took a kindly interest in my work. One was the philosopher F. S. C. Northrop, whose encouragement sustained and enlivened my activity in his field. The other was R. Bruce Lindsay, who, although but slightly older than myself, was already an established physicist and an excellent teacher with a profound interest in basic problems. He had come from Bohr's institute in Copenhagen and brought with him a fund of knowledge of the quantum theory. After a relatively short acquaintance he offered me the chance of collaboration with him on a book to be entitled 'Foundations of Physics', a text in theoretical physics with an emphasis on basic, even philosophical problems. The offer was eagerly accepted.

A year after my return to Yale Lindsay left for Brown University, accepting a prestigious invitation. Collaboration on the book therefore proceeded by correspondence, but the affinity of our thinking and the similarity of our styles narrowed the distance between us. However, the work went on under serious handicaps. The early 30's were a time when young scholars without tenure were insecure in their positions. The maxim, publish or perish, was applied with rigor, and many young scholars transferred from the larger, universities to smaller colleges. That imperative, furthermore, had a narrow interpretation: to publish meant to publish original research, not textbooks. Hence, while working on our book, I endeavored to keep my output of research papers aflow. Nor did I breathe a word about the book project to my superiors for fear of being dismissed.

There was also another strike against us. The combination of interest in physics and in philosophy was uncommon in the United States, and physicists who deviated from the narrow path were looked down upon by their colleagues. Philosophy of physics was definitely regarded as an aberrant, evasive sort of discipline, practiced by physicsts who 'could not make it' in their straight profession and, in somewhat greater numbers, by philosophers whose knowledge of physics was inadequate. Largely because of these considerations, Foundations of Physics did not appear until 1936. The statement in its preface, that in the previous decade there had been an increasing interest in the philosophical aspects of physical science, described our hope perhaps more than reality.

At any rate, the book was published by Wiley and Sons pretty much, if I recall their phrase, as a "high flyer" that was not likely to be a successful seller but perhaps an acceptable novelty in the science literature. For that reason they destroyed the plates after the first printing, and when some demand continued after the printing was exhausted, the book was reproduced by the photo-offset process which, at first at least, proved to be somewhat unsatisfactory. Twenty years after its original appearance, the copyright was returned to us and then sold to the Dover Publishing Company, who kept it on the market as a paperback.

Meanwhile, the benevolence of Professor Northrop and his most helpful advice and instruction convinced me that my interest in philosophy was not in vain. He made me feel that what some regarded as schizophrenia might, if developed, become a healthy trait. Himself trained in physics, close to Einstein in his thinking, he encouraged me to continue publishing in philosophic journals, and in his hospitable home I enjoyed many fruitful discussions on problems of common interest. Our meetings sometimes reminded me of the coffee hours with Sommerfeld in the English Garden of Munich.

During the 30's my research interests in physics shifted from spectroscopy to nuclear physics, in particular to the problem of the binding energies of light nuclei and the then current conjectures as to the nature and mathematical form of nuclear forces. In 1939 I obtained a fellowship in the Institute for Advanced Study at Princeton, where I enjoyed the very great privilege of working with Wigner and Wheeler. Meanwhile a few papers on philosophical topics, some of them included in this volume, made their appearance in print. A two-part article on the methodology of modern physics was the forerunner of two later books.

In 1940 I suffered a severe accident which kept me immobilized for nearly a year. This is the time when, largely confined to home and office, I was persuaded by a publisher to write a textbook on those parts of mathematics that had become most useful to the working physicist and chemist. George Murphy, a chemist friend at Yale, joined me in this effort. It proceeded rather slowly, was frequently interrupted by research after I gained mobility, and resulted in a book published in 1943. It hit a favorable market, and in retrospect I have sometimes asked myself why this should have been the case. The answer seems to be a dual one. First, there was a need for such a book because innumerable physicists and chemists were engaged in projects connected with the war effort and needed to know more mathematics than was taught in an undergraduate curriculum. Secondly, and this is perhaps more significant, the tenor of the book, the approach to problems, were different from those of the standard mathematics texts. The reason is probably that both Murphy and I were essentially self-taught amateurs (in the strict sense of the word, I hope) of higher mathematics and therefore more keenly aware of difficult points which elegant mathematics obscures. Being imperfect mathematicians made us perhaps better teachers.

I recall in this context an incident which may be worth reporting. A distinguished mathematician, in reviewing the book, not altogether unfavorably, complained of its lack of rigor. Since I knew him slightly I thanked him for the review, but also exposed myself to further criticism by noting that mathematical rigor seemed to me to be a standard changing with time; that proofs found in Weierstrass, for example, were no longer acceptable to some of my mathematician friends; and that rigor, if carried to an extreme, might become rigor mortis. This led to a friendly reply in which my critic said, in humorous candor, that one of the proofs in the book fell too short of rigor by any standard and was about as acceptable as the myth that babies were brought by the stork. To this I replied, with equal candor, that I had been brought up on that myth and that its later disproof had been pleasant and rewarding. I believe we remained the best of friends.

A Japanese translation of the book carried on its back the two names: Margenau-Murphy, and evidently came to be known by that designation. Years later I met a Japanese physicist who, after introductions and in an ensuing conversation, asked me if I was the author of a math book. On affirming this, I was repeatedly addressed by him as Mr. Murphy. I told him my name was Margenau. At that he bowed politely and said: "Sir, I believe I don't know you well enough to call you by your first name."

During the war years research in nuclear physics became in large part shrouded in secrecy. My foreign birth, I felt, excluded me from participation in war projects. Thus, while many of my colleagues were called away to do defense research, I remained at Yale, teaching both civilians and G.I.'s, sometimes at the rate of twenty hours per week. Understandably, my interest shifted from nuclear physics to other fields. I returned to problems in intermolecular forces, extending my previous calculations, which dealt with interacting atoms, to molecules. But at the same time a major swing of my interest back to philosophy took place, and this was largely the result of the arrival at Yale of Ernst Cassirer in 1941. He remained as visiting professor for two years, then, as emeritus, he became a research associate and died in 1944. His influence upon my thinking was decisive and my gratitude to him is deep.

One of the first courses he offered was a graduate seminar on Kant, and he asked Professor Northrop and me to join him in teaching it. Honored and delighted I accepted his request, and I learned more about Kant and Neokantianism than any of his students. Having read the 'Critiques' only in German my eyes were opened to the difficulties of translating Kant and of certain misconceptions that have arisen in the English literature on Kant because of the impossibility of rendering in simple English certain key terms, such as 'Vorstellung' and 'Darstellung.' Nor did I find Kant himself always consistent in his terminology.

Early in 1944 Cassirer asked me to collaborate with him in preparing an English edition, revised and brought up to date, of his 'Determinismus und Indeterminismus in der Modernen Physik,' at the time available only in German. I found a competent translator, the chemist Theodore Benfey, and set to work suggesting minor modernizations for Cassirer's approval. Our first step was to bring the bibliographical references up to date, the second to augment and amplify the text. But the latter task was never completed: my hero died before the work was done. I do not know whether he would have wanted me to finish it alone. Under the circumstances I felt that I should leave the text unchanged, add a preface over my name and append the approved bibliography. The book did not appear until 1956.

The epistemological view propounded in some of the articles of this volume, and expanded in a later book (the Nature of Physical Reality) was developed largely in consequence of Cassirer's and later Northrop's stimulation. It has been called constructionalism because it emphasizes that the elements of scientific theories, indeed the concepts of ordinary thinking, are not 'given' in the sense of the naturalism, positivism or any doctrine accentuating the process of induction; they are 'constructed' in accordance with specifiable, controlling metaphysical or methodological principles. Cassirer confirmed my initially hesitant disposition to use the word 'construct' in places where the reader might expect 'concept'. The reason is two-fold. First, it was intended to emphasize the origin of these noetic entities; secondly, to avoid the generic, collective implication of the term concept: a single electron, a single tree would in one of its aspects be a construct, though not a concept as that word is usually understood.

As the war went on I was finally asked to leave New Haven and join the Radiation Laboratory of M.I.T. in Cambridge. The problem posed to me was this. The radar devices then in use contained a so-called TR (transmit-receive) switch which allowed the use of a single antenna both for transmitting and receiving a signal. This switch was a gas filled tube which, when transmitting the strong outgoing pulse, was excited to incandescence. The electric discharge thus created, if it persisted, would have absorbed the returning signal so that it could not be recorded. Hence, the longer the discharge lasted the longer the dead time of the system, and the greater the minimum distance at which an object could be detected. My task was to design a way to shorten the discharge time of the TR tube.

The way in which this was done is not relevant here. The facts are being cited here merely to indicate why, for some years after 1945, my research in physics shifted to the theory of electrical discharges in gases, to a discipline now called gaseous electronics. This took the place of my earlier preoccupation with nuclear forces, the knowledge of which was growing behind closed doors. My stay at Cambridge lasted less than a year, but my interest in this new field continued.

The late 40's and all of the 50's were a period of frantic travelling, very often abroad. On one occasion I attended in succession two conferences in Europe, one dealing with gaseous electronics, the other on philosophy. When I returned and told a friend of the two events he asked: "Which one was the gas discharge conference?"

Before all that, however, an event which dominated my subsequent career took place in New Haven. I was invited to a party by the philosophy department. The attending professors were Blanshard, Fitch, Greene, Hendel, Northrop and Weiss. After dinner an intensive discourse on all sorts of philosophical problems arose. I felt drawn into it in a major way and sometimes hesitated to share my views. Finally I sensed that it was not my views but my knowledge that was being probed, with tact, to be sure, but also with thoroughness. Returning home that night I felt approximately as I did after my Ph.D. exam. A few days later, in 1946, I was asked to become a member of the philosophy department, to teach courses in both physics and philosophy. I understand that mine was the first joint appointment of this sort at Yale.

Since teaching was a crucial factor in my intellectual development as well as the testing ground of my philosophic views it seems appropriate that I should list here the titles of courses I have taught at Yale and, in a visiting capacity, in about a dozen other colleges, including Tokyo, Heidelberg and Fribourg. They are arranged in four categories: 1, undergraduate courses in physics; 2, graduate courses in physics; 3, undergraduate courses in philosophy; 4, graduate courses in philosophy.

1. Various introductory courses; Foundations of Physics (a historical development and careful analysis of modern theories, a course taught for more than 40 years); Electricity and Magnetism.

2. Mathematical techniques of Modern Physics; Quantum Theory I and II; Thermodynamics and Statistical Mechanics.

3. Philosophy of Science; the Role of Science in Human Experience.

4. Philosophical Presuppositions of Natural Science; Epistemology of Modern Science; Science, Epistemology and Ethics; Science and Ethics; Seminar on Causality; Philosophy of Space and Time.

Beginning in the middle 50's research in physics was sponsored largely by public agencies, in my case by the National Science Foundation, The Airforce Office of Scientific Research and the Office of Naval Research. The latter provided a contract calling for theoretical work in gaseous electronics with funds sufficient to finance the efforts of two graduate students. Their supervision and the writing of quarterly progress reports were burdens which curtailed the supervisor's own time for original investigations and led to numerous joint publications. The Airforce approached me with a request that I accept a contract and devote as much effort as possible to what seemed to be an outmoded field of research, the theory of spectral line broadening. This had been my preoccupation in the early 30's when I published a theory relating the intensity distribution within a pressure- and temperature-broadened line to the physical properties of the absorbing or emitting gas. The financing of the contract was so generous as to cause me to wonder about the Airforce's interest in a problem so singular and so far-fetched. There is perhaps a lesson in this, a lesson worth heeding by those who pretend to know the applicability of a theory at its birth and to judge its social usefulness or its potential for good or evil. The mystery was solved when, after 1954, I was hired as a consultant by the Lockheed Corporation which had been charged by the Airforce with a study of the physical processes taking place in the fireball of the first hydrogen bomb. There was no way of measuring temperature, pressure, ion density etc. directly; but it was easy to photograph the spectrum the fireball emitted during the various phases of the explosion. And here the innocent theory of the early thirties, which presented line structures in terms of the physical parameters of the source, could be put in reverse and made to yield information about source parameters from line structures. Never would I have dreamed in 1933 that the simple theory could be put to that kind of spectacular use.

In later years, consulting and lecturing took an increasing amount of my time. Among the most pleasant of these activities was participation in the work of the Navy's Underwater Swimmer's Committee and the Airforce's Radiation Weapons Committee. These met periodically in such interesting places as Key West and Cape Cod, and they brought me together with interesting people. Most enjoyable, however, were contacts with two institutions which activated my interest in philosophy. One was the Du Pont Corporation which offered seminars on various subjects to its executives. They were held in a beautiful resort hotel near Atlantic City, and I was a lecturer in philosophy of science. There I learned that an earnest concern with epistemology (perhaps not under that name), quantum theory and ethics is not confined to academe. The Brookings Institution conducted similar seminars in beautiful colonial Williamsburg, not for its own employees but for special professional groups, such as high military officers or union executives (which I remember with special vividness). I was on the Brookings list as lecturer on philosophy.

Lest the reader gain the impression that the concerns of the research branches of the Armed Forces were entirely pragmatic and military, I conclude by recording an embarrassing but perhaps not trivial incident. Our research contract with the Airforce called for work dealing with the topic of spectroscopy described above. At one point a problem close to the philosophical foundations of the quantum theory arose, and my philosophic interest was aroused. But the problem did not fit into the narrow field defined by our contract. Hence in my next appeal for a grant, I requested a small fund for work on a topic which was really philosophic but which I camouflaged to make it appear as physics. The reply was unexpected. It said in effect: We are glad to grant your request, but would you mind very much if we called it "philosophy of quantum mechanics"?

It is a curious fact, not often noted, that in the recorded history of human thought the physical sciences and philosophy are intertwined for at least the first 2 or even 3 millennia. The physical sciences are here taken to include astronomy, the oldest discipline, physics and chemistry. Their intimate connection with philosophy is evident not only in Western writings but also in the sacred literature of the East, where we find the Rig-Veda (written between 2500 and 1500 B.C.), the main source of Hindu philosophy, replete with discoveries and views regarding the physical world. Since this account will be limited, for reasons of space, to the better known facts of European history, I shall merely cite here very briefly a few of the physical speculations of Hindu philosophers which are usually ignored in American courses on the history of science.

In the Rig-Veda, one of the oldest known treatises on philosophy (and religion), it is suggested that the earth is spherical and suspended freely in air. One also finds the observation that "the sun has seven rays." What they meant precisely can hardly be known; but one authority (Dr. Dutt) maintains that it indicates the Hindus' knowledge of the spectrum. In the Brahamana one finds the statement: "The sun never sets nor rises. When people think the sun is setting, he only changes about after reaching the end of the day, and makes night below and day to what is on the other side. . . . In fact he never sets at all." The Rig-Veda says that "the moon shines by the borrowed light of the sun." But the most remarkable feature illustrating the parallelism between Eastern and Western thought, and one which I have never seen included in American text books on the history of science, is the atomic theory of the philosopher Kanada ($\sim$ 500 B.C.). He argues that there must be a smallest thing that excludes furthers analysis. Otherwise we should have a *regressus ad infinitum*, a most objectionable process in the eyes of Indian philosophers. These smallest and indivisible particles are held by Kanada to be eternal in themselves, but non-eternal as aggregates.

The intricate connection between philosophy and physics is apparent in the following quotation from the historian Colebrook (Trans. Roy. As. Soc, Vol I, 1930):

"Material substances are considered by Kanada to be primarily atoms, and secondarily aggregates. . . . The mote which is seen in a sunbeam is the smallest perceptible quantity. Being a substance and an effect it must be composed of what is less than itself; and this likewise is a substance and an effect; for the component part of a substance that has magnitude must be effect. This again must be composed of what is smaller, and that smaller thing is an atom. It is simple and uncomposed, else the series would be endless, and were it pursued indefinitely there would be no difference in magnitude between a mustard seed and a mountain, a gnat and an elephant, each alike containing an infinite number of particles. The ultimate atom is therefore simple.

"The first compound consists of 2 atoms: for one does not enter into composition... the next consists of 3 double atoms, for if only 2 were conjoined, magnitude would hardly ensue, since it must be produced either by size or number of particles; it can not be their size and, therefore, it must be their number. Nor is there any reason for assuming the union of 4 double atoms, since 3 suffice to originate magnitude. The atom then is reckoned to be the sixth part of a mote visible in a sunbeam."

Kanada, by the way, was approximately contemporaneous with the Greek originators of the atomic theory.

Whatever the force of these arguments may be, whatever their esthetic appeal, they do illustrate the point made in the beginning, that even in Eastern thought philosophy and physical science were fused in the significant early literature of the Orient. Later, to be sure, science fell behind in that quarter, but it was never separated from its sister discipline.

The history of science and philosophy in Europe is better known to Western readers. Hence, to illustrate their intimate relation, we shall perhaps somewhat arbitrarily select sporadic illustrious figures and mention their contribution to either field. It might also be mentioned that college courses in the history of philosophy and the history of sciences cover approximately the same ground up to about the time of Newton.

But just a word of explanation for our indulgence in this theme of fusion. I shall attempt to show that the close connection existed approximately until the beginning of the last century. Then there occurred a radical separation: physics and philosophy became different departments of study in our colleges, contact between them loosened progressively. Or, to be more precise, a single, somewhat superficial philosophy, vaguely called empiricism or positivism, dominated the scientific scene and suppressed the more elaborate metaphysical views of earlier thinkers. Then, during the first 3 decades of the 20th century, interest in philosophy revived among physicist. The main purpose of this preface is to show forth the reasons both for the 19th century split and, in somewhat greater detail, for the 20th century reunion.

First, then, a casual glance at Western history of science. Beginning with ancient Greece, the oldest known philosopher was *Thales* ($\sim$ 600 B.C.), whose philosophy was a form of animism and whose physics attempted to base all phenomena on the existence of only one substance: water. *Anaximenes* ($\sim$ 550 B.C.) chose the fundamental substance to be air, and *Heraclites*, another member of the Ionian School, took it to be fire, likened reality to flickering flames and urged upon his followers a thorough skepticism concerning sensory perceptions. Only reason was to be trusted.

His contemporary, *Parmenides*, developed a philosophy which carried the metaphysical postulate of simplicity (cf. papers in section I) to an extreme, paying minimal attention to empirical confirmation. His physics can be summed up as follows. Beginning with the primary postulate: There exists something because it is the object of our thought, he argues that being and filling space are synonymous. Hence vacant space does not exist. Ergo: there can be no motion, since motion takes a thing from the space it occupies to the space where it is not. Since the latter (in his view) does not exist, motion can not take place. Neither can there be a separation of the world substance. Continuing this reasoning he finally concludes that coexistence and succession are non-entities, fallacies of the senses. The "existing" is eternal, uncreated, indestructible, unchangeable, homogeneous. Invoking a principle of simplicity or beauty he pictures it as a sphere. Here we find a form of physics embedded in perhaps the most extreme form of rationalism or metaphysics ever proposed.

Let us take a jump to the Greek philosophers who are often regarded as founders of the atomic theory. *Leucippus* ($\sim$ 450 B.C.) introduces his *atomoi*, imperceptibly small entities. He endows them with reasonable physical properties like shape, size and motion in space. But to account for motion he turns philosopher and invents a special agency which forces atoms to move.

His follower *Democritus* elaborated upon the atomic theory, accounted for the existence of different states of aggregation of matter, but his writings also contain the statement: "By necessity are foreordained all things that were and are and are to come." We find here the first version of a principle of determinism or causality. *Pythagoras*, mathematician, physicist and philosopher, discovered the connection between pitch and length of a vibrating string, proposed a theory of vision and assigned simple stereometric forms to different atoms. But he also proclaimed numbers to be the only reality, taught recurrence of events and the transmigration of souls.

*Plato*, (427–347) whose philosophy we shall not review, also produced a theory of vision, displayed intimate knowledge of air pressure, magnetism, capillarity and the formation of crystals.

We need not take space here to show the consanguinity of our two disciplines in the writings of the most influential philosopher-physicist of antiquity, *Aristotle* (384–322), whose doctrines dominated practically all of the Middle Ages, a period in which theology kept its constraints upon both disciplines.

No great discoveries were made until the times of the Copernican revolution, the movement connected with the names of *Copernicus* (1473–1543), *Kepler* (1571–1630) and *Galileo* (1546–1642). All of them, especially the latter two, were oriented toward and creative in philosophy. Kepler, in his early writings, gives a strong impression of worshipping the sun. His philosophic leanings become explicit in such metaphysical utterances as the following: "Natura simplicitatem amat." "Amat illa unitatem." "Nunquam in ipsa quicquam otiosum aut superfluum existitit." "Natura semper quod potest per faciliora, non agit per ambages difficiles."* And in his early erroneous theory of the solar system he was strongly influenced by esthetic considerations: he regarded the orbits of the 6 known planets as circles inscribed in and circumscribed about 5 spheres.

*Galileo*, (1546–1642) known chiefly for his contributions to mechanics, his construction of the telescope and his sacrificial advocacy of Copernican doctrine, also argued about the philosophical issue of primary vs. secondary qualities. On the other hand *Descartes* (1596–1650), the author of the dualism between *res cogitans* and *res extensa* (mind and matter), also invented one of the fundamental tools of physical theory, the Cartesian coordinate system, and developed a theory of vortices.

*Newton's* (1642–1727) work in physics is too well known to require exposition here, and no biographer fails to record his deep philosophic and later religious convictions. But we include here another bit of information, not so often recalled: There was no name for physics, no professorship of physics at Cambridge University. Newton's title there was professor of natural philosophy — a title given today to college teachers who do not quite fit into a conventional physics nor a philosophy department.

---

* Nature loves simplicity. She loves unity. Nothing useless or superfluous exists within her. Nature always does what she can by simple means, she does not make difficult detours.

Most discussions of the history of physics either stop or change their approach with Newton. For his scientific breakthrough was so successful, so immense that a mere survey of facts will not do it justice. His work in the foundations of mechanics, now called classical mechanics, along with his contributions to optics made possible the solution of innumerable detailed problems, each important in itself, so that physics not only became a science in its own right but began to dominate all other sciences and to loosen its allegiance to philosophy. Great philosophers following Newton, notably *Leibnitz* and *Kant* (cf. the Kant – Laplace theory of the origin of the solar system) still made important contributions to physical science, but the center of gravity shifted toward analytic and highly accurate accounts of physical phenomena. The foundations of mechanics seemed wholly secure. They were amplified and refined by men like *D'Alembert, Largrange, Gauss* and *Hertz* without any change in Newton's laws.

Brilliant successes illuminated other areas of physics, notably those related to the electromagnetic field, which yielded almost miraculously to *Maxwell's* (1831–1879) mathematically elegant equations. The story of successes is long, and it is not surprising that two important developments took place: first, the classical physics of Newton and his successors invaded other, initially unrelated fields, especially biology, which by its own emancipation – primarily Darwin's theory of evolution – was brought closer to the camp of physics. And secondly, philosophy, impressed by the stunning advances in physical science, adapted itself to the most convincing features of the physics of the time: to the universality of mechanistic processes, which illuminated simultaneously the study of sound, of heat, even of light, which was regarded as a vibration in a mechanical ether. All this, together with the expansion of knowledge in the field of electromagnetism, led to the dominance of a philosophy based on the ideas of matter, which is the substance of mechanics, on the absolute certainly, universality and invariability of physical laws, on experimentation as the most important principle of science, as taking precedence over reasoning. Thus, at the end of the 19th century, philosophy had been pushed aside by physical science, the two had parted company, and the prevalent philosophies were empiricism, positivism and materialism. The biologist Haeckel was able to announce, in the late 1800's, that materialism was the only reasonable philosophic view. All reality consisted of matter, whose properties were hard and definte and clearly understood. Idealistic philosophers were fools; no new discoveries that would revive their claims could be expected. Marx and Engels inhabited this milieu.

The solidity of these convictions was destroyed primarily by two discoveries early in the 20th century: relativity and quantum mechanics. They shook the foundations of Galilean-Newtonian physics, of materialism, the principle of clarté Cartésienne which insisted on a clear, visually conceivable explanation of all phenomena. Along with these two revolutionary discoveries came a return of interest in philosophy among some physicists, the urge to ask questions not tolerated by the standard brands of positivistic methodology. Many of these questions, and the answers to which they led, compose the content of this book. The remainder of this preface will be used merely to sample the exciting novelty of the two mentioned discoveries which motivated the return of philosophic interest mong physicists.

Relativity gave rise to the so called twin paradox, discussed in numerous places. I shall illustrate it by an example of space travel which embodies features similar to those of the twin paradox.

Could a person, in scientific principle if not with means available at present, visit a star 1000 light years away within a human life time? Einstein gave an affirmative answer and the details are as follows. The astronaut leaves the earth on a rocket ship travelling with a speed approximately 3 miles/sec. less than the speed of light. He will arrive at the star after 20 years according to the clock in his ship or any other device (e.g., his own pulse, if it is sufficiently regular) that travelled with him. Turning around may be a problem, but if he manages it alive he will return to the earth on the same rocket ship after 40 years. But while he will have aged by only 40 years, he will find the earth 2000 years older. While in transit, our traveller's mass increased to about 5 tons, and it took the energy of approximately 10,000 hydrogen bombs to send him off with the requisite speed.

Relativity theory has other astonishing consequences: it permits the existence of black holes, indeed of massive bodies whose energy is zero. St. Thomas' doctrine of *creatio ex nihilo* (creation out of nothing) could be justified by an appeal to relativity.

This theory applies mainly to the macrocosm. But in the microcosm, too, the laws have changed in fascinating fashion.

Our sense organs do not permit us to observe directly what goes on in the domain of atoms, molecules and electrons. We do, however, have instruments capable of registering their behavior. Let us imagine an intelligent observer, perhaps as small as an atom, to be equipped with such devices and ask what, in accordance with the laws of quantum mechanics, would he see? He would find it difficult, for instance, to verify Galileo's law of free fall. For he would at once observe that there are no continuously visible small objects such as electrons, or even atoms. And if he collected a lot of them, i.e., succeeded in concentrating numerous intermittent patches of luminosity — his equivalent of a body — into a small space he would find that they do not remain densely packed nor stay at rest. They would jostle and bounce about, and there would be no way of quieting them down.

Assume that this unruly collection is dropped — perhaps from a microcosmic leaning tower — and their progress is recorded. It will be impossible to trace a single entity; the record of the collection will be a multiplicity of individual light emissions and reflections, each of them unpredictable and random. Yet there is some semblance of lawfulness and coherence, especially to an observer far away who is unable to distinguish the light flashes from individual atomic entities. To him, the falling group has the appearance of a swarm of fireflies (at night), a swarm which is clustered fairly tightly at the beginning but diffuses into a larger and larger assemblage of luminous dots as it falls. The behavior of the individual dots is regulated only by chance, by the laws of probability; but the center of gravity of the entire cluster follows Galileo's law.

These few, arbitrarily selected examples illustrating some of the consequences of relativity and quantum mechanics, by their defiance of common sense, have revived, indeed necessitated a concern with philosophic problems among physicists. The contents of this volume respond to it and attempt to clarify some of the answers given by modern physics to the questions inherent in these examples and related problems. The reader will doubtless sense that such ideas as probability, cause and effect, a new meaning of reality, of observation and of measurement, the connection between a sense experience and the scientific concept which corresponds to it, the flow of time — all these will play major roles in the subsequent discussions.

The titles of the 4 sections are meant to convey general descriptions of their contents. The sections themselves are closely related.

# ACKNOWLEDGEMENTS

The Author and Publishers gratefully acknowledge the permission given by the Editors, Proprietors and Copyright Holders of the following Books and Journals for permission to reproduce the articles mentioned below in this work:

'The Problem of Physical Explanation', The Monist, vol.39,321,1929.
'Probability and Causality in Quantum Physics', The Monist, vol.42,161,1932.
'Meaning and Scientific Status of Causality', Philosophy of Science, vol.1,133,1934. Copyright 1934, The Williams & Wilkins Co.
'Methodology of Modern Physics', Philosophy of Science, vol.2,1935. Copyright 1935, The Williams & Wilkins Co.
'Metaphysical Elements in Modern Physics' Reviews of Modern Physics, vol.13.
'Is the Mathematical Explanation of Physical Data Unique?', reprinted from 'Logic, Methodology and Philosophy of Science' E. Nagel, P. Suppes and A. Tarski (eds.), Stanford University Press. Copyright 1962 by the Board of Trustees of the Leland Stanford University.
'Probability, Many-Valued Logics and Physics', Philosophy of Science, vol.2,1935. Copyright 1935, The Williams and Wilkins Co.
'The Frequency Theory of Probability', Philosophy and Phenomological Research, vol.6,1935.
'Can Time Flow Backwards', Philosophy of Science, vol.21,1954. Copyright 1954 by The Williams and Wilkins Co.
'Causality in Quantum Electrodynamics', Diogenes, vol.6,1954.
'Relativity: An Epistemological Appraisal', Philosophy of Science, vol.24,1957. Copyright 1957, The Williams and Wilkins Co.
'Philosophical Problems Concerning the Meaning of Measurement in Physics', Philosophy of Science, vol.25,1958. Copyright 1958, The Williams and Wilkins Co.
'Bacon and Modern Physics: a Confrontation', Proceedings of the American Philosophical Association, vol.105,1961.
'Western Culture, Scientific Method, and the Problem of Ethics', American journal of Physics, vol.15,1947.
'Physical versus Historical Reality', Philosophy of Science, vol.19,1954. Copyright 1954, The Williams and Wilkins Co.
'The New View of Man in His Physical Environment', The Centennial Review, vol.1,1957.
'Science and Human Affairs', Discourse, 1958.
'The New Style of Science', Yale Alumni Magazine, Feb. 1973.
'Phenomenology and Physics', Philosophy and Phenomenological Research, vol.5,1944.
'Physics and Ontology', Philosophy of Science, vol.19,1952. Copyright 1952, The Williams and Wilkins Co.
'Faith and Physics', The Yale Alumni Magazine, May 1957.
'The Pursuit of Significance', The Wooster College Magazine, 1966.
'Religions Doctrine and Natural Science', Twayne Publishers, 1962.

ACKNOWLEDGEMENTS

Above all the author wishes to thank his distinguished colleague, Mario Bunge, for his suggestion that this volume be published, and Mr. I. Priestnall of the D. Reidel Publishing Company for his competent editorial advice and his success in easing the author's mind about the selection of essays.

# PART I

## METASCIENCE: PHILOSOPHICAL ANALYSIS OF SCIENTIFIC TRUTH

# PART I

The papers in this section deal with general philosophical problems concerning the role of theoretical principles *vis à vis* experimental discoveries. The first few were written before the basic ideas of quantum mechanics were fully understood and while they were still under discussion by the theory's originators.

Physical examples used to illustrate philosophical points reflect the state of the science (and the author's understanding) at the time they were written. I wish to call particular attention to the erroneous expectations contained in the last paragraph on page 87, which was written before the experimental discovery of the neutrino.

Some of the arguments contained in part I were later synthesized and published in coherent form in "The Nature of Physical Reality" (McGraw-Hill, N.Y. 1950; Oxbow Press, New Haven, 1977).

CHAPTER 1

# THE PROBLEM OF PHYSICAL EXPLANATION

Statements of the meaning and purpose of physical explanation are numerous. Every philosopher who has written on "Inductive Logic" has in some manner described or circumscribed the problem of physical explanation. The present paper is not a review of existing knowledge or opinions. It was written with an acute consciousness of the incapacity of most theories of explanation to provide grounds upon which modern hypotheses may stand, and it attempts to formulate a concept and an understanding of physical explanation which is, as far as logical principles permit, in conformity with *modern* views.

Scientists often satisfy themselves, and dispense with the complexity of the problem, by stating that explanation is nothing but a search for causes. But the semblance of truth which attends this assertion proves, on closer analysis, to depend on a delusion about the meaning of "cause." Cause, in the parlance of those who are accustomed to precise manners of speech, is an observable fact which, in a series of sense perceptions, invariably occurs as the antecedent to a definite set of other observable facts, called consequents. Hence, when in the course of this inquiry I speak of the cause of any phenomenon, I do not mean a cause which is not itself a phenomenon. A necessity of transcending into a metaphysical world of nonperceptible entities does not exist for the purpose of scientific investigations; nor will any doctrine concerning the transcendental nature of phenomena be found in the following remarks. Causes are definite and discernible components in the course of natural events, data of nature which need not be inferred by speculation. However, explanation deals with entities which are not, in general, observed, and can therefore not properly be called causes.

The term "nature" will occur more frequently than any other in the subsequent discussion; hence it should be more clearly defined, and stripped of all colloquial implications which do not belong to its strictest meaning. Nature, in the most noncommittal sense, is the totality of sense impressions, this latter term including observations of a systematic and scientific character. It comprises nothing external to the human mind and postulates no things in themselves. Nor does it involve any assertions with regard to the non-existence of things in themselves. Data of nature will be understood to be parts of our consciousness, and should imply no suggestion as to their origin. The very word datum is misleading. For if our perceptions be gifts we might expect to find a giver, which some recognize in an objective nature, others in a divine agency. But inductive experience teaches nothing about a giver, nothing about the origin of sense impressions; it leaves us in the bare state of *having* cognitions, without even an indication of their whereabouts. The term "habita" instead of "data," the use of which was advocated by a philosopher whose name I have forgotten, would appear preferable in view of this condition. But its introduction would constitute too radical a break from current terminology to warrant its usefulness. The word datum will be employed in this specified sense.

We have seen that nature is the sum total of data. However it is a common practice of the mind to project these data into an external space and time, to endow them with substantiality and construct an objective nature which is a thing in itself. This process is so universal that a few critical remarks need be made concerning its legitimacy. Let us analyze it briefly and consider some of the steps. Certain experiences cause us to form inseparable associations between sense perceptions, persuade us to become convinced of their existence independently of our perceptions, which we finally accustom ourselves to look upon as accidental. We have invested the possibility of sensation with the quality of permanence. So far an objective nature is a phychological structure. Belief in an external world is simply the belief in a permanent possibility of sensation. It is important to recognize that we need not go farther than this to establish a ground for science. But the realist will refuse to stop at this point. His next step will be to class these permanent possibilities as existences distinct from our sensations, and this may be proper if "existence" is not understood in the medieval ontological sense; but if he thinks of them as generally distnct, as causes of our sensations, he is deluding himself by a faulty use of what is called causation. The law of causality may not derive its unconditional binding force *from*, but certainly manifests it *in* actual experience, and every conceivable case to which it is applicable consists of two sets of data of which one may be inferred if the other is known. Now an objective nature is not a determinable factor of experience, and we have no evidence that it comes within the range of entities over which causality has legislation. It is for this reason that the argument fails—that causality can never form a bridge between the known and the ultimately unknowable, and that an objective nature, if it exists, is forever unintelligible. Another feature which strengthens the fallacy of an objective nature is the tendency of our mind to ascribe greater immediateness, not to use the word reality, to notions which present themselves more readily than others. For instance we are tempted to consider matter as more immediate and real than, for example, energy or an electric field. But it is evident that this propensity is caused by the constitution of our senses, or mind in general, and must therefore be regarded as accidental. Indeed it will appear that the algebraic relation $pq - qp = (h/2\pi i)\epsilon$ may be just as essential a component of reality as is the notion matter. But this consideration must be deferred until later. The recognition that the realistiic appeal of certain natural data is largely dependent upon our modes of comprehension deprives naive realism of a considerable portion of its persuading force.

If nature has no intelligible objective counterpart, and if its data have no generic agency, how can the uniformity of nature be accounted for? I am using the term uniformity in John Stuart Mill's original sense as referring to the prevailing regularities encountered in the order of data. The occurrence of laws in nature is indeed one of the strongest arguments against the ideal immanence of physical science. Although the validity of numerous laws proves nothing by itself about the objective reality of nature, we do find it less plausible to assume a great number of independent uniformities in our experience than to assign to an objective nature equally many *properties* from which the uniformities in experience follow generically. The simplest and most elegant way out of this difficulty is that taken by the extreme idealist who interprets all laws as categories projected into an imaginary external world from within. But we do not wish to make any postulate with regard to an external world as a thing in itself, not even its non-existence; we merely wish to record our absolute ignorance concerning it. The only thing we have shown is that there is no causal connection between it and natural phenomena, and that, if there be any link, it is unintelligible. Moreover, we wish to avoid the dilemma wherein the idealist finds himself when he attempts to understand the separate existence of conscious beings. It remains to be seen whether or not the separate uniformities impressed upon us by the very data of nature can be resolved into few simple assumptions. If it is possible to reduce them to one or two axiomatic principles our surprise at finding nature so peculiarly obedient to a multitude of laws will be much diminished, and the tendency of assisgning independent properties to an objective thing in itself behind nature, in order to produce these laws, will have lost its power. This reduction, I hope to show, can be performed.

In any group of sense perceptions there will, in general, be definite features of comparison. The group, first chaotic, displays similar traits, allowing the intellect to begin its ordering action. Such resembling features may be termed synthetic properties of the complex of sense awareness. They themselves constitute no laws; it is the regularity of their occurrence that forms a law. This regularity, or uniformity of behavior, imposes a very definite restriction upon nature, which is not inherent in its concept; it makes nature subject to a form of pre-determination which permits us to describe its unknown data with some expectancy of correctness. The character of this restriction might be of great complexity; the predetermination might exist in a multitude of ways. Instead, we find that the regularity is of the simplest imaginable type, namely, that a given set of natural events determines a definite set of others whenever the elements composing the primary set combine to restore it. This fundamental axiom, it will be observed, is more general than the principle of causation, for it establishes the possibility of numerical relations and geometrical intuitions which could be deduced only improperly from the law of causation. It will be referred to in the course of this discussion as the principle of consistency of nature, and shown to include the principle of causation. I find it difficult to formulate this latter in a manner that is universally acceptable, in view of the circulation of such vague and divergent notions as are expressed by: *"Causa aequat effectum;"* "Everything has a sufficient cause;" "Everything has a beginning in time." The general recognition seems to be the determinateness, the impossibility of arbitrary intrusion into the course of events. This, however, is precisely the content of the principle of consistency of nature if the relation between the primary and the determined set of events be one of succession. It may be remarked parenthetically that this same principle yields numerical and geometrical laws if both sets are coexistent. With more particular reference to physics the principle of causation may thus be stated: No matter at what time an experiment be performed, its result is the same if the original conditions are identical.

Perfect identity of causal conditions can never be achieved. Hence, if the principle of consistency be regarded as an *a posteriori* datum it can be established only as an ideal abstraction from an approximating, ever recurring experience. As such, it would exhibit all the defects of an experimental discovery, would carry with it no assurance of universality or the necessity of its continual recurrence.

On the other hand, it may be supposed that it is a category of *a priori* rank, a fact which lies at the very basis of experience as the condition for its possibility. The decision has little to do with the application of the principle; the physicist, in attempting to understand the nature of his explanations, need not commit himself on this point.

A difficulty which the last mentioned form of the principle of causation makes no attempt of concealing lies in the rôle which "time" plays in it. It would seem at first that the abolition of absolute time and the dissolution of the concept of simultaneity by the theory of relativity confront us with a grave problem concerning the truth of the principle. This matter does demand attention. However, since it is but loosely connected with the topic of this paper, I shall not deal with it in detail, and merely state that the principle is in conformity with relativity considerations as far as I have been able to extend the analysis.

Bridgman, in his *Logic of Modern Physics,* refers to a fundamental postulate which he calls the "principle of essential connectivity" and phrases in some such way as this: If two apparently similar isolated systems start from the initial conditions, and differences develop in their behavior, these differences are evidences of other previous differences. It is clear that this follows as an immediate consequence from the principle of consistency of nature; and it appears to me to be less general than the latter. Moreover, he appears to make the validity of the principle of causation dependent upon the existence of linear differential (or difference) equations to describe the phenomena. It is certainly instructive to point out that in case of non-linear equations we may be unable to carry out the causal analysis because of the unmanageable mathematical complexity of the problems. That may even be true in the case of linear equations. But the principle still applies if our analysis is impossible; the effects of the partial events need not be additive, the equations need not be linear. In general, a complex effect may not be representable as the superposition of elementary effects—yet it is still unique, and the principle of causation requires no more.

The preceding remarks are likely to appear out of place, since they have but little to do with the topic of this discourse. Yet I must contend that an analytic inquiry into the meaning of "nature" and "natural law" is essential in any attempt to rise to as general as possible a view of the problem of physical explanation. It is indeed my effort to outline its widest aspects, for any restrictive definition would exclude much that is valuable in the field of modern hypotheses.

The first step in our constructive program is to acquit ourselves of the accusation of doing unnecessary labor when we frame explanations. Let us, then, pause for a moment and contemplate the reasons for the necessity of explanation. If physics were merely a science of measurements there would be no room for explanatory investigations; the theoretical physicist would lavish his energies in diluting concrete facts. But it is the characteristic of a true science that it proceed in close association with philosophy, that it be conscious, in every step of its development, of its relation to the fundamental principles of logical reasoning. Why do we not content ourselves with a mere description of nature, renouncing all attempts to explain it in terms which it does not itself offer without the intervention of reason? The answer is that the context of sense awareness is unrelated, and that we are compelled by the very constitution of our organs of recognition to conceive of events as continuous, as related. The unsophisticated mind experiences considerable dissatisfaction when it observes that gradually varying causes produce effects completely out of harmony with their gradual changes. Charge two spheres by constantly increasing their potential. During the first part of the process nothing directly observable occurs; then, suddenly, a violent discharge appears. Heat a substance slowly, and the effects will be consistent only up to a certain limit; then an unexpected change in phase is witnessed. Physics is full of instances of unrelated variations. If it were not, the desire for explanation in notions other than purely descriptive would probably never be felt. Again, there is the transition from one phenomenon to another which appears too abrupt to gratify our longing for coherence. Let an ordinary piece of steel act upon a bit of iron and there will be no effect. Another piece of steel, similar in its external properties to the first, will attract the iron. The unsophisticated observer is puzzled. The principle of consistency of nature, which he unconsciously embraces, compels him at once to formulate theories with regard to the internal properties of the second piece of steel: he is indulging in physical explanation. To investigate further this basic urge of human nature, which causes us to work for unity and coherence, to yield uncritically and without reserve to its universal appeal, would lead us into interesting speculations. But for the purposes of this discussion it is sufficient to state its existence, and to acknowledge its constraining power.

We are now about to realize the function of physical explanation. It is clear that its purpose is more fundamental than to reduce the number of different concepts needed in the description of nature, or to serve as a guide in experimental investigation. For it falls short of accomplishing the former task; it introduces new concepts in addition to the ones required to comprehend our experience, which still remain necessary. The latter purpose, that of guiding the experimenter, it incidentally achieves, but its mission is far nobler than its capacity as a handmaid of observation. In the remainder of this discussion I shall try to analyze the process of explanation into its elements, starting with simple forms of explanation, determining their special features. Then it will be my aim to widen the meaning of the term, and finally to lay down some requirements which it must always answer. In the end it will be desirable to survey the field of physics and to determine in what respects existing theories conform, or do not conform, to these requirements. In following this course we shall observe that the meaning of physical explanation has undergone distinct modifications, the most radical of which is taking place in modern physics.

Whenever we make any attempt to explain an occurrence, that is a set of observable events or phenomena, we associate with each phenomenon in nature some definite concept, or relation between concepts. This statement sounds very abstract and almost inane. Yet it is necessary to make it in such a general form in order to avoid the unpleasant obligation of qualifying it later. To be sure, it means that we establish a correlation between nature and a certain system of thought, called the explanation of nature. The elements of this system of thought, to which I have referred as definite concepts, or relations between concepts, I shall take the liberty of calling explanatory symbols, or simply symbols. Up to this point we have discovered nothing, either in nature or in the constitution of our mind, that would restrict our choice of possible symbols. Nevertheless we find that in most explanations the symbol is chosen to be some imaginative process, mostly capable of visualization. For the sake of simplicity we shall permit ourselves for the present to think in terms of such processes, although it will appear later that there is not only no necessity for such restriction, but that, under certain conditions, it may actually be desirable to abandon it. As examples of such correlation we have the measurement of heat on the one hand, rapid motions of idealized spheres—called molecules—on the other; or the appearance of a spectral line on the side of nature, and the electron transition as its physical explanation. Our problem will be solved when the nature of the correlation in question is determined. The common misconception that it be a causal relation has been guarded against at the beginning. It is here to be noted that an ill defined notion of causality has done much to confuse the minds of scientists, and has unhappily suppressed the appreciation of the function of physical theories and hypotheses. Motion of ideal entities is *not* the cause of heat, if the term "cause" is to have any meaning at all. Not even the invention of such a vague and misleading phrase as "hidden cause" or "underlying cause" enables us to make this statement in the affirmative. The cause of the heat, or, better, the body's being hot, is the process of heating it, a phenomenon with which nature presents us directly. The assignment of the symbol (motion of molecules) to the observable phenomenon of heat is an arbitrary act of reasoning, the justification of which we have to discuss and establish.

Another very common view is that the correlation is one of identity. The imaginary process is said to be the same as the occurrence explained. In fact it seems as though this were the universal credo of physicists today. Almost every experimenter believes sincerely that the emission of a spectral line *is* the same as the transition of an electron. An analysis of this situation would involve the problem of objective reality, which I am carefully avoiding because of its complexity, and, fortunately, we need not consider it to perform our task. One thing is certain: namely, that the indicated attitude is extremely useful as a working hypothesis and probably much to be encouraged. Yet the logician would assert that the imaginary process and the natural event are not identical; at least we should have no means of proving the identity. And if we still insist on calling heat and kinetic energy of molecules identical, we should be open to the accusation of using the word identical in a sense differing from its original and rigorous meaning. Granting, then, that it is advantageous to think in many instances, of our explanation as the event to be explained, we must maintain from our analytical point of view that identity fails to designate the correlation we are seeking. We are forced simply to call it, without pretense of further specification, a definite correspondence.

When the illusion has been destroyed that explanation stands to natural event in a causal or identical relation, the explanatory symbol loses much of its substantiality and its immediate appeal, but not its value. Disregard of objectivity causes us to feel more free in selecting symbols: we need no longer have scruples in exploring the regions of abstract thought in our search for suitable symbols. It is true that visual processes are desirable for their convenience, but it may happen that we are driven to rid ourselves of the shackles which they impose. Then it will be imperative to abandon convenience and to accustom ourselves to more arduous methods of reasoning.

The wider the domain from which we may chose symbols for the explanatory representation of experience, the more necessary becomes a criterion for their selection. There is evidently an infinite number of imaginative processes that can be associated with any event. Why do we prefer a single one to all others? And, after having selected one, whence do we draw the evidence for the correctness of our choice? One fact is recognized at once: Whenever a new field of science, presenting but a single, or few observations has been discovered, there are a multitude of conflicting systems of explanation. An increase in the number of known phenomena is always accompanied by an elimination of hypotheses. It is safe to say that if in physics we had knowledge of but one single fact, its explanation would be meaningless, impossible, and no one would feel the need of it. Physical explanation has meaning only if there is a group of observations to be explained. The symbols used to construct a theory covering this group of observations must in the first place be limited in number, and secondly, they must be related. The significance of these two restrictions will be more clearly understood if we analyze the example already referred to. Kinetic energy of molecules is a combination of symbols, or, in this case, the term that calls to our minds a set of imaginative processes designed to explain heat. We consider it a satisfactory explanation. Suppose we attempt to find a few reasons of a general character why the kinetic theory is preferable to the old caloric theory, for instance. We find immediately that there are in our experience with heated bodies certain changes, such as freezing, fusion, and evaporation which demand the introduction of symbols other than are involved in the existence of a caloric, and entirely unrelated to it. The kinetic theory, however, has a sufficient number of related symbols to cover these phenomena. It may be said that the kinetic theory is more consistent, which is but another way of stating that its symbols are related. Moreover, if any theory needed the assumption of moving molecules, of the existence of a caloric, of the attraction between molecules, and perhaps a dozen other assumptions, we should reject it without test for it is wasteful of symbols.

These remarks are still much too general to be of assistance in determining which individual symbol within a group is to be associated with a given phenomenon, and before this has been specified there is little meaning in other considerations. Again let us look to the kinetic theory for elucidation. To be more specific, let us consider a gas. Under certain controllable conditions it rises in temperature, that is, a change occurs that can be measured. To this phenomenon we assign a symbol, tentatively that of "added swiftness of molecules." If the gas be enclosed in a vessel, another phenomenon may be observed and measured at the same time: that of increase in pressure. There is distinctly no similarity in the outward appearance of these two effects; they seem perfectly incoherent. Yet they have the same cause. The explanatory symbols, by their very reason for existence, must preserve the relatedness of the two events which nature obscures. Now we know from elementary observations that a moving particle has momentum, and that a change in momentum produces a force. Such change in momentum takes place at the boundary of the vessel. Thus we are led to associate with the observation of change in pressure the symbol change in momentum of molecules. On the part of nature we have two occurrencies, quite different in character, but joined by the similarity of their causes. To them we assign two symbols so related that one is implied in the other. Take now two different gases and allow one to pass into the other. We are still dealing with gases, hence at least part of the causes are the same as those previously considered. We observe the phenomenon of diffusion, an effect which no longer resembles any of the other two. Yet, I repeat, one of the causes is the same, for we are using the same gas as before. But diffusion is explained by simply assuming the properties of molecules belonging to different gases to be different. We are introducing a new symbol, or what amounts logically to the same thing, modifying slightly an old one and we employ it to account for the effect of diffusion. Thus we might continue. We discover more and more phenomena, completely different in appearance, but related by the similarity of their causes. Our system of explanation preserves the continuity or relatedness of the causes; it comprises only symbols of such character that a transition from one to a related one corresponds to the step from one occurrence in nature to another one which has a similar cause.

The situation reminds one indeed of the process which mathematicians call mapping. In choosing this process to illustrate the meaning of physical explanation I am fully conscious of the inadequacy of the use of pictures to represent abstract facts. But this illustration is so formal and devoid of accidental details that is not likely to distort our analysis. We shall henceforth frequently refer to it. Think of the data of nature as points in nature space. They will be dense in certain regions, sparse in others, depending on the care and thoroughness with which the region has been investigated experimentally. Each point is a known fact, its coordinates may be descriptive properties. Now experiment reveals but single phenomena, never any relations between them. Hence, however densely the points may be clustering in a certain domain, their distribution will always be discontinuous. As Bridgman expresses it: every experimental discovery has a penumbra of uncertainty about it. The extension of this penumbra may be reduced by increased precision of measurement but it can never vanish completely. This is quite true in nature space. Yet while Bridgman satisfies himself with the existence of this uncertainty as a deplorable but inevitable fact, *we demand that physical explanation abolish it.* We desire a representation of empirical knowledge which is, in terms of our graphical illustration, continuous. Therefore, we require a correlate to nature space, which we may call explanation space. In this explanation space, each point is a symbol in the previously defined sense, and each symbol corresponds to a point in nature space by the condition that a transition from one point in nature space to a nearby point be represented by a small and continuous variation in the associated symbols. A certain set of symbols, such as that involved in the kinetic theory, may, or may not, suffice to explain the entire domain of physical data. This matter will be discussed a little later. For the present we shall only point out this two-fold possibility, and note that such a determined set of symbols defines in our explanation space a certain mode of representation, which may possibly have to be changed as we proceed to explain another field of physical data.

At this point it becomes necessary to remark on the origin of the symbols. I fear that I have given the impression that they are, or might be, entities entirely distinct from natural phenomena. The extreme idealist would have it so. Surely, it would be beautiful and conservative of much mental effort if we could split the totality of human knowledge into two distinct categories, one comprising nothing but natural data and the other one our speculations about nature unstained by the admixture of sense impressions. Unfortunately, however, we can not even gain natural data without the aid of intellectual processes, and we can not think about nature in terms completely free from reference to elementary external events. The greatest problems in the theory of knowledge arise from the diffusion of these two classes of things. We are compelled, then, to choose our symbols of explanation in accord with this situation. A group of purely logical postulates is never to be selected as a set of symbols, for there would be no intelligible connection between the explanatory theory which they constitute, and nature's data. Hence, in every mode of representation of our discoveries about nature, there must be found some symbols which are directly derived from nature by observation. It is to be observed that even the most abstract theories of modern physics conform to this requirement. This statement is not really a restriction upon the choice of symbols; it is merely the condition that they be intelligible, that they have meaning.

Perhaps the grandest and most successful attempt at physical explanation is the work which has led to the mechanistic view of the universe. It will be interesting to devote our attention to it for a little while. First of all we must understand that the mechanistic view is by no means the same thing as the tendency to design mechanical models. The latter has not been confined to the mechanistic theory; it is to be found even in theories most remote from all attempts of mechanical explanation, and merely reflects a prevalent habit of thought. If we wish to find the mechanistic view in its purest form we must look to Helmholtz, I believe, who expressed it in a classical manner in his famous lecture "On the Conservation of Force," held before the Physical Society of Berlin in 1847. He designates as the aim of physical science the explanation of all natural phenomena in terms of motions of discrete material particles subject to the action of attracting or repelling central point forces, whose magnitudes are functions of the distance between the particles. *There* is a most clearly defined and coherent set of symbols, which exhibit the properties of which we have already spoken. Their number is small; they are related; and at least some of them, such as motion, material are immediate natural data, whereas others, such as particles, point forces, are idealizations which are not found directly in nature.

Suppose that we investigate the range of validity of this system of explanation. Mapping the points in nature space we begin with mechanics and find that they are all represented by one continuous curve in the associated explanation space. We pass from one phenomenon to the next by changing slightly the combination of our symbols. The same curve carries us without a break into the regions of sound and heat and a good distance into electricity. Points in nature space can be connected, that is, experimental data can be conbined, in an unlimited number of ways, each particular way corresponding to a curve in explanation space. The number of possible curves in a certain portion of the latter space will depend upon the density of points in nature space; but we postulated that explanation space be continuous, hence there must be points between the curves. And it frequently happens that, noticing an interesting point somewhere between two curves to which a point in nature is not yet known to correspond, we look for it in nature space and find it there.

When we realize that our curves stop somewhere within the region of electricity, we are about to make another interesting discovery. There are phenomena in nature that can not be mapped as long as we use the same simple mode of representation. Incidentally they have to do with propagation of electric disturbances. Here, our symbols fail to be applicable, for they would mislead us into expecting that the energy transmitted should be associated with material particles. (I am professing at present no knowledge of the quantum theory). But to this conclusion, which is a point in explanation space, there corresponds no point in nature space. Now it is within rights to combine the symbols in such a manner as to invent an ether, although some might doubt that this can properly be called a *small modification* of the elementary symbols. Yet even then we do not find the complete correspondence which we are seeking. To be sure, we can construct it after the fashion of McCullagh and Lorentz, endowing the ether with rotational elasticity. Indeed, it is interesting to note that Henri Poincaré has shown that every natural phenomenon may be explained by the mechanistic theory, even in an infinite number of ways. Why is it, then, that we feel strongly disinclined to accept as a physical explanation an ether with connected molecules containing three mutually perpendicular gyrostats in uniform angular motion with little angels flying around and keeping them properly lubricated? At first one would probably answer that the set of symbols involved is too artificial and complicated. Yet we must reject that answer if we wish to save some of the most widely accepted theories of to-day. It is highly unscientific to reject a thought because it is inconvenient. The correct answer is that there is nothing in nature to correspond to the added explanatory symbols, except the single phenomenon of propagation of electric waves. We are introducing into our explanation symbols capable of variations for which nature has no independently variable attributes. This condition may be roughly designated by saying that the system of symbols has more degrees of freedom than its natural correlate. Here we recognize another essential requirement to be imposed upon any satisfactory system of explanation, a criterion for the validity of a physical theory. In the instance under discussion, a defender of the mechanistic view might help himself out of his difficulty by fixing rigidly the axes of the molecular gyrostats, and by postulating that they be forever free from mechanical defects and inaccessible to experimental change, thus reducing the degrees of freedom permitted by the explanatory symbols. But if such practice were considered legitimate, it would violate the first requirement, namely that our symbols be few in number; for its uncontrolled use might increase their number without bounds. Consequently we are forced to admit the failure of the mechanistic theory of explanation in the domain of electricity.

Now we are in possession of several criteria for the fitness of a theory to explain occurrences in nature. They were derived by analyzing the meaning and purpose of physical explanation. Let us muster them in review, before we go on. At first we found that the explanatory symbols must be limited in number, and kept so by all permissible procedures of extending and elaborating on the system of explanation. At the same time we recognized that the symbols had to be related. Then there is the requirement that the set of symbols have no more variable attributes than the phenomenon it represents in nature. Last, but most important, perhaps, is the peculiar correspondence between events and the symbols used for their explanation, a correspondence which was made clear by reference to the properties of the curves in explanation space, which had to be continuous, and were made to cover completely a certain domain of space.

This latter correspondence will occasion some further scrutiny. The breakdown of the mechanistic theory shows that a system of explanatory symbols, which has been very successful in certain fields of physics, may fail to be applicable to other groups of phenomena. In other words, our points in explanation space cover completely and continuously a definite portion of it. Outside of this part of space we can not go, unless we are willing to use a different mode of representation, that is, a different set of symbols. This situation is unsatisfactory indeed, and it caused considerable alarm at the time of its discovery. Even now, most of us are still inclined to view it with some sort of apprehension. It is my personal belief that this attitude is not merely the result of our surprise at learning a few novel and unexpected aspects of nature, but that it contains a justifiable feeling of doubt as to whether the possibilities of finding an explanatory system covering all phenomena have been exhausted. Considering how wide a field of choice we have for our symbols, and how few efforts have been made to extend a theory over all ranges of natural events, we should certainly be convinced that we need not yet commit ourselves to the necessity of explaining nature in terms of different, non-consistent theories. There is much opportunity for profitable and constructive work in theoretical physics, and some interest in achieving unity and coherence in the various systems of explanation has fortunately arisen.

However, it is not at all certain that such unity is possible. Does, in that case, physical explanation lose its meaning? If it is to retain its meaning, the correlation between the points in nature space and the points in explanation space must be definite, that is to say, the event corresponding to a certain variation in the symbols must be definable and unique. This is equivalent to a point to point correspondence between the data of nature and the points in explanation space. Two ways are imaginable in which the correspondence might fail to be definite. First, one explanation could cover two or more phenomena. This case may be dismissed without further consideration, for it will always appear that, whenever this difficulty exists, it is due to a fallacy, caused by an insufficient analysis of the symbols involved. The other possibility is more serious: one phenomenon might be represented by several distinct sets of symbols, or, in terms of our favorite simile, one point in nature space might correspond to more than one in explanation space. This situation can not be tolerated, for in every conceivable form of this dilemma it is found that, in explanation space, points in the immediate neighborhood of those under discussion can be selected which designate conflicting phenomena in nature. To refer to one instance: both the kinetic and the caloric theory explain certain observable facts about heat. But the first would demand that heat be always associated with matter, whereas the latter would carry with it the possibility of its isolation, two things which can not both be true.

If the reasoning that has led to this particular conception of physical explanation is at all conclusive, we may formulate some ultimate inferences which are interesting because of their immediate applicability to modern theories. There is a universal desire that a system of symbols be found which will serve to represent all physical phenomena. This may be said to be the aim of scientists. But it may be beyond our power to design a system of this character; the final verdict in this matter is nature's. Even then physical explanation has meaning as long as there is a point to point correspondence between explanatory symbols and natural events. If this correspondence breaks down, physical explanation is impossible. Fortunately, we have as yet no cause to believe in the ultimate failure of this correspondence, although there are instances in physics at the present in which it does not exist.

Such conclusions as these might be regarded and applied as clearing and ordering agencies among modern physical theories. It would be possible and interesting, perhaps, to base upon them a critique of scientific hypotheses. However, I have not the space to survey the whole field of physics, nor would

I be capable of cataloging exhaustively all the theories that would enter into the discussion. Permit me to criticize, from the point of view of this analysis, only one outstanding theory, the quantum theory of popagation of light. Can we consider it as satisfactory, or not? To be sure, the quantum theory, in its more general form, does not satisfy our desire for unity, but we saw that this was not absolutely necessary. There is a sudden break in the mode of explanatory representation when we pass from the emission of light to its propagation. Two distinctly different sets of symbols are required for these naturally related events. Yet this is permissible since we postulate continuity of explanation space only in discrete domains. Hence if the application of the quantum theory were confined to explain problems of emission and the wave theory to account for those of propagation, both would be perfectly legitimate theories, in spite of the fact that we know of no connecting link between them. However, as soon as the quantum theory is called upon to explain the phenomenon of the propagation of light, and it functions, it ceases to preserve its character as a satisfactory system of explanation.

From our present point of view, what is the difference between a theory and a hypothesis? The answer is easily given: a hypothesis is a set of explanatory symbols valid over only a small range of natural observations. It may be obliterated by another hypothesis which expands and covers its region, which we should then be inclined to call a theory. As the range of application determines the value and power of a set of symbols, we are quite justified in attaching greater importance to a theory than we give to a hypothesis. Yet there is no essential difference in the nature of these two constructs.

Modern physics abounds in theories of limited ranges of application. In reflecting upon this condition, and observing the insistence with which each investigator elaborates upon and extends his theory of explanation, we might feel that science is no longer inspired by the hope of finding a single, consistent theory to cover all of nature's data. But there is a very promising feature about most modern theories: most of their explanatory symbols are purely formal. This does not make them better *per se*, of course, yet it entails a tremendous advantage, namely the possibility of merging one theory with another. Had the $\psi$-function of Schroedinger's theory represented something very definite, its meaning would have precluded all chances of fusion with the theory of Heisenberg. Page's success in uniting the theory of special relativity with that of electrodynamics might not have been so complete if the relations of relativity—its explanatory symbols—were less abstract and general. It was indicated before that there is no reason why the ultimate system of explanation should not be of an abstract and formal character, as long as some of its symbols relate to actual experience. Abstract explanations are to be encouraged because of the greater ease with which they combine.

The reason why many physicists seem averse to accept an abstract theory is quite apparent, and has been alluded to before. They find it less convenient, more elusive, than their accustomed combination of workshop models. Some would emphatically protest, pray to their little idol called truth or reality, and recite their well learned creed denouncing abstract things as unreal. But I have not yet heard of any definition of explanation in terms of what is real; moreover, if we knew reality, we should not bother about explaining it. Reality is not primary to explanation; it can only be said to be identical with the most satisfactory system of physical explanation. We shall discuss this matter again shortly.

*Anschaulichkeit*, picturesquencess, as Biggs wants it translated, is not a necessary attribute of explanatory symbols. The question concerning the role of time and space in modern theories has become quite acute, in particular their continuous, or discontinuous nature. It is clear at the outset that every system of explanatory symbols will involve time and space in some manner, for if it did not, it would not be intelligible, and a relation to actual experience could not be found. One sometimes hears the assertion that the modern quantum theory has proved time, or space, or both, to be discontinuous. This I wish to show to be a fallacy, at least an incorrect statement of the case. I know that I run the risk of being ousted from the association of fashionable physicists when I confess the heretical belief that time and space are not wholly facts of experience. I shall not go so far as to say with Kant that they are forms of intuition of pure reason, completely independent of being influenced by empirical recognitions. Nor can I follow Einstein and others in their rather uncritical acceptance of the dogma that they represent relations we have learned from nature. I hold that they contain an element which is uncontrolled by natural data, without which experience would be impossible.

It is certain that there must be traits in our knowledge which mark the peculiar organization of our mind, notions that would be different if our modes of comprehension were different from what they chance to be. As slight a change in our mental make up as a shift of the sensibility of the eye from light to X or $\gamma$-rays would occasion a complete change in our conception of nature; what would happen if our ways of reasoning were altered can not be foreseen. You will probably object that our mind is but a datum of nature. Yes indeed, but you will forever be unable to understand the possibility of experience unless you assume the existence of a mind prior to experience. There are other reasons, but this is the most pertinent one for assigning to the mind a position unique with regard to occurrences in nature. Time and space are among the elements whose form is conditioned by psychological predetermination. No experience can change their property of continuity—among others which do not interest us here. Try as hard as you like to imagine discontinuous time or space; you will find it as impossible as to visualize time of two dimensions, or space of four dimensions. Furthermore, there is no physical experience that forces us to abandon our conception of time and space. The theory of relativity simply suggests a change in the units employed to measure them, and a rearrangement of the order in which observable phenomena appear in time and space. That is perfectly compatible with our views. So is the suggestion that so-called "physical things" may not have a continued and identifiable existence in time and space. There is nothing in the concepts of time and space that would prescribe the behavior of phenomena with regard to them, although both carry with them inseparably the property of continuity. In fact, the very statement of the discontinuous existence of things implies a continuous background for judgment. Thus we find that this seemingly paradoxical and monstrous consequence of Heisenberg's theory is not to be rejected on general grounds, but is consistent with the requirements of a good explanation. However, this is no proof for its correctness, nor a reason why it should be advocated, and I should say that if the suggestion has no stronger a support than Heisenberg's much too famous uncertainty argument, it is interesting only because of its novelty.

A treatment of the nature of physical explanation is incomplete unless it deals with the function of probability and statistics. It is understood that until very recently the theory of statistics has been a powerful method of describing aggregate phenomena, used discriminately by careful investigators with a proper sense of its limitations. Of course Boltzmann introduced the postulate that the elementary motions of molecules are chaotic in order to provide a basis for the application of probabilities. But he regarded his assumption as an approximation which derived its validity from the existence of a large number of molecules. With this understanding, statistical concepts are symbols of description rather than of explanation, a concession of ignorance rather than an assertion of definite knowledge. Yet, why should probability relations not be incorporated in a system of explanation? There is nothing in our requirements about explanatory symbols that would brand them as objectionable. Clearly, it is the assumption on which they are founded. If the postulate of chaotic motion be detached from the condition that there are very many molecules, that is, if causal relations controlling the motion of a single molecule are dissolved and annihilated, then we are violating the principle of consistency of nature upon which the possibility of explanation rests. To be quite accurate, I should have said that this violation is incompatible with the theory of explanation which I have attempted to present. Whether or not an alternate theory can be constructed, independent of the axiom that nature is consistent, I am unable to say at present. This can, however, not be denied on *a priori* grounds.

With the reader's kind indulgence I shall conclude this discussion by a few remarks on the relation between explanation and reality. Reality has not yet been defined in this essay, nor anywhere, as far as I am aware, in a satisfactory manner. Every attempted definition makes an assumption about nature that is unwarranted by sufficient evidence. But we are still in the process of learning about nature, and that process will continue so long as there remains a science. Hence it will forever be hopeless to determine reality in terms of an existing nature of which we have no cognition. Moreover, the concept of reality has nothing absolute and unconditional about it, neither in its logical nor in its philological implications. A study of the various, sometimes conflicting, notions which the word reality has been used to convey throughout the recorded history of the race, is exceedingly interesting. It would show convincingly that the meaning of the term is variable, subject to arbitrary fixation. And this fixation must be performed with closest reference to our knowledge of nature. Since we can not define the latter in terms of the former, the argument seems cogent to me that reality must be the best system of explanation, the ultimately satisfactory one. Thus reality has been taken from the dusty store-house of logical categories and removed to the lofty heights of ideals. Its substance has vanished before our eyes and reappeared as the light of a shining star, guiding us in scientific labors. Reality does not exist as a discoverable something in the phenomena of nature, called into being by the single act of a designing spirit of the universe. We do not *discover* but continually create it in our efforts of designing *the* ultimately satisfactory system of explanation.

CHAPTER 2

# PROBABILITY AND CAUSALITY IN QUANTUM PHYSICS

In attempting to formulate the logical structure of quantum physics it is necessary to refer frequently to probabilities; even the basic axioms of this system of thought which progressively pervades the various branches of physical science involve an explicit appeal to probability concepts. Now it has been customary in mathematical and physical investigations to examine very critically and with utmost care all consequences of a set of postulates when such a set is given, but the postulates themselves, in particular the questions with regard to their consistency and their relation to other fundamental suppositions, often receive unduly little attention. The mystifying power of the term axiom, applied to any set of assumptions, whether basic or not, appears to insure its immunity against inquisitive attacks from the quarters of undesirable sceptics. It is only too natural that such a situation should exist, for one science always starts where another leaves off, and axioms mostly have their place in the boundary region between two, in a domain, hence, over which neither can claim complete jurisdiction and in which neither is perfectly competent to operate. Probability, upon which quantum mechanics is axiomatically founded, lies precisely at the triple junction of mathematics, physics, and philosophy, and partakes of the uncertainty and diffuseness that is likely to affect the concepts arising in such neglected, yet much disputed, boundary fields. The mathematician originally designed the theory of probability in a rather playful fashion, enjoyed his paradoxical position of predicting despite the absence of laws, and finally settled to the more serious business of molding it into an analytical scheme of hypothetical character, which, if you grant its premises, will permit you to make definite predictions. But within this science he found no criterion for the validity of these premises. The physicist, in turn, frequently unaware of the precise epistemological status of the probability scheme, applied its formalism to some of his problems and found it very useful. His chief concern was with the results, and little did he care about the origin of the analytical tool with which the mathematician had so kindly presented him. Meanwhile, mathematicians gained confidence in their cherished product, lost sight of its hypothetical nature and began to feel that physicists had really proved the probability calculus to be applicable to nature.

During this dvelopment the philosopher has by no means been an idle spectator. He has felt very keenly that his territory was being partly invaded and that his comment ought to be respected. Probability to him appeared to be largely a secondary issue, he viewed it as a manifestation of exact laws in nature whose precision was blurred by factors not within control of the observing individual. As its importance in the natural sciences increased, philosophers became more and more probability-conscious; they grasped the concept in its widest implications and analyzed its faintest shades of meaning. In this endeavor they became interested in various non-scientific ramifications and frequently lost touch with applications; their opinions became diversified, and finally physicists and mathematicians no longer heeded their remonstrant voices which were, in part at least, conflicting. It is mainly for this reason that philosophy has failed to imbue with its profoundly critical spirit the attitude of the scientist toward his own fundamental problems, that the endeavor of working physicists had been for the most part pragmatical and mathematical, particularly in its recent phases. But the circumstance most harmful to the logical structure of present physical thought, though extremely fortunate in another sense, has been its spectacular material success which not only caused philosophers to look with reverent awe upon the physicists' achievements, but impelled physicists to move on and on in quest of new discoveries, forgetting about possible flaws at the very bottom of their science.

One result of this development is the view, prevalent today among mathematical physicists, that all predications about nature are probability statements. Moreover it seems to be commonly understood that the probability concept which thus enters on the scene, may not be further analyzed, and that its validity excludes the working of exact natural laws. This state of affairs is usually summed up in the categorical remark that the causality principle can no longer be adhered to.

The author has attempted to show in a previous paper[1] that such a position, while it is not definitely self-contradictory, may not be regarded as one to which quantum physical reasoning inevitably leads; that it has considerable disadvantages from a metaphysical point of view, and that is is desirable, in view of current theories, to retain the causality postulate in its most general form (principle of consistency). The present note is designed to furnish additional evidence for this point of view, to clarify the manner in which probablity concepts and theorems are used in quantum dynamics. In one very important respect it goes considerably further than the paper quoted, for we shall attempt to demonstrate, by analyzing the very methods of quantum mechanics, that the entire formalism presupposes the existence of transcendent, continuously and uniquely varying states of physical systems, whose changes are precisely those which causality requires.

The term probability admits of a colloquial meaning which is never considered in any of the exact sciences, and will therefore be rejected at once. This is a sort of likelihood, based upon individual experience or judgment, referred to in phrases like "the probability that it will rain tomorrow," or "the probability that someone's statement is true." It is evident that no calculus is available to settle such matters, that they rely wholly on subjective estimation. To be sure, each of the two phrases might be given a statistical sense which would render the use of "probability" scientifically accurate; for instance, in the first example one could think of a certain locality and inquire, without placing any particular emphasis upon the day tomorrow, what the chances of rainfall are. Or in the second, the person in question may have made a great number of statements which were either true or false, so that the chance of any one of his statements being true can be determined. In these instances and in all others, scientific meaning is attached to the concept of probability when it involves the possibility of recurrence for the event in question. This recurrence may be either in time, as in the examples just cited, or in space, as in physical problems of statistical ensembles, where one system is repeated many times in space. This fact excludes from scientific consideration all probability theories which are not frequency theories. Hence, although we do not wish to detract from the general philosophical value of a priori probability conceptions, we find that they are without application in all strictly scientific schemes.

The subsequent discussions will be seen to follow very closely the important work of von Mises,[2] both as regards terminology and general outlook. It is felt to be out of place here to present the logical reasons for accepting his views; we merely desire to state that his doctrine of the foundation of the probability calculus appears not only to be most defensible and clear cut, but also to link up quite directly with the physicist's mode of thought. It is desirable at the beginning to fix the terminology to be used. The probability calculus is applied to a large number of observations, each of which will be called an *element*. The total sequence of elements is designated as a *probability aggregate*, a term meant to be synonymous with von Mises' *Kollektiv* (first used, but with a somewhat different connotation, by Th. Fechner). It is clear that not every set of observations is of a character which justifies the application of probability arguments, in other words, a probability aggregate must satisfy very definite conditions, which will be mentioned later. Every element is regarded as having one of a limited number of mutually exclusive *properties* that can be observed. If $n_i$ is the number of times which the $i^{th}$ property is observed and $n$ the total number of observations or elements, then the quotient $n_i/n$ is the *relative frequency* of the $i^{th}$ property. Supposing finally that $n$ becomes very large, i.e. that the probability aggregate contains an infinite number of elements, in which state it will be considered as an ideal one, then the limit of the quotient $n_i/n$ will be defined as the *probability* of the $i^{th}$ property. It is clear that $\Sigma_{i=1}^{s} n_i/n = \Sigma_{i=1}^{s} \lim_{n\to\infty} n_i/n = 1$ For a given aggregate, the limit of $n_i/n$ depends only upon the index $i$ and may be regarded as a function $w(i)$. The totality of $w(i)$'s is known as the *distribution* of the particular probability aggregate.

As an illustration one may refer to the aggregate formed by a large number of throws of one die and the corresponding observations of the numbers which appear on the upper face. Each observation subsequent to a throw is an element; there are 6 properties, the appearances of the numbers 1 . . . . 6. The probabilities in this case are known to be equal to 1/6 for every property $i$, so that the distribution consists of 6 numbers $w(i)$ which are all equal.

In general it is not necessary, of course, that the number of properties be finite; examples, such as the determination of the coordinates of the die after its throw, where the number of properties is not finite and the distribution continuous, are easily adduced. In these cases no essential modification of the definition is required, except that summations are to be replaced by integrations. It is to be remembered, however, that in a strict sense no physical observation is continuous, for every measurement is performed by means of an apparatus which yields as the value of any determinable property an integral multiple of some fundamental quantity; but this recognition is by no means new and has no bearing upon the postulates of quantum theory which definitely admits continuous distributions. If the discrete character of measurements is felt by some physicists to call for a theory which takes adequate account of this discreteness, the recent mathematical developments do not satisfy their desires.

A difficulty which has caused some mathematicians to reject the foundation on which the present notions are built is involved in the definition of probability. It has been shown that the limit of $n_i/n$, taken in a mathematical sense, does not exist.[3] For no matter how large the number $n$ becomes, the probability calculus furnishes a finite probability that the quotient $n_i/n$ be different from $w(i)$ by more than a preassigned small amount. This fact is contradictory to the rigorous mathematical concept of a limit. The difficulty connected with it appears to have been satisfactorily removed by Hohenemser,[4] who substitutes for the relation $\lim_{n\to\infty} n_i/n = w(i)$ another one which avoids the transition to the limit but allows in a somewhat modified sense all operations which derive their justification from von Mises' definition. For the details of this analysis we wish to refer to the paper quoted.

The probability calculus is incompetent to provide a priori information regarding the distribution in a given aggregate. The fact, for instance, that in the game of dice the probability of throwing a 5 is equal to that of throwing a 2, or that either is 1/6, could not have been deduced by arguments peculiar to this calculus. The distribution depends in all cases on extraneous, here physical, conditions, as is seen from the circumstance that it will change if the die is loaded. Hence the probability theorist must content himself with an experimental determination of his distribution by proceeding to observe throws for a very long time, and all concepts involved in his dealings must be based on these a posteriori data alone. It may be, however, for reasons not implied in the calculus, that there exists an extensive correlation between a given aggregate and the character of its properties, a correlation which is conditioned by the constitution of nature and hence is equivalent to, and of the type of, a natural law. In the example of the die a connection between probabilities and the physical properties of the object on which observations are made is at once seen to exist, but it is not sufficiently typical to be generalized. If the coordinates of the die after any throw were measured, the resulting probability distribution would again show a definite dependence on physical properties, for instance the nature and shape of the surface on which it is thrown. But the dependence is of a different kind. It is mainly for this diversity of dependence on extraneous conditions that the attempts of founding probabilities upon anything but relative frequencies have failed. This point is of considerable importance, for it will soon be shown that quantum mechanics is unique in the sense that it does assume a *general*, though very abstract correlation between the state of a system and the various probability aggregates generated by observations on its measurable qualities.

The only aim which the probability calculus can properly achieve is to produce the distribution of a derived aggregate when that of the primary one is known. Given, for instance, that all $w(i)$'s in our previous example of throwing one die are equal to 1/6; what is the probability of throwing any particular number between 2 and 12 with two similar dice, is a question which the calculus can legitimately answer.

There is a definite group of ways to form a derived aggregate by combining primary aggregates, or by changing the elements or properties of a given one. To each of these changes there corresponds an equally definite operation for finding the distribution. However, these normal operations of the probability calculus are of very limited occurrence in quantum mechanics and hardly justify detailed discussion. More important is the construction of mean values. To illustrate, let there be attached to every property $i$ (the number from 1 to 6) in the game with one die a certain gain $f(i)$. Then the initial aggregate whose properties were numbers has been changed to one of gains, but with the same probability distribution. Now a typical question with which the calculus is often concerned, regards the expectation of gain. The answer is evident; the expectation, or as it is usually called the *mean value* of the gain, is $\Sigma_i f(i).w(i)$.

This concept is capable of generalization. Let us introduce for the expected mean the symbol $\bar{f}$. Then, as was just stated, $\bar{f} = \Sigma_i f(i).w(i)$. If the aggregate in question has a continuous distribution and the symbol $x$ is used for the continuous sequence of properties, this relation is to be modified into $\bar{f} = \int f(x)w(x)dx$, the integration being extended over the entire range of $x$. Finally we may drop the explicit reference to the gain in our example and consider $f(x)$ merely as a function of $x$ which is defined for any value in the range of $x$. Then $\bar{f}$, constructed by the rule just given, will be the expected mean of this function with respect to the probability distribution $w(x)$.

These discussions place us in a position from which the symbolism of quantum mechanics can be conveniently surveyed. The exposition to follow is not entirely conventional, it differs in its starting point from the usual treatment and shows some of the newly developed concepts in another setting. The analytical consequences of our point of view must, of course, be identical with the well established current theories.

As in classical science, we define a physical system as an object upon which measurements can be performed, but contrary to classical convention, we do *not* define the state of a physical system by the results of possible measurements alone. This has often been attempted. A well conceived and logically clear pursuit for this type was the famous work of Heisenberg, Born, and Jordan, which led to the destruction of its own premises in the transformation theory. For this analysis, as well as all others, introduced eventually a continuous function akin to, but somewhat more general than, Schroedinger's $\psi$-function, which is utterly foreign to measurements. To minimize the annoyance of such undesirable elements they were given a name that would make them appear entirely sub-ordinate to measurements, and would assign to them the purely analytical rôle of intermediaries between observations; they were called transformation functions. But anyone who is at all familiar with the methods of quantum mechanical calculation knows that this name has not done away with their importance. Hence we shall dignify them by regarding them as significant, in fact the $\psi$-function is taken to be the abstract definition of a state.

Postponing for a moment the question as to how these $\psi$-functions are to be determined, which is a mathematical one, we find ourselves confronted with a very unaccustomed situation: the state of a physical system is an entity removed from the realm of measurements. To be sure, it will be shown to govern all possible measurements on the system, but it is not to be expressed in terms of them. It becomes necessary to separate clearly the realm of measurements from the realm of states of physical things. The latter is an absolute domain, in which description remains highly abstract and relations can not be visualized. To it we shall now confine our attention.

Instead of describing a state by the positions and velocities of hypothetical parts of the system in question, as was customary in classical physics, it will now be characterized by a function of all space co-ordinates and the time. Specific reference to parts of the system is abandoned. The changes which occur are changes in this function and are not intelligibly related to changes in the intuitive parts of the system. Any objections against the abstractness of this scheme are overruled if this mode of description proves to be of value in making physical predictions. Not every function of space and time co-ordinates may be chosen as a $\psi$-function; certain mathematical restrictions naturally arise from the rules for finding them.

To continue the discussion it is necessary to introduce the notion of an operator, which will take us partly into the realm of experience. Following Dirac, we shall call quantities capable of physical measurement (position, momentum, energy, etc.) observables. In measurements, they simply appear as numbers. With each observable in the abstract sense will be associated a mathematical operation (multiplication by a variable, taking a derivative, taking the Laplacian, etc.), and this operation is symbolized by an operator. If, for instance, $P\psi$ is written, this implies the result of performing the operation defined as $P$ upon the function $\psi$. There are, at present at least, no unique rules for determining the form of an operator corresponding to a given observable: this has to be ascertained by repeated trials, just as in classical physics the determination of the proper function for kinetic energy, or momentum, was a matter of experience. The forms of the most important physical operators, however, such as energy, momentum, angular momentum, position, are now well known.

Retaining the use of the symbol $P$ to denote an arbitrary operator, and designating by $p$ a number, the equation

(1) $\qquad P\psi = p\psi$

makes mathematical sense if it is understood that $\psi$ has as its arguments those co-ordinates upon which $P$ operates. It is an ordinary or a differential equation for which solutions can be found. But, in general, solutions do not exist for any value of the numerical parameter $p$, if these solutions are to satisfy certain simple conditions not of interest at present. There will then be a set of $p$'s, denumerable or non-denumerable, such that to every $p_i$ there corresponds a solution $\psi_i$. The $p_i$'s are known as the eigenvalues of the operator $P$, the $\psi_i$'s form a special set of $\psi$-functions, each characterizing a particular state of the system under consideration. This system, it is true, is nowhere specified, but its physical character enters into the determination of the operator $P$ and thereby makes this scheme sufficiently definite. Thus we have arrived at a set of $\psi_i$'s, from which any other $\psi$ can be constructed, for it is a well known mathematical fact that the solutions of (1) form a complete set of functions, that is, any function (with due limitations) can be written as a linear combination of the $\psi_i$'s, thus:

(2) $\qquad \psi = \Sigma_i a_i \psi_i$

where the $a_i$'s are numerical coefficients.[5] Let us refer to the states defined by the various $\psi_i$'s as "pure cases with respect to the operator $P$," for the present only as a matter of nomenclature. Every operator $P$ will then generate a set of pure cases or states, which, so far, have only mathematical meaning. (The term "pure case" was introduced by H. Weyl, and will later appear to be a most suitable one.) Relation (2) may then be expressed by saying that any state of a given system, described abstractly by a function $\psi$, may be considered as a superposition of pure cases with respect to any operator $P$.

So far we have not proceeded beyond our starting point, which was fixed by the postulation that states, in themselves, are described independently of measurements by continuous functions. It is now necessary to construct a bridge with experience. The act of acquisition of knowledge involves an interaction of the subject and the object of experience which can not be accurately described in mathematical terms. We are not referring in this connection to the physical disturbance which an apparatus used for measuring will produce in the state of the system if the latter were conceived in a classical way, but of the much more fundamental interplay of knowing and being. The former hybrid conception, which intermingles classical and quantum mechanical elements of thought, is responsible for a great deal of confusion in current ideas. As a consequence of this interplay it will be supposed that experience does not partake of the completely determined character of states, or $\psi$-functions. Instead, the realm of measurements consists of various *probability aggregates*.

To every operator $P$ there corresponds one probability aggregate whose elements are successive observations of the observable to which $P$ belongs, and the eigenvalues of $P$, as given by (1), will by hypothesis—subsequently justified by experience—be taken to be its properties. This implies that a value of an observable does not exist unless it is one of the set $p_i$. Now our remarks regarding states would be of little merit if they did not admit of a connection between the $\psi$-function and the experimental probability aggregates, since states in themselves were said to be unobservable. Indeed it will be shown that the $\psi$-function regulates the probabilities within every possible aggregate. This means that, when the state of a physical system (its $\psi$-function) is given, the probability distributions of all possible measurements are at once determined. The rule for finding them is simply this: When a system is in the state defined by $\psi$, then the expected mean of the observable corresponding to the operator $P$ is

$$(3) \quad \bar{p} = \int \psi^* P \psi \, d\tau$$

where the integration is to be performed over the range of all coordinates of which $\psi$ is a function.

There is no possibility of deriving this relation on a priori grounds. It is a supposition to be made and tested. But it is at the basis of all quantum mechanical analysis and allows this entire mathematical system to be developed if it is combined with the notion that the properties of the aggregate of measurements are the eigenvalues of the corresponding operator. If the system is in a state $\psi$, let the function be expanded according to (2). Then it is seen that

$$\bar{p} = \int \psi^* P \psi \, d\tau = \int \sum_i a_i^* \psi_i^* P \sum_j a_j \psi_j \, d\tau = \sum_{ij} a_i^* a_j \int \psi_i^* P \psi_j \, d\tau = \sum_i a_i^* a_i p_i$$

This result shows that the definition (3) is well chosen since it has the form of an expected mean as previously given; but it also offers additional information. If the numbers $p_i$ are the properties of the $P$-aggregate and $p$ their expected mean, then the quantities $a_i^* a_i$ or $|a_i|^2$ must be probabilities. Hence $|a_i|^2$ is the probability of measuring $p_i$ when the system is in the state $\psi = \sum_i a_i \psi_i$. Occasionally a statement to this effect is claimed to be the axiomatic foundation of quantum mechanics, or it is asserted that quantities like $|\int \psi_i^* \psi \, d\tau|^2$ determine these probabilities. It is easily verified that the last statement is an analytic consequence of the former, which we have shown to follow from the mean value relation (3). We are therefore inclined to regard (3) as more fundamental and prefer to start with it.

To determine the meaning of the set $\psi_i$ which appeared to play a special part in our analysis, one may suppose the system to exist in one of the states $\psi_i$. Then, by (3),

$$\bar{p} = \int \psi_i^* P \psi_i \, d\tau = p_i \int \psi_i^* \psi_i \, d\tau = p_i$$

There results the interesting fact that the mean value within the $P$-aggregate is one definite property $p_i$ appearing with a probability 1. This in turn means that a measurement will yield the value $p_i$ with certainty, the probability aggregate has degenerated to a single property. The adequacy of the term "pure case with respect to the operator $P$" is now apparent. Whenever the state of a system represents a pure case, that is whenever the $\psi$-function defining it is the $i^{th}$ eigenfunction of a certain operator, a measurement of the observable corresponding to this operator will yield with certainty its $i^{th}$ eigenvalue.

If relation (3) is to be useful it must be possible to evaluate it, the integral must exist. This consideration furnishes the conditions which any $\psi$-function must satisfy: If it is combined with *any* physical operator $Q$ in the manner $\psi^* Q \psi$, this combination must integrable over the complete range of all the arguments of $\psi$. From the point of view here taken this condition appears very natural and simple, a fact not unworthy of note in view of the numerous conflicting proposals regarding the "boundary conditions" of $\psi$, etc. In particular it is seen to be beside the point and insufficient if it be postulated that $\psi$ as well as its first derivatives must vanish at infinity, as has sometimes been done. It is important to observe, also, that equation (2) places no further restriction on the $\psi$-function, especially that it need not be related in any manner to a set of pure states. Indeed the variety of states in quantum mechanics, which is identical with the variety of functions satisfying our conditions, is much larger than that of the possible combinations of positions and velocities which define a state in classical physics.

Let us now pause for a moment and review the logical procedure up to this point, which in the foregoing considerations was of necessity encumbered with mathematical details, often beclouding the view. We have started by drawing a clean line between the domain of being and that of experience, showing that one is describable by continuous functions and the other by probability aggregates. The continuous functions are *not* intuitively suggestive of the physical properties of "parts" of the system, indeed the concept of "parts" is the result of an unfortunate retention of classical habits of speech. Our knowledge of $\psi$-functions is purely formal and governed by mathematical rules. The probability aggregates, on the other hand, are directly defined in terms of experience and regulated by an hypothesis containing a priori elements and regarding the selection of possible properties of the aggregates. In a sense, quantum mechanics confronts us with a form of transcendental idealism separating sharply the realm of experience from that of things, but instead of renouncing all facilities for dealing with "things in themselves" in conformity with the Kantian doctrine, it allows us to operate with states in an abstract manner. The analogy to which we are alluding is a fascinating one, and appears capable of enlarging the philosophical aspect of modern theories. However, we are too poorly equipped at present, on account of the vagueness in some of the quantum mechanical concepts, to pursue it fruitfully. We hope to return to it at another occasion.—The correspondence between the domain of being and that of experience is formed by a relation which fixes for any state of being all probability distributions within the aggregates in which that state may manifest itself. This is our mean value relation (3).

At this point we must not forego answering those who insist that the separation of the two domains here emphasized is illusory or at least unnecessary, since $\psi$-functions have no physical meaning that may be tested by itself. It is argued that the description of states in terms of continuous functions is nothing more than a convenience, that the $\psi$-function is merely a suitable bag in which to carry all the information supplied by measurements. This view, however, is definitely fallacious for it is impossible to construct a $\psi$-function out of experimental data, no matter how numerous they may be. One can not start with a number of measurements and arrive at a $\psi$-function except by an extrapolation which surpasses in arbitrariness by far every idealization which is necessary, for instance, in order to pass from experimental data to the statement of a physical law. It is true that the only intelligible use to which $\psi$-functions can be put relates them to measurements, but it is equally true that finite measurements never suffice to determine their form, or to explore all their implications. But the main argument against the thesis which we are proposing to discard does not even touch the validity of these considerations. Even if it were possible to proceed in a unique manner from measurements to an abstract description of states, the fact that the reverse process is feasible would demand attention. It was shown that, when the symbolic representation of a state (any function satisfying certain conditions) is given, the outcome of all possible measurements is at once determined in a statistical way. This reciprocity would place states and measurements— $\psi$-functions and probability aggregates—at least on an equal footing. These remarks lose none of their weight if we admit, as we must, that in preceding paragraphs, where the general status of $\psi$-functions was discussed, frequent reference to measurements was made. The attitude behind the objection in question is easily seen to be governed by the tendency, widely shared at present, of defining all concepts in terms of experimental operations. This has been dealt with elsewhere,[1] and we shall here dismiss it, except to repeat that it is utterly destructive to the notion of a science and annuls itself it carried through in a logical manner. An appeal to reality, a very problematical issue which receives meaning only by statutory definition, can not be made in this connection since it is no fixed basis on which a science can stand, but rather a result of scientific investigation. The most consistent view at present is to regard abstract states, and the functions which are their representations, unreservedly as *real*.

It has been stated that these functions depend in general on space coordinates and the time, and when in the course of this analysis there occurred an integration over the coordinates there has been no explicit committal as to whether the time was to be included in the integration or not. This attitude was mainly one of caution, since the question involved is not yet analytically clear. Very probably in the final form of the quantum theory time and space coordinates will appear in a symmetrical manner, certainly in order to satisfy the requirements of the relativity theory if for no other reason. To be in keeping with current practice the integrations in (3) and the subsequent expressions should be understood to be performed only over the space coordinates, but if it is thought desirable for the sake of symmetry, as we feel it is, one may well integrate over a period of time without introducing essential changes in the results (provided that the $\psi$-functions be properly normalized with respect to time). We are far from believing, however, that this would remedy the weaknesses of existing theories. The ultimately satisfactory manner of dealing with the time dependence of $\psi$-functions will involve more radical refinements. The success of the present methods in quantum mechanics in spite of these defects is of course to be understood. In most problems of interest physicists are confining their attention to the state of a system at a given instant of time. These are then expanded by superposition of pure cases with respect to the energy operator, and temporal changes are afterwards treated by means of a perturbation calculus, for instance Dirac's method of "variation of constants." Such methods do not yield an explicit expression of the $\psi$-function which involves the time as an argument except in the manner peculiar to stationary energy states (where it disappears when $\psi^*\psi$ is formed). Nevertheless, although the usual treatment focusses attention mainly upon the spatial part of the description of states, their change in time is governed by a law which is well known.

This brings us to a point of great importance. The law in question is a differential equation:

$$H\psi = \frac{h}{2\pi i} \frac{\partial}{\partial t} \psi$$

in which $H$ is the energy operator, $t$ the time, and the remaining quantities, except $\psi$, numbers. One sees that as long as $\psi$ itself is finite it can undergo only *continuous* changes in time. Also, if the complete form of $\psi$, as a function of space co-ordinates and time, is given for any instant, the $\psi$-function for all future times may be calculated, at least in principle. This is tantamount to saying that in quantum theory *we describe stares in causal manner*. The assertion that systems "jump" from one discrete state to another without passing through intermediate stages is wholly due to a confusion of concepts. To the continuous changes of the states there correspond continuous changes in the probability distributions of the properties of every aggregate of measurements, changes which are perfectly compatible with the circumstance that the properties themselves may remain discrete and that no intermediate values between these properties can be measured. We wish to uphold this statement also with regard to the changes caused by measurements. It is frequently supposed that, when a measurement is performed on a system which exists, with respect to the observable to be measured, in a state describable only by a superposition of pure cases, the state will be discontinuously changed into one of the pure states with respect to the observable. This feature is sometimes regarded as one of the inexplicable anomalies of quantum physics, endowing measurements with a very mysterious character. The matter is cleared up, and seen to be a consequence of previous considerations, if it is remembered that every measurement of an observable requires and involves an *isolation* with respect to that observable. But an isolated system will, *without external influence*, transorm itself to one of the prue cases, so that, if the measurement is merely an isolation without further disturbance, a pure case with respect to the operator in question must result. Since this point is often overlooked we desire to give a physical example. Starting with an atom in its normal state, one may first irradiate it and then attempt to measure its energy. After the radiation has been allowed to interact with the atom for a small time the latter will no longer represent a pure case; its $\psi$-function will be a linear aggregate of its stationary energy states. Now it may easily be shown that, as long as this "impure" state exists, the energy operator involves the time, for otherwise Schroedinger's equation would have a stationary solution. This merely means the existence of energy interactions, or lack of isolation. As soon as these interactions are removed, the energy operator becomes constant in time and the atom will very naturally settle into one of its stationary states. A measurement of the energy in this case might be made by stopping the radiation and determining the frequency of the spectral line emitted by the atom; at any rate it is essential that the perturbing radiation be screened off at the instant of observation in order to make the state definite. It is this isolation which brings about a rapid but continuous establishment of a stationary state, and it is by no means surprising that a measurement *may* leave a system in a state characterized as a pure case.[6]

The form of relation (3) offers a strong temptation to interpret $\psi$-functions in a more concrete way. The integral $\int \psi^* P \psi \, d\tau$ might itself be regarded as a formal mean of the operator $P$, independently of the analysis which leads to $\Sigma_i |a_i|^2 p_i$. $\psi^* \psi$ would then take on the meaning of a continuous probability distribution in space and time and indicate, for every set of arguments of $\psi$, the corresponding chance for measuring the observable belonging to $P$. But one will readily convince himself that such a simple interpretation is not generally possible. For whenever the operator $P$ involves a differentiation (more properly, does not commute with $\psi$) this supposition becomes meaningles, unless one is willing to extend suitably the definition a mean. Such an extension, however, appears rather speculative and may even be misleading. To be sure, there are simple operators, like the "charge operator" (multiplication by the total charge $q$) and the "co-ordinate operator" (multiplication by coordinate $x$), for which this interpretation happens to be permissible. The usefulness of calculating mean "dipole moments" by such a scheme has been recognized in some of the earliest papers on the subject and has for a time led to the view that $q \psi^* \psi$ was to be considered as a virtual charge density. But it is now generally conceded that the meaning of the $\psi$-function is not to be confined to such specific formulations, and that they receive their justification only from the larger context in which a more abstract formalism has placed them.

The term "pure case," when it occurred in this discussion, has always been qualified by reference to a certain operator or observable. Let us inquire as to the possibility of a pure case in an absolute sense, that is one with respect to all possible observables. In classical physics, where the complete description of intuitive parts of a system entered into the picture of a state, there was simultaneous certainty with regard to all measurable properties at every instant. Quantum mechanics does not permit this situation. Let us determine whether a pure case with respect to the operator $P$ is also a pure case with respect to the operator $Q$. The state with which we start is then $\psi_i$, a solution of (1). If now the integral $\int \psi_i^* Q \psi_i \, d\tau$ results in a degenerate probability aggregate for the observable $q$, that is if only one $q_i$ appears and with the probability 1, we shall know that the state represents a pure case with respect to $q$; if, however, there appears a "spectrum" of probabilities, this is not true. We must, therefore, expand $\psi_i$ in terms of the set of $\varphi_i$'s, defined as solutions of $Q\varphi_i = q_i\varphi_i$, thus: $\psi_i = \Sigma_j b_j \psi_j$, and insert this expression in the integral, which will yield, in general, $|b_i|^2 q_i$. This tells us that there cannot be simultaneous degeneracy of the $p$- and $q$-aggregates, in other words, that a state cannot at once be a pure case for both observables. Only when all $b_i$'s vanish except one, so that $\psi_i = \varphi_i$, is this possible. In that case it may be shown that $P Q \psi = Q P \psi$, a fact which is expressed technically by saying that the two operators commute. Now, as it happens, certain pairs of operators, belonging to so-called canonically conjugate observables, do not commute, a fact which implies that there cannot be simultaneous certainty with regard to the outcome of measurements on both observables of a pair. This is the meaning of the famous uncertainty principle which has been illustrated almost ad nauseam by thought experiments whose main purpose seems to be to obscure the issue. Not infrequently they involve legendary apparatus and prove trivialities by achieving the impossible, in spite of their authors' insistence upon close touch with experience. Their usual aim is to show that the measurement of one observable, by direct physical interference of the measuring apparatus with the state of a system *conveived classically* produces an uncertainty in other observables of an order of magnitude expressible in terms of a universal constant.[7] Such a demonstration, in this author's opinion, proves nothing; for the existence of such a disturbance has never been doubted in classical physics, and its relation to the universal constant is invariably brought into the discussion by arguments not pertaining to the thought experiment itself and always presupposing the validity of some quantum theoretical postulates. Moreover, they place the emphasis upon measurements, where it does not properly belong, and sometimes allow measurements to be considered as the only cause of uncertainty. The foregoing discussions have shown that the type of uncertainty peculiar to quantum mechanical predictions, while always manifesting itself in measurements, does not necessarily depend on the physical interference of a measuring device, that it is inherently conditioned by the state of a system, i.e. by its $\psi$-function, and that it is present as an abstract situation whether an experiment is performed or not. For instance, when it is known that the function $\psi_i$ describes the state, then it is definitely settled that a measurement of $p$ will give the only possible result $p_i$, and that the observation of a canonically conjugate observable $q$ can only be predicted in terms of a probability distribution, independently of whether the observation of $p$ has actually been carried out or not. To be sure, the existence of pure case $\psi_i$ may be the manner of ascertaining a distribution, may now be looked upon as an a result of a previous measurement of $p$, or it may not. The question: how is one to know that this state is present, except by measuring it? is easily answered, for we may have found, by preliminary experimentation, a method of producing the state and may now rely on a principle of uniformity in predicting its existence.—Uncertainty has its origin in the dual manner of describing physical entities; it appears when we descend from the causal regularity of states to the domain of probabilities connecting the elements of experience.

The relation between causality and probability in quantum physics is not one of antithesis or mutual exclusion, but of coordination. In particular, the causality principle has not been abandoned, since it governs the behavior of states. This harmonious juxtaposition avoids all the difficulties with which the postulation of an autocratic and exclusive rule of probability relations in nature is beset. The precariousness of such a doctrine was exposed in another paper,[1] where the implications of the so-called probability postulate were analyzed.[8] If we claim that experience presents us with a probability aggregate comprising all possible observations, without at the same time invoking the principle of consistency of nature (causality), the question as to the meaning of that assertion is highly perplexing. For it involves an examination of the conditions under which a set of data is a probability aggregate. These contain one known as invariability of the aggregate with regard to arbitrary selection rules, a condition extremely difficult to establish except with the aid of a principle similar to that of causality. The fact that many physicists are willing to content themselves with the blunt assumption that probability governs the world can be explained only as the result of an insufficient analysis of these matters. As stated in the introduction, probabilities were introduced into quantum physics in an axiomatic way, which produced the feeling among physicists that they were relieved of any obligation of critical analysis with regard to its non-physical meaning.

The new situation created by the quantum theory may be viewed from another angle. It was pointed out and exemplified that the probability calculus establishes its distributions by performing a great number of observations on an aggregate, because no general correlation between the physical nature of a system and the probability distributions to which it gives rise is available. With an important reservation, quantum mechanics can be considered as having provided a correlation of this type through its mathematical nexus between states and probability distributions. In so far as the $\psi$-function is characteristic of a state, and hence certainly of the system itself, probabilities are regulated by the abstract properties of that system, and what appeared, from the standpoint of the probability theorist, as the primary manner of ascertaining a distribution, may now be looked upon as an a posteriori verification of consequences flowing from the character of things. But the reservation in question is this: it is incorrect to think that through the ingression of a formal element into the probability calculus this discipline has lost its purely empirical status. For by starting with a $\psi$-function one can only arrive at hypothetical distributions, that is, distributions which exist if the state belonging to that $\psi$-function is present. If this limitation were removed it would be possible to construct the elements of experience out of a priori data, which is certainly an untenable position.

In one sense, however—and this point we wish to emphasize in concluding—the introduction of quantum mechanical states has extended the regularity of experience. If we were bound to a *completely empirical determination* of the constituents of our physical knowledge, it would be necessary, with the present probability scheme, to observe by means of a great number of measurements the probability distribution of all observables. The $\psi$-function, together with the interpretation of relation (3), makes this process superfluous. For its form is determined if only the distribution of the properties in one aggregate is empirically known, since we can then certainly expand it by superposition of pure cases with respect to the observable in question. Its expansion in terms of pure states of other observables is then merely a matter of analysis, and relation (3) tells us, of course in a statistical way, the outcome of measurements on all other observables.

It is our hope that the present discussion will serve in some small measure to clarify the exact logical and metaphysical position of recent physical theories. The author is aware that some of his developments do not strictly coincide with current ideas, which are divergent among themselves, but feels that his intended departures improve either logical completeness or coherence. He also believes he has exposed the inadequacy of the supposition that causality has vanished from the physical description of things. It has merely receded, in part, from the realm of physical experience.

## NOTES

[1] *The Monist*, XLI, 1, 1931.
[2] *Wahrscheinlichkeitsrechnung und ihre Anwendung in der Statistik und theoretischen Physik*. Franz Deuticke, 1931.
[3] *Cf.* Sternberg, *Angewandte Math. u. Mech.*, 9 (1929), 501.
[4] In *Naturwissenschaften*, 19 (1931), 833.
[5] The functions $\psi_i$ are completely determined by (1) except for a constant multiplier. This will always be so chosen that $\int \psi_i^* \psi_i d\tau = 1$. The integration here is extended over the

complete range of all arguments of $\psi_i$, and $\psi_i^*$ represents the complex conjugate of $\psi_i$. Moreover, every $\psi$ function will be supposed to be "normalized" in this manner. The $\psi_i$'s are also taken to be orthogonal, i.e. $\int \psi_i \psi_j d\tau = 0$ if $i \neq j$. This property either follows at once from (1), or can be produced by choosing a linear combination of solutions of (1).

[6] My use of the term "pure case" in this early publication differs from that of Von Neumann in his famous *Mathematical Grundlagen der Quantemechanik*, Springer, 1932.

[7] The use of the term "uncertainty," or any one of its synonyms which have become current, is in a sense deplorable. In our phraseology it is to designate a *limited range of properties* over which the probability distribution is known. The term uncertainty is particularly unfortunate, because it appears to suggest to many minds the entire absence of knowledge regarding the distribution within the range, a totally erroneous supposition. Moreover, it is necessary to define the *limits* of the range by some (arbitrary) mathematical criterion in order to introduce iniformity into the discussion. Heisenberg fixes the range in terms of a constant appearing in the error function; another feasible way would be to take as the range that part of the total domain of properties within which the probability is greater than a certain amount $\epsilon$. There are many other possibilities. It is easy to show that the constant $c$ is in the relation $\Delta p . \Delta q = ch$ depends on the definition of range, or, popularly, the definition of "uncertainty."

[8] An inconsequential modification of a statement there made becomes necessary if we adhere to Hohenemser's probability definition; the number of elements in a probability aggregate need then not be finite. But the condition of invariability of probabilities with regard to "blind" selections of elements must still be imposed.

CHAPTER 3

# MEANING AND SCIENTIFIC STATUS OF CAUSALITY

The disagreement with regard to the validity of the principle of causality, existing to-day among scientists, has its roots in the diversity of definitions of the principle itself rather than in a problematic scientific situation.[1] As far as the formulation of quantum theory is complete its bearing upon philosophical questions can be fixed with precision provided the questions are phrased intelligibly. But a question is intelligible from a scientific point of view only if it satisfies two conditions: (1) the meaning of its terms must be fixed; (2) it must be in accord with the conventions of the science to which the question is put.

The necessity of the first requirement is at once once evident; if it were not satisfied the question would have several correct but self-conflicting answers, such as those to which the undisciplined discussion of philosophical problems usually leads. The second requirement, however, reflects a particular weakness of philosophy. Within the domain of the latter, words have retained a variability of meaning which, to be sure, makes for beauty and flexibility of expression, but impairs precision of speech. So persistent was the tendency toward figurative flourish that even in cases where science had standardized the meaning of a term in a very systematic and useful way philosophers continued to use the term with its former diffuseness. When accused of this procedure they said they were speaking non-technically and deplored politely the scientist's manner of degrading terms into technical ones. Indeed the difference between philosophy and the specialized sciences began when the latter caused their concepts to crystallize and agreed to name them universally and with care. Even now the distinction between science and philosophy is best described in saying that science proceeds by making the acceptance of its terms obligatory for all its pursuants, while philosophy allows its advocates to coin largely their own phraseology. This individualism frequently causes confusion and indefiniteness of philosophical attitude. To illustrate: whether causality is a category or not, is chiefly a matter of definition and nomenclature and may be answered correctly by yes and no, in fact the question may be meaningless; proponents for each of these three answers are to be found among modern philosophers. However, the question: is mercury an element has but one correct answer. Similarly, the term energy which has a perfectly definite scientific meaning is constantly used in phrases such as "mental energy" which signifies nothing unless ignorance of the laws of physics on the part of the speaker. Science has avoided such ambiguities. Hence we can effectively guard against them by using technical lingo wherever it is possible. In discussing causality it is necessary to formulate the problem in terms of physics as far as they are available.

Considerations like these are common with all scientists and many philosophers today; they constitute no unsympathetic critique of philosophy, or of the methods employed in philosophical investigations. For the solution of numerous problems the methods of science, which, speaking figuratively, is nothing but the crystallized part of philosophy, are not, or at least not yet, at hand. It is also true that the philosophical terminology is more strongly subjected to popular misuse than scientific language, and therefore prevented from being standardized. None of these arguments is sufficient, however, to justify the use of vague philosophical phrases where definite scientific terms present themselves.

Kant's formulations of the causality principle are to be rejected mainly for these reasons. When he says in the first edition of his "Critique of Pure Reason:" "Alles, was geschieht (anhebt zu sein), setzt etwas voraus, worauf es nach einer Regel folgt," this statement is not scientifically clear. "Regel" is entirely undefined; it is possible to find a rule for everything that is susceptible of description. On the other hand, if the word is to be interpreted as a means of knowing subsequent events in advance, there arises the difficulty of who is to know and employ the rule, together with all the other inconsistencies which will be encountered shortly in connection with similar formulations. Another possibility of interpretation places the emphasis upon the first part of the sentence and neglects the last as an inessential explanatory phrase. But then, if "voraussetzen" is taken in its temporal sense, Kant's statement amounts to nothing more than the assertion that the universe has no beginning in time, which is plainly not identical with the causality postulate.

Kant's modified formulation, as it appeared in the second edition of the same book, reads: "Alle Veränderungen geschehen nach dem Gesetz der Verknüpfung von Ursache und Wirkung," and is subject ot the same criticism. Physics knows of no such law; as a matter of fact there is no plausible way of defining cause and effect. No laws of physics, if properly stated, involve reference to either of these concepts, and if the distinction of cause and effect is artificially impressed upon the phenomena which these laws regulate, then the laws do not even allow us to differentiate between the two. This fact follows at once from the well known property of reversibility possessed by most natural laws. Newton's law of gravitation, for instance, sets up a relation between an observation on the rate of change of the radial velocity between two masses on the one hand, and the distance between them on the other. But it contains no criterion to determine the causal status of these observations. There is no law of connection between cause and effect known to science; moreover, these concepts are foreign to physical analysis. Nor is it of any avail to inject them externally, for the meaning usually conveyed by the words in question is expressed more adequately and precisely by technical terms like boundary condition, initial and final state.

Most of the difficulties discussed so far are avoided in Laplace's statement of what he and many later scientists consider to be the essence of causality. He postulates the existence of a universal formula according to which all happenings take place, and expresses this state of affairs as follows (Théorie analytique des probabilités): "An intelligence knowing, at a given instant of time, all forces acting in nature, as well as the momentary positions of all things of which the universe consists, would be able to comprehend the motions of the largest bodies of the world and those of the smallest atoms in one single formula, provided it were sufficiently powerful to subject all data to analysis; to it, nothing would be uncertain, both future and past would be present before its eyes." This is certainly an intelligible proposition; it is excellent in its clarity and precision. All of its terms are well defined; the word force is to be understood in its accurate physical sense as the product of mass and acceleration, and "knowledge of a force" means knowledge of the differential equation which relates this product to a function of position, this function being also known. This spirit of Laplace's proposition pervades all of classical physics and has been eminently fruitful in the development of that science. Is the proposition true? In answering this question we shall find reason for abandoning this particular formulation of the causality principle.

Of course Laplace's statement is true. Imagine if possible a perfectly arbitrary universe with a god agitating it according to his ever-changing desires. Although we are still searching for a suitable definition of causality, it seems clear that this would constitute the model of a non-causal world. If we further suppose the happenings in this chaotic universe to be discernible and describable, then it must be possible also to represent them by means of equations. These equations will not necessarily contain analytic functions only, nor will they be differentiable. However, if things take place with reasonable smoothness and not too sudddenly, if "natura non facit saltus," the functions will possess the property of differentiability. It is then clear that differentiation will, in general, simplify the equations, for it will cause additive constants to diasappear. An intelligence powerful enough to know all these differential equations together with the values of coordinates and derivatives at a given time, and able to solve them, would have a complete survey of all events, future and past. In Laplace's world this survey must be possible on the basis of a knowledge of all forces, i.e. differential equations of the second order. This, in itself, constitutes no further restriction upon our arbitrary universe, since there is nothing to prevent us from differentiating the equations twice; but it expresses a preference which should manifest itself in a universe satisfying this particular causality postulate. In such a world laws should take on an especially simple form if they are stated as second order differential equations. Thus it is seen Laplace's postulate is not a stringent one, it is true for almost any imaginable universe. It imposes nothing upon a world which by itself runs smoothly, and is certainly a valid approximation to the course of events in the arbitrary, non-causal universe. In that sense the statement in question is true, but it does not seem to characterize causality. Whether or not nature is conveniently describable in terms of second order differential equations is an entirely different issue and must probably be affirmed—although there are cases of physical analysis where description by equations of higher order is customary.

The fact that the postulation of a universal formula, as phrased by Laplace, involves the hypothetical existence of an omniscient demon has been considered unsatisfactory by numerous investigators. An excellent critique of this point is to be found in Ph. Frank's book *Das Kausalgesetz und seine Grenzen*. The appeal to a higher intelligence is certainly unscientific and to be avoided if possible. But it is not intrinsically bad if it merely serves to clarify the proposition. The question is: does Laplace's supposition of a superior intelligence constitute an essential point of his statement? Evidently not, for if the reference to this intelligence were omitted the assertion, weak as it may be, that nature be conveniently describable in terms of second order differential equations would still remain. But this can be progressively tested by experience and admits of definite verification. Hence it is not legitimate to say that the demon does not exist or is impossible and therefore reject the proposition. A real criticism should attack its meat and not its form. It may well be observed in this connection that a very common argument against causality fails for the same reason. This argument appeals to the uncertainty principle which does not permit all the simultaneous data necessary to integrate the universal formula to be known. Hence, it is concluded, the causality principle can not be valid. Here, too, it is forgotten that the criticism is directed against an inessential point in Laplace's formulation which, as we have seen, fails to express the characteristics of a causal universe anyway.

The differential equations governing the processes in the arbitrary, lawless world which has been imagined will in general involve the time explicitly.

Consequently the forces (which are always to be defined in terms of accelerations, not popularly as "pushes and pulls") will change with time in an essential manner, and not only through the co-ordinates on which they depend. In order to predict the future, Laplace's demon would have to know not only the instantaneous values of all forces, he would require their complete form as functions of the space co-ordinates and of time. An entirely different situation arises if we postulate that the forces be functions of space co-ordinates and possibly their time derivatives only. This would constitute a very definite limitation upon natural processes, a limitation indeed which the world agitated by a god would fail to exhibit. For now the forces have the same instantaneous values whenever the co-ordinates and their derivatives assume a given set of values; this endows nature with a regularity which, it appears to us, is very nearly what causality is meant to convey. But this important feature hardly follows from Laplace's formulation and may, at any rate, be expressed more directly, as we shall see presently.

Before continuing this trend of thought it seems necessary to deal with a definition of causality which is most widely accepted at present. The argument runs: introducing a superior intelligence is not permissible because it is man who makes his science and it must be he who is to judge whether his world is causal or not. A universal formula without an individual knowing it is a vague phrase. Why not modify the proposition last considered by substituting man in place of the demon? The causality principle is then valid if it is possible for the scientist, on the basis of known laws, to reconstruct the past and to project the future when the present state of the world is completely known. We shall term this conception of causality, in which a clearly anthropomorphic attitude combines with utilitarian considerations, the positivistic one for want of a better name. It is quite in harmony with the modern trend of eliminating things that have merely logical status but no concrete meaning in terms of physical operations. Nevertheless it pays to examine it closely.

First it is defective from the point of view of intelligibility. The term "possible for the scientists," or, if it is preferred, "possible for the human mind" is objectionable. If it conveys its popular connotation, who is to decide the capabilities of the scientist or the human mind? Even though a present physical theory denies the possibility of knowing all data upon which a detailed analysis of the future depends, this possibility may be restored by later developments. It is true that this fault can be remedied by modifying the positivistic formulation of the causality principle and stating: causality exists if no confirmed physical theory contradicts the acquisition of data by which a determination of future and past events can be made. However, this is very specific and has never been proposed, nor is it exempt from the criticism that follows. In line with the present argument we also note the occurrence of words like "law" and "theory" which, though they figure prominently in physical discussions, still lack universal definitions and are very far from being technical terms. Direct statements about nature are always clearer and less involved than terms like these which refer to our reasoning about nature, and are therefore better suited to define causality. If such considerations seem pedantic one must answer that their neglect has caused more difficulties than has the actual solution of philosophical problems.

Yet they do not touch the principal weakness of the positivistic causality formulation. We propose to show that the latter may be satisfied by a non-causal universe. The point is simply this: causality has nothing to do with the question whether future events may be known in advance, its prerequisite is not that the scientist turn prophet. Suppose that the god who agitates his universe according to his inscrutable desires and without restrictions by law or order should give the scientist exact forewarnings of his actions, so that the latter is able to prophesy with accuracy. Would this make his playland a causal universe? Of course there is no rigid answer since we are still searching for a suitable definition of causality. But if we interpret correctly the universal implication of this concept regardless of its various formulations we feel that one must answer: no. It is commonly conceived, for instance, that the occurrence of miracles contradicts causality; in fact disbelief in miracles is usually justified by the causal constitution of the world. The circumstance that many miracles have allegedly been predicted is hardly sufficient to restore a sceptic's belief in them, and hence to reconcile their occurrence with the causality postulate. We conclude: the positivistic formulation with its main emphasis upon human ability to know in advance does not express the nucleus of what is understood by causality. Besides, it is anthropomorphic and reflects distinctly the present utilitarian color of our science.

Let us now put an end to criticism and select a definition that will satisfy the outlined requirements more widely and state the central part of the concept in question. It will be granted that the crucial feature which makes the arbitrary world non-causal is the irregularity arising from the whims of the god, whether this irregularity may be known or not. To be more specific it is the fact that in a non-causal world the force between two electric charges varies, say, as the inverse second power of the distance today, but possibly as the inverse tenth power tomorrow. Or even, while they attract each other today, they may repel each other tomorrow. We feel that causality is violated when a given state $A$ is not always followed by the same state $B$. A definition of the word "state" is certainly necessary, but we may reserve it until later. At present the customary intuitive meaning will suffice. This property by which any given state has associated with it a unique consequent state has previously[2] been called "consistency of nature." To avoid circumlocutions we shall continue to use this term. In more adequate phraseology, and without the unsatisfactory reference to states, consistency of nature may be characterized by saying: As a result of the constitution of nature, the differential equations by means of which it is described do not contain explicit functions of the time. This statement is less definite than it seems because it does not contain directions as to the choice of variables appearing in the equations, a vagueness which is the counterpart of the indefinite meaning of "state." The existence of variables must of course be supposed since they form the condition under which description of nature is possible at all. But then, if nature is consistent in this sense, the integrated equations will not depend on time in an absolute manner; more specifically if $x = f(t)$ is a solution, then $x = f(t-t_0)$ is another, so that the motion in question has an arbitrary beginning in time which becomes fixed only if accessory conditions are known. Furthermore, to use a previous example, the exponent in Coulomb's law of attraction will be invariable in time; if it is $-2$ today it will be $-2$ forever. The same should hold for all parameters appearing in the differential equations of physics if nature is consistent or causal.

It is to be remembered, however, that the occurrence of equations which do not satisfy this requirement, is not at once a proof against causality. In fact we often encounter such equations when a problem is not completely analyzed, for example in the case of forced oscillations. But here, as well as in all similar cases, the explicit time dependence could be eliminated by including in the analysis those parts of the system which produce the varying force, i.e. by "closing" the system. "Impressed forces" always indicate that the physical system in question is an open one. In fact a closed system is simply one to which causal analysis can be applied, that is one which can be described by differential equations not containing the time explicitly. Another point of importance is this: An equation which originally satisfied the causal requirement will, after a single integration, no longer do so. We express this state of affairs by saying that we now have an "equation of motion" and not a "law." (Incidentally it would be very desirable to standardize this particular meaning of the word law in natural science.) Nevertheless this introduces an uncertainty into our formulation of the causal principle, but one which cannot be avoided. It will occupy us when we discuss the analytic character of causality, its property as a non-tautological proposition.

Consistency, the central issue of the causality postulate, banishes absolute time from the description of nature (equations of motion) by eliminating time explicitly from its *essential* representation (differential equations, laws).

It is the only formulation which does so. The elimination is necessary because absolute time has no physical meaning. Another advantage is the simplicity of the consistency formula, which nevertheless involves everything implied in the usual conceptions of causality, such as the existence of unique laws, strict determinism as far as it is a concomitant of causality. (It does not imply, however, that future events be actually known or knowable by an individual.) Theorems of conservation (energy, momentum), generally felt to be in some way connected with the causality principle but unaffected by its customary formulations, follow at once as analytical consequences from the fact that the differential equations do not involve the time explicitly.

The principle of consistency, and hence of causality, has no meaning if it can be applied only to the universe as a whole. For in that case the number of describing variables would probably be infinite and the description in terms of differential equations loses its sense. Using the more intuitive definition according to which a state $A$ is always followed by the same state $B$ the failure is evident when we realize that state $A$ may occur only once. Causality is then an empty phrase unless the universe is periodic. But the same would be true of all the laws of physics, in fact of physics as a science, if its statements were inapplicable to small domains of nature. Hence the condition that causality shall have meaning is the same as that for the existence of science. The philosopher who argues that the consistency definition be void forgets that his argument nullifies science as a whole. As a matter of experience the universe *is separable* into smaller systems to which differential equations can be applied, and if these equations are of the type here postulated then nature is causal.

An observation of this kind carries little weight with those who feel too keenly that the processes in the universe are separable only to a rough approximation. One is entitled to discuss this attitude in earnest only if he is willing to accept all of its consequences which include the proposition that science is an illusion. But we note in the first place that it is by no means necessary for an arbitrary nature to be *roughly* or *approximately* separable into systems whose fates are independent; to say that it is imposes a definite restriction even if the separation is not completely possible. But let us consider more exactly how it is performed. Measurements on the force of attraction between two electric charges will not in general verify Coulomb's law. We observe that the force depends in some peculiar way upon the position of external charges, which suggests to us that the measured effects are not entirely due to the system in question, namely the two test charges. This is expressed by stating that the system is not completely separated from its surroundings. Next we remove our system farther and farther from surrounding bodies and notice a gradual improvement in the consistency of the meausrements. Now the situation would be very simple and satisfactory if progressive isolation produced better and better agreement between the observations. For then the condition of complete isolation, i.e. a closed system, could be defined by a simple limiting process much in the same manner as limits of functions are defined in mathematics. However, the situation is here of greater complexity, though still manageable. The agreement is improved only up to a certain point, and then further isolation fails to make itself felt. We have reached the limit of precision of our measurements. This limit of precision is a very definite thing which scientists have always considered very carefully. Quantum mechanics emphasizes it greatly and even renders its value in some instances calculable by means of the uncertainty relation. Nevertheless there is a well known method of dealing with the ever present divergence of observations: the theory of errors allows us to compute the most probable value of a measurement from any group of observations. It is the limit of this most probable value upon which we base the derivation of Coulomb's law and not the limit to which actual observations tend. The latter does not exist, but the former does. Hence it is possible to define a state of separation, not by actual physical operations but by blending experimentation with reasoning. We wish to emphasize in this connection that physical concepts need not—and can not—be defined solely in terms of experimental operations or observations; it is both customary and proper in scientific investigations to characterize an abstract state of affairs by its logical consequences if they are more simply expressible, whether they are observable or not. For example, when we state the second law of thermodynamics in the form: the entropy of the universe increases, then there is no way of testing the law as such; its evidence arises from considerations formally similar to those which have convinced us of the existence of closed systems to which differential equations are applicable.

By its definition, a closed or independent physical system is a causal one, because we call it closed when the laws governing its behavior do not involve the time. But strictly speaking only closed systems are accessible to physical analysis. Thus it would seem that physics can never inform us of a failure of the causality principle. This brings us to a point of importance: Is the causal principle a tautology? Here we are forced to make a large concession; an unbiased investigation must not fail to recognize its character as an analytic proposition. Kant, who thought of it as an *a priori* synthetic judgment, did not formulate it in a way in which its analytic character became apparent. It is certainly true that whenever a physical system does not appear to be closed, that is when the differential equations describing it contain the time explicitly (if it does not behave consistently), we conclude that the variables determining the state in question are not completely known. We then look immediately for hidden properties whose variation may have produced the inconsistencies, and whose inclusion in the analysis would eliminate them; moreover if we do not find any we invent them. This procedure is possible because in the consistency formulation of causality the term "state" is undefined. The corresponding indefiniteness in the more abstract formulation lies in the absence of specifications as to the number of variables entering into the differential equations on the one hand, and of the order of the equations on the other. If the number of variables increases indefinitely, or if an indefinite number of differentiations is permitted, time dependence can be ultimately eliminated no matter how complicated the processes of nature, provided only that the equations of motion can be differentiated a sufficient number of times. But this requires no more than a certain smoothness and continuity which nature certainly satisfies.—On the whole it seems, then, that the causality postulate reduces to a definition of what is meant by "state." It is an agreement to consider those quantities as composing the state of a system which enter into a time free differential equation describing its behavior.

This line of reasoning leads at once to the inevitable question: Why retain the proposition if it is merely tautological? Tautologies, as everybody knows, add nothing to the knowledge of nature; they expose a property already included in the term they are to explain; their careless use often produces vicious circles. All this is true in a sense; but to suppose that tautologies are always useless and to be avoided is a very common fallacy. Every definition is a tautology except the first time it is stated. To the writer the word tautology conveys something more objectionable than the good old "analytic judgment." Really the two are synonymous, and one must free himself from any intuitive bias that may cling to the former term. The principle of conservation of energy is a tautology in the proper sense of the word, yet nobody doubts its fruitfulness, and it is even customary to speak of its validity as though it were an actual proposition about the world. The point in question is this: The general definition of the principle contains no restriction as to the number of different kinds of energy which may be transformed into one another and whose sum is constant. If this number became indefinitely great as a consequence of invention whenever a new type was needed the principle would hardly be applicable, it would merely define energies. As a matter of fact, however, nature permits us to get along with very few different types of energy, and its description in terms of energies is exceedingly useful and convenient. Therefore, while the energy principle in its logical formulation is a tautology which amounts to a definition of energies, the *analysis of nature in terms of this definition is advantageous*. Precisely the same is true with regard to causality. Its logical formulation is inevitably tautological and leads to a definition of physical "states," or a selection of variables in the differential equations. But this selection is useful and applicable, so that the causality principle, though tautological in its abstract form, does amount to a statement about nature. Moreover it makes sense to say that nature is not causal, for this would be true if, in an attempt to describe nature in accordance with our definition of consistency, the resulting differential equations were found to contain very many variables or to be of very high order. One might, of course, arbitrarily restrict the latter and permit only equations of the second order, but we feel that this would do violence to the common conception of causality.

We have seen that the tautological character of the consistency postulate is no particular fault. Indeed the postulate may be transformed into a synthetic statement if this be desired, but somewhat at the expense of its precision; it might be phrased: Nature is so constituted that its description in terms of differential equations which do not involve the time explicitly (such description is admittedly possible!) is *convenient*.—We merely state that Laplace's causality formulation is also tautological, while the one identifying causality of nature with human power to prophesy is not. But it stands to reason that the latter modification, missing, as it does, the central point of the casuality concept, pays too large a price for its non-tautological form.

In formulating the principle of causality we have almost solved the question as to its validity. Classical physics was based upon it and therefore presents no argument against it. The influence of quantum mechanics upon its status has been considered in detail in two previous publication,[3] whose results may be stated briefly as follows: Classical analysis had come face to face with experiences which its usual methods failed to describe; in fact the treatment of certain problems threatened to become non-causal. At this very instant quantum mechanics achieved a revolutionary feat of great importance: it redefined the concept of physical states in a more abstract manner (in terms of mathematical functions satisfying certain requirements) and *thereby restored the causal character of physical analysis*.[4] It was the conviction that the causality principle must be retained which inspired quantum mechanics although some of its creators have not been altogether conscious of this fact. The impression that quantum mechanics violates the principle now arises whenever the older classical conception of states (positions and velocities of the component parts of a system) is carelessly carried over into the new field of description in which it has no meaning. The uncertainty principle forbids ultimate extrapolation to the quantities defining a state in the classical sense, but it does not prevent an ultimate extrapolation to $\psi$-functions. To be quite impartial one should add that the trustworthiness of quantum mechanics even in questions of ultimate extrapolation to classical states is not entirely evident, for it is precisely in the very small domains of space (structure of the nucleus, structure of the electron, and its trembling motion) where its present axioms break down. Quantum mechanics does not constitute an argument against causality.

However there comes news from other quarters which may upset the validity of the causal principle. If, as has been reported, the velocity of light, which is an essential parameter in the differential equations of physics, undergoes a slow variation in time, then a revision of the postulate may be in order.

Another significant objection has long been known although it is ordinarily overlooked. It has to do with spatial continuity of the universe. If the structure of nature's elements is continuous in space, and hence infinitely detailed, the equations representing the behavior of any of its parts will of necessity contain an infinite number of variables. Hence causal analysis, as it proceeds into finer and finer details of structure, will meet the same obstacle which prevented it from exploring an inseparable universe as a whole. Two means are available for avoiding this difficulty; both have been employed. One is to adopt a field theory which fixes minutely the values of a physical quantity at every point of space, but fixes it by means of simple functions so that the scientist is enabled to dominate in one grand sweep all the intricacies of spatial structure. There are reasons, however, why this procedure is unsatisfactory. Unless the field functions used in physical theory are periodic they imply singularities which, while they are insignificant as far as many properties of nature are concerned, certainly do not exist. The other and probably the better way to escape the difficulty is to eliminate its roots, that is to abandon the conception of a continuous universe. This is done in quantum theory by assuming the discrete existence of electrons, protons, neutrons, energy quanta and the like. The magic formula here was to reduce the number of variables appearing in causal description by endowing finite parts of space with homogeneity, so that this finite portion requires no more elements of description than does a point. Here again, quantum theory comes to the rescue of the causality principle.

## NOTES

[1] For a careful survey of the many meanings of the words causality and determinism see M. Bunge (*Causality*, Harvard Press, Cambridge, 1959). He does not confine the meanings of cause and effect to states of isolated systems but uses a much wider definition which tolerates such statements as "force is the cause of acceleration." Although common language sanctions this usage, we reject it as inconsistent with modern physics. (Note added in 1975). On the same subject, see also Bunge's *Myth of Simplicity*, Chap. 11, Prentice Hall, Englewood Cliffs, 1963).

[2] H. Margenau, *The Monist*, Jan. 1932.

[3] H. Margenau, The Monist, Jan. 1931; *ibid.* April 1932.

[4] R.B. Lindsay makes this point by saying that quantum mechanics abandons determinism but retains causality.

CHAPTER 4

# METHODOLOGY OF MODERN PHYSICS

## 1. SURVEY

Methodology might be understood to mean a description of various individual procedures which have led to the successful solution of specific problems. In studying the subject of physics from this point of view, i.e. with special emphasis on method, one would naturally turn his attention to the traditional divisions of experimental and theoretical physics, the former with its measuring devices and the latter with its mathematical techniques. In no other sense than this does the term methodology make any direct appeal to the working physicist, and if you would ask him to define his methods he would probably answer with a description of experiemental technique or the methods of setting up and solving differential equations. His answer would tell you *how* he solves his problems, but hardly how he *finds* them and *why* he solves them.

Nor does he assume this restrictive attitude from ignorance of the larger setting of his problems. His reasons for it are partly those of the artist who refrains from comment upon the origin of his artistic conception; they arise in part from his systematic tendency to put first things first, for he is afflicted with a keen realization of the futility of attempts to discuss scientific methodology in its fundamental implications before the incidental details of his technique are understood. If he is very conservative he may even maintain that his field is entirely foreign to extra-physical speculations such as those necessarily involved in a basic evaluation of his technique, and that any mingling of his physical notions with such speculations debases the former. But he is more likely to admit the justice of philosophical circumspection, professing, however, neither interest nor competence with regard to the latter. Again, his lack of interest is well founded, for is it not true that the late brilliant scientific advances have been least co-ordinated with conventionally sound philosophic judgment?

Finally, there is some scientific modesty in the typical physicist's attitude toward "larger settings." In all his operations he prides himself on the accuracy of his judgments, he takes care not only to minimize his errors, but even to ascertain their probable magnitude. This he cannot do when he indulges in appreciative reflections upon the ultimate meaning of his work or upon its relation to the more speculative sciences. Hence he feels lost, being unable to use his customary tools, and he consequently avoids this awful territory.

In spite of this well founded preference of the typical physicist for his own technical domain, it is not the description of special scientific procedures to which this paper will be devoted. For, after all, there are cogent reasons, reasons indeed which should appeal to physicists, for surveying the special methods applied in any science from a vantage point which places them in view simultaneously. Unless this is done we shall be forever studying local currents and eddies in the larger stream composing our science, without discovering the general direction in which the water flows. True enough, we shall never know exactly where a boat, adrift on the stream of physics, will ultimately land, but we can, by taking a larger view and determining the main direction of flow, decide whether it will be worthwhile to set out at all. In less metaphorical language, what I shall try to undertake is to abstract from all specific and diverse physical procedures their common background of method, to analyze their residual features and to estimate their epistemological significance.

This task in itself is not new, and many attempts of carrying it through have been partially successful. I shall forego the opportunity of criticising them in detail and merely point to a defect that has been very common. The diversity of valid physical theories is extreme, and it is easy in designing a general methodology either to forget entirely, or to underrate the importance of, apparently stray tendencies. It will be found that most writers, especially those whose main interest lies in philosophy, are using *mechanical* theories and laws as the sole basis for their abstractions, leaving out of account the non-mechanical ramifications of thermodynamics and electrodynamics. Indeed a methodology of physics is incomplete if it ignores the more formal reasoning which characterizes these two fields. In fact if we were permitted to limit ourselves to the experiments and theories dealing with large scale bodies our task would be a simple one. Difficulties arise, as we shall see, when we attempt to juxtapose highly formal methods and those perceptibly evident inferences which seem to govern large scale mechanics. It will also be necessary to *bridge*, not merely to contrast, the use of probability notions in some parts of physics with the idea of constrained evolution of mechanical states. These goals have not often been attempted, nor could they have been reached until fairly recently, for the very meaning of statistical theories, especially in connection with quantum mechanics, was not clearly understood; it seems to have crystallized only during the last few years. To bring the latest quantum mechanical developments within the range of a uniform methodology is perhaps the most difficult part of our task. It would be simpler if we could regard as a correct solution the interpretation of those who see in the new theories merely a rejection of causal description. But I believe that this extreme point of view, which on the surface seems to harmonize so well with the mathematical technique of the matrix theory, must be ultimately unsatisfactory and is due to an insufficient analysis of the basic concepts involved. It is necessary to admit that physical experience has in a sense become indeterminate, but the methods by which we describe physical states remain as causal as they were before. To put it crudely and in a manner which will receive its fuller meaning later, the world of physics has retained its causal consistency, while its connection with the perceptible world has been placed on a statistical footing. The recognition of the dichotomy between the two worlds just mentioned is an essential feature of the doctrine which we are here proposing.

To meet more definitely the attitude of the staunch realist who sees no possible value in any extra-physical speculations about physical methodology one may appeal to the following consideration. Such an attitude can be based only on one or the other of the two suppositions: (1) The domain of physics is entirely secluded from all other fields of human interest, each field of thought being self-sufficient and independent of cross-fertilization. (2) Physics is the science *par excellence* and will, in time, pervade all other types of endeavor; the reason why we acknowledge a distinction at present is the imperfect state of analysis prevailing in the latter. Neither of these assertions can be said to be flatly false. But the first is clearly unsatisfactory in the sense that if it were to be ultimately true we should be sorry, for reason is unwilling to stop its inquiry in front of an array of pigeon-holes of which the contents are unconnected. If we maintain the second we are, I believe, indulging in an optimism for which there exists at present no sound basis. Perhaps the hope of such a possibility has vanished with the materialism of the last century. But if we wish it to be true, that is if we claim for physics the prestige of being the final touchstone of truth, then it is evident that we must enrich its structure by principles and methods not at present embodied in its constitution. We are safe in maintaining, therefore, that it is no disgrace for the physicist to speculate about the meaning of his science.

A methodology grows and changes with the expansion of knowledge. The impressive system of analysis outlined in Kant's "Critique of Pure Reason" was adequate for the science of his day, but his transcendental esthetics broke down when Einstein discovered the interrelatedness of time and space. More and more of his methodological principles crumbled away in the face of newly discovered facts. We are referring to this example because it shows so clearly how the individual tenets of a speculative system were invalidated by individual physical discoveries, a process which in general takes place more obscurely. Whatever claims may be made for any methodology of physics, finality can not be among its merits.

The question of method is very intimately bound up with the purpose or aim of scientific activity. It is only if we face squarely the somewhat annoying question: What is the physicist trying to do, that we come to grasp the nucleus of our problem. Let us, then, determine with some care the general object of the physicist's investigations.

## 2. DIFFICULTIES IN CURRENT CONCEPTIONS

It would be naïve to settle the matter by statements of the sort: The physicist explains nature, or he discovers reality. For in what sense does he explain? Does not such an utterance merely push the entire mystery of the physicist's activities into the word "explain" and seal them up, hidden from popular view? And then again: in what sense is his reality real? Does he not rather invent than discover? To such questions we desire to find an answer.

The object of physical investigation is not merely an accumulation of a certain type of knowledge, or of all the "facts regarding matter and energy," as elementary textbooks occasionally put it. The distinctive character of scientific systematization which pervades this knowledge does not exhaust itself in mere classification of facts. The existence of satisfactory physical theories which go beyond facts is sufficient proof for the inadequacy of the view that regards physics as a mere aggregation of knowledge.

But perhpas we are dealing with this argument in too summary a fashion. For there is, after all, a more painstaking group of analysts who will admit that a mere collection of facts does not make a science, but that aggregation must be supplemented by abstraction, synthesis and induction, processes which transform the collection into a logical array of facts. This claim, it must be observed, is usually made by those who wish to bar transcendental elements from the realm of physics, and this wish is, perhaps unwittingly, the father of the thought. Hence to answer this assertion in toto one must destroy two illusions: first that physics is based solely upon the logical principles of deduction and induction; second that it is unscientific to harbor transcendental notions, that is notions which do not spring directly from experience. We have already dealt with the latter thesis in the preceding chapter and shall meet with it again. All of modern physics compels this view. The question of the intrusion of intrinsically non-empirical concepts into our science has ceased to be an academic one; only by taking proper account of them can we understand the recent formalization that has taken place in physics. But this is a point to which we must return.

As to the contemporary structure of physics, even with the exclusion of the latest development which to some appear still slightly problematic and of doubtful security, this is certainly not a logical array of facts. For, though fact be taken in its widest possible meaning there can be no way of saying that the changing symbolism of atom clouds, electron waves, vacancies in the distribution of electrons can represent facts. I am not trying to belittle these physical notions; if anything, I am belittling facts. Facts are far too crude to fit into the delicate texture of physical theories; it is to mitigate their crudeness that the latter exist. It is true that logical principles, like deduction and induction, are used in physical as in all other reasoning, but their use is by no means characteristic. Induction, in particular, is used only as a tentative procedure; that is, the propositions to which it leads are always subjected to empirical verification and never proclaimed as certainties. For this reason, whilst the physicist employs inductive reasoning, he need not trouble himself with the profound problems involved in a logical analysis of induction.[1]

Next we come to the view, widely held at present, which identifies the object of physical investigation with the acquisition of facilities to predict. It can not be denied that physics provides such facilities in an ever increasing measure and that its practical value largely resides therein. But is not this advantage rather a by-product than the sole end of research? Scientific knowledge is power and will forever be power; nevertheless there are many who claim that they would pursue their scientific endeavors even if this were not true. They call their motive, somewhat enigmatically, love of research, much as though it were a natural instinct whose satisfaction gave them pleasure, and this pleasure is not peculiar to the act of prediction; it is present whenever a problem is solved. Among these, we believe, would be some of the most successful physicists. Thus it seems, at least from a psychological point of view, that the driving force in physical investigation can not be exclusively the fun or the profit of prediction.

Then, too, there are numerous physical theories which, thus far, have failed to make any noteworthy predictions. The theories of molecular binding and of ferromagnetism are among them. In these two fields experimental knowledge is so far ahead of "explanation" that a long time is likely to elapse before any significant purely theoretical predictions of new phenomena will be made. Yet both of the theories are regarded very important contributions despite their remote chance of predicting. The value of physical theories is clearly not fixed in proportion with this particular merit. Nor does a physical theory lose its value when its range of application is almost closed and it is hardly in a position to enable further predictions.

The misconception that scientific research be essentially a matter of predicting new phenomena has produced a most insidious practice not altogether uncommon among contemporary physicists. The scheme is to publish, from time to time, a sufficient number of predictions without giving detailed reasons for them, so that at least a few of them will come true by the laws of chance. Since valid predictions are very impressive and the memory of blunders tends to fade away, this scheme is sometimes regarded a vehicle to fame.

Evidently the physicist's business involves something more than prophecy, something that makes his experience peculiarly coherent and produces an internal fitness which facility for prediction alone does not convey. We know a great deal, for instance, about the phenomena attending the flow of an electric current through a wire at very low temperature and we have a strong desire to understand them. In spite of this experimental knowledge, in spite of the fact that we can predict many of the things which will happen when similar experiments are tried, it is felt that this subject is in an unsatisfactory state, and physicists would be grateful for a (correct) theory which will do no more than "explain" these facts. The meaning of this term, explain, is the object we are seeking.

Is it the exposure of causal relationships? This is indeed frequently asserted. To decide the issue conclusively one must look more closely at the meaning of the word cause. An occurence or event is said to be the cause of another if it precedes the latter invariably in time. Let us not worry too much at this point about the phrase "invariably," for we are not attempting here a critical discussion of the matter but only to find the customary use of the term. Moisture in the air, we say, is the cause of rainfall, or economic maladjustment the cause of war. It is to be noted that the causal relation connects two situations both of which are perceptible or, in a sense to be defined later, both are data. But this is perhaps the most primitive use of the word and clearly one that has little to do with physical research. As we have seen, the physicist does not satisfy himself with a record of facts, not even if they are placed in their correct temporal order. It is true, of course, that causal relationships of this type are the basis on which he prospers, but it is equally clear that he elevates himself far above them.

We can also say: A certain thought is the cause of a question; or a given mathematical condition is the cause of a special solution, without running any risk of being misunderstood. Here the word is employed in a different way, for the situations which are connected are no longer perceptible, and the time element is pushed slightly in the background. To be sure, the physicist often uses cause exactly in this way, as when he speaks of mass as causing a gravitational field, or of conditions causing constraints. In fact all imperceptible concepts occuring in physical theories are embedded in causal connections of this type, but to place concepts in such relations does not constitute a physical theory. Again we see that the statement "the object of physics is to set up causal relations" aims at the target but misses the mark. For, besides being related among themselves, these concepts must be in some way connected with experience.

Obviously now, there is a third manner of using the term cause, exemplified by the statement: an electromagnetic field of rapidly fluctuating intensity is the cause of the sensation of light, or the chair which I kicked has caused me pain. Here a relation is set up between something that is perceptible and something that is not. The time element has nearly dropped out of the picture and its place has been taken by a curious notion of enforcement. The use of the word has become very hazy indeed, as probably even those will admit who attach little weight to the difficulties attending the relation between stimulus and perception. Philosophers sometimes argue that this use of cause is unfortunate and should be abandoned. To this conclusion I should like to subscribe, not so much out of philosophical convictions, as because it is unwise to use one term in too many different ways. However, it is precisely this latter sense in which most advocates of "causal explanation" wish the word cause to be understood. They would suggest that physics supplies unobservable causes for the events of experience, thus rounding off the universe in a satisfactory fashion. Statements of this kind can not be spared the criticism of indiscriminacy of speech.

But there is a more serious reflection upon this view. People who hold it are not always careful to keep the *notions* distinct. Thus, by using the term cause in the last sense, the force of gravity may be said to be the cause of a body's fall. By using it in the first one could say: the force of my arm (muscular exertion) is the cause of the stone's flight. In both cases the antecedent is a force, the consequent a type of motion, the relation is said to be causal. Nevertheless every physicist realizes that the term force in the last case is not understood in its true physical sense, but one of sensory perception. Thence has sprung the confusion in the meaning of force, and the resultant difficulty, known to every teacher of physics, of conveying to the beginning student a properly abstract and non-anthropomorphic conception of this physical quantity. Force is not something that can necessarily be described as a push or a pull.

Let us now, after finding such an immoderate amount of fault with supposedly current notions, set out to determine the object and the course of physical investigation.

## 3. PHYSICAL RESEARCH AS SYMBOLIC CONSTRUCTION

Physical inquiry in its essential form moves along a peculiar cycle: it starts with definite, perceptible matters of facts; proceeds from there into a field in which at least some of the elements of operation are not directly perceptible, where there is greater freedom from empirical constraints; and finally it emerges again in the realm of perceptible facts. To illustrate: we observe a falling body, or many different falling bodies; we then take the typical body into mental custody and endow it with the abstract properties expressed in the law of universal gravitation. It is now no longer the body which we originally perceived, for we have added properties which are neither immediately evident nor empirically necessary. If it be doubted that these properties are in a sense arbitrary we need merely recall the fact that there is an alternate, equally or even more successful physical theory—that of general relativity—which ascribes to the typical bodies the power of influencing the metric of space i.e. entirely different properties from those expressed in Newton's law of gravitation. We are here clearly operating on a plane where there are fewer constraints than in the domain of perceptible facts. But having specified these properties, we at once return to the latter and say that the planets should revolve about the sun in elliptical orbits, or that the perihelion of mercury should advance through a specified angle per century. To be sure, a great amount of mathematical reasoning is involved in this simple return, but this represents only a formal expansion or modification of the properties produced in the former act of speculative creation. We say that our "explanation" of the original observation, the falling of bodies, is in error if the latter propositions are empirically false. In case of their verification it may be either correct or false. Our procedure begins and ends among matters capable of perception, the totality of which we shall call *nature* for the sake of definiteness.

Let us choose another example. The physicist observes, as part of nature, the deflections of an ammeter placed in an electric circuit. In particular does he notice the changes in the position of the pointer which take place as he introduces more or less wire into the circuit. He then invents, in the privacy of his peculations, entities of which he has never made the acquaintance through his senses although he may feel that he has perceived "similar" things in other connections. these entities were named electric current, resistance, electromotive force. If you press him, however, he will admit that no stock is to be placed in this similarity aside from its helpfulness in visualization, and that these entities are characterized by the properties which are assumed for them, that is, the relations in which they can enter. In the present instance, one such relation is Ohm's Law. This, of course, does not exhaust their definitional properties, for they may also enter relationships with other constructs (heat generated, etc.) not here introduced. From Ohm's Law we can then predict other experiences, which, if verified, lend some support to this mode of explanation. It is important in this connection to see that the return trip to nature is made possible by placing the speculative elements themselves, i.e. current, resistance, electromotive force, in definite correspondence with experience. We must know how to identify in a quantitative way the experimental counterpart of a current, resistance or E.M.F. The physicist expresses this by saying that he "measures" current strength, resistance, or E.M.F.

The same phenomenon permits an alternative explanation. The physicist can make his notion of an electric current more definite by supposing it to be a stream of entities known as electrons. To these electrons he assigns properties such as being charged, or of producing certain effects when set in motion, effects which are observable only through their experimental consequences. The electrons themselves are never objects of perception despite the intriguing vividness of Wilson's cloud tracks or the moving oil drops in Millikan's famous experiment. They can not be said to constitute nature in the sense in which we wish to employ the term. (Cf. also section 5.) But from the notion of electrons we can, by virtue of definite rules of operation, go back to nature and make specific statements about perceptible matters of fact.

There is not of necessity any conflict between these two explanations of the same phenomenon. For we may satisfactorily assume as one of the properties of electrons that their motion constitutes a current of the former kind, thereby rendering the first explanation reducible to the latter. The electron theory of current flow has the semblance of greater concreteness and makes, perhaps, a more direct intuitive appeal. But, we repeat, the construct of an electron is just as remote from the source of physical inquiry, nature, as is the less easily imaginable notion of a current of electricity, with electricity intuitively undefined. This semblance of difference has led to the distinction between phenomenological explanation—the type which satisfies itself with constructs of little or no imaginative appeal—and, causal explanation, such as the electron theory. It is clear from the foregoing considerations that the use of the latter term is very much to be discouraged, and that, on the whole, the distinction is illusory. This last point, however, is one to which I shall return later.

The full course of physical explanation, as we have now seen, begins in the range of perceptible awareness, swings over into what we shall now term the field of *symbolic construction*, and returns to perceptible awareness, or, as we have said, nature. The field of symbolic construction is populated with many entities, such as masses, electrons, electromotive forces, which we shall call *constructs* of explanation, or, more briefly, constructs. These will be subjected to closer scrutiny as we proceed. The essential feature of a physical explanation is evidently the transition from nature to the realm of constructs, and the reverse. These transitions are governed by rules of which we must gain a clearer conception. But before we consider the relation between these two classes of concepts it it well to direct attention to several points which can easily give rise to misunderstanding.

I have not been very specific in the exposition of the meaning of "nature." The term is obviously not used with its customary significance. We have defined it as the totality of all matters capable of sensory perception, including thereby hallucinations as well as bona fide sense impressions. The former are, of course, of no interest to the physicist, and we shall suppose that there are means of eliminating them from the remainder of nature. Again, not all the rest of perceptible situations have hitherto become the starting point of physical explanations, but they must be regarded as holding such potentiality for the future.

But now arises the question, so vexing to many philosophers: Are we going to deal with nature merely as a complex of awareness, or must we attribute to it the status of transcendental objectivity? It is very fortunate, I feel, that physics is not forced to make this decision. In fact we can, and shall in these considerations, pursue the most cautions and possibly the wisest, course of regarding nature to be *merely* the aggregate of our perceptible experiences. I confess to the belief that the reader will not doubt the existence of such an aggregate in his own mind, and that this aggregate is largely similar to that which I have accumulated myself. If I am wrong in this supposition there is nothing further to say, for physics, as well as any other science, is founded upon this miracle. On the other hand, if any one wishes to go farther, as most physicists do, and postulate an objective nature he is free to do so at his own risk; no science can compel this step. This paper is written from the point of view of one who is unwilling to sacrifice a grain of certainty for a pound of common sense evidence. Accordingly, we should replace the common term data, which occupies so prominent a place in the physicist's vocabulary, by *habita*, implying that there may exist no external agency to which we are indebted for its gifts. (I will add that on my part this attitude is purely academic.) But to avoid the coining of a new and strange term we shall continue to speak of the elements of nature from which the reasoning of physics starts, as physical data.

Every datum has the attributes of extension and duration. To determine the precise logical sense in which space and time may be said to be attributes of nature is a problem which need not be solved for the present purposes; the only point to be observed is that they are very essential attributes, so essential in fact that a datum not in time and space not only fails to be realized but is even unimaginable. It is for this reason that all constructs, which as we have seen, must be accessible from physical data (in a manner to be discussed in section 4), partake of temporal and spatial qualities in order to provide a possibility for correspondence with nature.

It is sometimes asserted[2] that all physical data are pointer readings. A statement of this sort is likely to emphasize the quantitative precision which the physicist enforces within his nature—there is no difficulty in the fact that we can arbitrarily enlarge and modify our perceptible experience—and this emphasis is certainly well placed. But as a statement of fact it seems in error. While pointer readings form by far the largest class of physical data, there are at least two others. The first comprises experimental decisions between two alternatives: Does a certain phenomenon occur or not? The very appearance or absence of a band in the spectrum of a given molecule may be a perfectly significant datum. The other is that group of experiments in which the investigator relies on counting individual events, e.g. the clicks of a relay attached to a Geiger counter. Many of the data of nuclear physics, recently acquired, are of this type. Thus pointer readings, while of very great importance among physical data, possess no peculiar distinction and do not exhaust the possibilities. This remark is not idle, for no satisfactory methodology can be constructed on an oversimplified basis.

The physical concepts belonging to the class called nature, such as position, matter,[3] spatial coincidence or distinctness, motion, etc., present no greater difficulty of definition than do other properties of so-called concrete things. Some of them, e.g. position, are usually taken as primary concepts of which the meaning is intuitively evident, and the others, like spatial coincidence, may then be defined in terms of them. But a far greater complexity arises in conncetion with the other class of concepts which we have termed constructs of explanation. For it is often true that here a single construct admits of many different definitions, of which two are typical, and it is important that these be distinguished. To be specific, let us select a few illustrative constructs at random: mass, temperature, electron, magnetic field strength, magnetic field. Let us admit at the outset that no definition is an internally closed affair; it always presupposes certain terms to be understood or situations to be realized. We shall find this to be true when we attempt to define these constructs. Now then, how can we define mass? One possible procedure is to select a standard physical operation which can be performed upon every object to which we desire to ascribe a mass. We may then define mass as a definite functional relation between certain results of this standard procedure.[4] This definition is equivalent to the assignment of a perceptible, quantitative counter part to the symbolic construct "mass." It is, however, not unique, for there is a great number of processes which might be chosen as standards, e.g., gravitational interaction of two bodies and observation of accelerations, attachment of a body to an elastic spring and observation of the resulting acceleration, etc. Logically, each of these defines a *different* mass.

An alternative definition is available if, for example, we suppose the concepts of force and acceleration to be understood, for we can then say in general that mass is the ratio of force to acceleration (speaking non-relativistically, of course) without specifying a single operation. It is true that from a physical point of view this definition is equivalent to the former because it happens that the mass of a single body is independent of the choice of force, so that standardization does not detract from the generality of the definition. Indeed if this were not so the concept of mass defined in the present manner would be devoid of physical interest, but not of meaning.[5] Viewed logically, the definition of the present paragraph, which is not itself in terms of experimental operations (the auxiliary concepts involved may be thus defined!) and which I will term the *constitutive* definition, is different from that in the preceding paragraph, to be called the *epistemic* definition. Constitutive definitions are largely frowned upon by physicists, and epistemic ones are often considered the only admissible type. Nevertheless the interdependence of the two types is plain, and the fact is, I think, that in virtue of the epistemic definitions of our constructs we can measure, and in virtue of the constitutive definitions we are enabled to reason about the constructs. Epistemic definitions alone would leave physical constructs without coherence, moreover, there are constructs which do not admit of this type of definition as we shall see. The present example is perhaps not very well adapted for exhibiting the necessity of constitutive definitions; the following one is more suitable to this end.

Let us look at the construct temperature. We are well acquainted with its epistemic definition in terms of the efficiencies of reversible heat engines (Kelvin scale) or of thermometer readings. In this case, an important constitutive definition happens to be on record also, namely the mathematical one which characterizes Gibbs' modulus of distribution. This modulus, the reader will probably recall, is essentially identical with temperature.

In the case of the electron we find the epistemic definition missing, since the electron is not a construct to be "measured."[6] Any definition by means of operations would fail to define it. It is not difficult, however, to define it constitutively by reference to other constructs of explanation, or by certain mathematical relations which describe its properties. Magnetic field strength again is capable of both definitions: If we accept the validity of Maxwell's equations we can describe it fully in terms of measurable accelerations of magnetic poles, as is cutomary; otherwise, if we grant independent meaning to all symbols except $H$ in Maxwell's equations we can, if we desire, define $H$ by reference to these equations. The last of the constructs listed, the magnetic field, can not be defined epistemically; we must designate it as the region of space surrounding a magnet or current, or in some equivalent way.

All constructs can be divided into two groups: Those which can be defined only constitutively (electron, magnetic field, atom, molecule, wave, photon, etc.), and those which are susceptible of both definitions (mass, temperature, field strength, force, energy, etc.). Constructs belonging to the former group will be classed as *physical systems*, the others as *quantities*. Through their epistemic definitions all physical quantities are measurable; moreover, they acquire variability of amount. A combination of quantities with a system will be called a *state*. Later (section 7) we shall examine more carefully the ways in which states are formed; the simplest possible way is clearly a direct assignment of a certain number of quantities, each of given amount, to a specified system. But, as we shall see, the combination may be less direct.

To summarize briefly with the use of the conventions now set up: For every set of data the physicist creates a set of constructs, in which systems are somehow combined with quantities, forming states. Both data and states are capable of temporal variation.

We must now consider by what rules, if any, the correlation between data and constructs is made.

## 4. GENERAL REQUIREMENTS REGARDING CONSTRUCTS OF EXPLANATION

Physical explanation would be a useless game if there were no severe restrictions governing the association of constructs with perceptible situations. For a long time it had been supposed that all permissible constructs must be of the kind often described as mechanical models or their properties, but this view is now recognized as inadequate. The only apparent justification for it arises from a false interpretation of the significance of constructs, namely that which sees in them a composing part of some mysterious "reality." Our analysis has shown that they are different from what one should plausibly call nature; hence they need not satisfy the intuitive requirements of any fictitious reality. They may, in fact, be as abstract or insensible as we please (Cf. section 6).

*A.* But, while no restriction can be made as to their choice, their use is subject to very strong limitations. It is easy to find a set of constructs to go with a given set of data, but we require that there be a *permanent* and *extensive correspondence* between constructs and data. I hasten to illustrate the meaning of this unintelligible label. Let us choose two examples from extreme parts of physical theory, the first from ordinary mechanics, the second from quantum mechanics.

1. The initial set of data consists in the momentary perception of an apple in the process of falling to the ground. The corresponding group of constructs of explanation includes: two masses, those of earth and apple, each with properties described as mutual attraction according to the inverse square law. Many other combinations of constructs might be devised to correspond to the momentary set of data. But we now insist that the correspondence originally set up be maintained in time, in order words, that the changes in the state implied in the properties of our constructs correspond to the changes in the data by the same rules which defined the original correspondence. In this particular case, the rule of correspondence is that the measurable acceleration of the apple was to be proportional to $1/R^2$; we postulate that this shall be true at any later time. (This means, of course, that the acceleration shall be nearly constant, since $R$, the distance between center of earth and apple is practically constant.) Also, this correspondence shall hold whenever the apple is dropped again. This is what we mean when we postulate a permanent correspondence.—But we require more. We wish to apply the same rules, not only to every other apple, but even to every other body. Thus by assigning masses to the planets and then applying the same rules of correspondence we may arrive at Kepler's Laws, and we demand that the entirely different set of data, motion of planets, conform to our expectation. We have extended the range of applicability of our initial set of constructs by making it correspond to a wider range of nature. These matters are almost trivial; our recapitulation of them has the only purpose of showing how they fit into the scheme of a general methodology.

2. Let us pass to the second example. The group of data in question is the measurement of the frequency of a spectral absorption line. Its constructional counterpart is a multitude of atoms characterized by mathematical quantities such as Hamiltonian operators, definite assignments of degrees of freedom (number of electrons). Out of these we can build variable states ($\psi$-functions) which carry in themselves the rules of their temporal changes.

This assignment of constructs is to be permanent in the sense that a) to every state there must correspond a perceptible situation,[7] and b) whenever the measurement is repeated the same group of constructs must be applicable. The correspondence in question has to satisfy the requirement of extensibility, which may be seen from the following example. The Hamiltonian operator is modified if the absorption takes place in a gas under pressure, for then, by the properties of the constructs under consideration, it acquires interaction terms. Their influence, again in accordance with the rules of correspondence peculiar to this system of explanation, is a definite change in the frequency of the spectral line which is actually observed. Many other instances are imaginable, and most of them are experimentally verified.

There is no sense in which the correspondence with which we are dealing can be said to be a one-to-one correspondence. For in the first place, the smallest part of nature is not a datum, but a collection of data. An isolated element of perception, without relation and relata, is an absurdity. Secondly, every set of constructs contains some for which there exists no single perceptible counterpart as one may easily verify by scanning the foregoing examples.

As far as the permanence of the correspondence between constructs and data is concerned the matter is fairly clear, for whenever it fails the theory, i.e. the set of constructs, must be abandoned. The extensive character of the correspondence, however, presents a somewhat different aspect. It is not true that a theory is discarded unless it is uniformly applicable to all parts of physical nature. The set of constructs used in the first example, the theory of gravitation, is not expected to correspond to all physical data; electrical charges, for instance, are denied the property of (gravitational) mass, and this procedure is simply an arbitrary means of stopping the original correspondence. We say that the theory has a limited range of application. (That we can define mass electromagnetically and thereby widen the range is true but incidental; we could have chosen a different, but probably less familiar example.) Nevertheless we feel that there are enough data which are "explained" by this theory to warrant its retention.

There is no exact criterion to fix the domain within which the constructs involved in any specific mode of explanation must correspond to experience. This indefiniteness is reflected in the difficulty of distinguishing a theory from an hypothesis. The latter is simply a theory which has, at present at least, a limited range of application. It is promoted to the status of a theory if and when its range is deemed sufficiently large to justify this more commendatory appellation; there is no sharp distinction.

It may be of interest to examine a few physical theories from this point of view. First, there is the type whose constructs, together with their relations to nature, can be extended to a coherent range of data, but fails curiously to explain a particular set of empirical facts. Lorentz' theory of the electron is an illustration. Its central construct was a small spherical agglomeration of charge furnished with a number of abstract properties, all correlated with definite data. This symbolism explained, for a time at least, all known experiences collected under the heading of electrodynamics, except that it gave no account of the stability of the electron itself, which is demanded by its rules of correlation. The theory was not rejected on this score, but it was felt to be defective. In fact, this defect has not been removed at present.

Secondly, there is the type of theory which can not be successfully extended for the opposite reason: instead of explaining too little it explains too much. Here Schrödinger's wave mechanics in its original form, that is, before it was supplemented by Pauli's famous exclusion principle, is a suitable illustration. According to the rules of this theory, the states of individual systems such as atoms or electrons can be combined in any arbitrary manner (the resulting $\psi$-function may be symmetrical, antisymmetrical, or mixed). But no correlate on the part of nature could be found for some of these constructs. It was only by limiting drastically the permissible combinations (Pauli Principle) that correspondence with nature was restored.

Finally, it is possible for a single mode of explanation to fail on both of these counts. The early form of Dirac's theory of the electron, so successful in many respects, was guilty of this fault. It demanded the existence of negative kinetic energy electrons, constructs with properties so paradoxical that they could not possess an observable counterpart, and on the other hand it left the occurrence of electrons with positive charges unexplained. Both these difficulties are now eliminated by a brilliant modification which we shall discuss later (Cf. section 9).

We have seen that in trying to extend the correspondence between a given set of constructs and a given set of data we often encounter obstacles of an essential kind. These have always been annoying to physicists and have prompted them to generalize or modify their constructs in a way that would avoid them. Great achievements have resulted from such efforts; long strides have been taken towards ultimate unification of the constructs of explanation. Indeed the successes, especially those attained in the general theory of relativity, are so sweeping as to suggest the question: Will it be possible to explain *all* of nature by means of one great symbolism, forming a hierarchy of constructs of which the basic ones are completely extensive without break? It is the hope and the inspiration of the theoretical physicist that it can be answered affirmatively, for this prospect alone makes his efforts in the deepest sense worthwhile. But like all ideals, it provides him with no assurance that it will ever be realized—we shall probably be forever approaching the object of our dream. Nevertheless there is fun in the approach.

B. The requirement of permanent and extensive correspondence between constructs and data has an important supplement, namely that of *simplicity*. It demands that the number of unrelated constructs in any theory be small. Again, this requirement is not perfectly definite; but this lack of uniqueness is no lack of precision in our formulation of physical methodology: it corresponds to an actual indefiniteness in our choice of theories. There can be rival theories explaining the same range of data with equal success. If one involves as few constructs as the other it is difficult to decide between their merits. But faith in the ideal of an ultimate uniform explanation of nature then leads the physicist to assert that only one of the rivals can be tenable, and his usual procedure in these embarrassing circumstances is to wait for the emergence of new data which will encumber one theory with additional and unrelated constructs while merely extending the other. The former is then said to violate the requirement of simplicity and is abandoned on these grounds.

For a long time the caloric theory and the energetic theory of heat were competitors. The former assigns to the complex of data "hotness," the construct "caloric," which is to be understood as a large collection of extremely small and almost massless particles, much in the way in which the atomic or fluid construct corresponds to matter.[8] Thus the construct caloric is not entirely unrelated to others already in use; the quantities distinguishing it (compressibility, extension in space, etc.) i.e. the further constructs involved are familiar from the accepted explanation of matter; their number is small. Much the same can be said about the energetic theory, which assigns to hotness the familiar construct: kinetic energy of moving particles. The advantage of the latter was not at once apparent; we recall that Lord Kelvin was able to derive his absolute scale of temperatures on the basis of the caloric theory, Carnot explained his well known cycle and proved the theorem regarding the maximum efficiency of a reversible heat engine, and Clapeyron calculated without much error the value of the important constant (subject to a different interpretation) now known as the mechanical equivalent of heat, by using the caloric construct.

It was not until the experiments of Rumford and Davy were performed that a clear decision became possible. These experiments demonstrated that a body can become hot without having any matter added to it. In terms of the caloric theory this means that caloric can be created. Thus the construct assumes a property which sets it aside from the constructs by means of which we are accustomed to account for matter. If it is to be retained it must be counted as a *new* construct added to the others. In this respect the energetic theory is better; kinetic energy can be communicated without transferring matter, hence no further constructs need be introduced. Thus the caloric theory is exposed as inadequate because it violates the requirement of simplicity.

If we were at liberty to ignore this requirement the physicist's ideal of a uniform system of explanation could easily be attained. For we could populate the world of constructs with mechanical models capable of simulating all possible data. Poincaré, I believe, was the first to call attention to this fact, which can be demonstrated without trouble. But the number of independent constructs in this scheme would be so great that physicists would at once reject it. There are other difficulties in the use of mechanical models of which we shall speak later.

C. In part A of this section, when discussing the permanent character of the correspondence between constructs of explanation and data, we spoke of changes of state implied in the properties of constructs. State, we recall, is a purely constructional term and is not in general identical with a perceptible situation, although, of course, it must be capable of correspondence with data.[9] (Cf. remarks near end of section 3) In what manner are these changes of state implied? The answer is simple: They are enforced by our use of the *causal mode of description*. I do not wish to enter again upon a detailed discussion of the broadest sense in which the causality principle is applied in physics.[10] The usual, though not the only possible way of satisfying the causal requirement in connection with the changes of constructional states is to subject the quantities whose changes are responsible for the modification of the states, to differential equations. The order of these equations in the independent variables, co-ordinates and time, depends in general upon the number[11] and the kind of quantities (i.e. constructs!) chosen to define a state. In mechanics, where the systems are particles and the associated quantities are positions and momenta, the equations are of the second order in the time. The same is true in a great number of other fields where the symbolism of explanation has been fashioned after the pattern of mechanics.

In general the physicist can *predict* as a result of the causal nexus between successive states. What he predicts is another state of his constructs, but from this new state he may proceed immediately to data using the specified rules of correspondence. In practice the process of prediction usually takes a standard course: One seeks a general solution of the differential equations, adjusts this solution to the known state at any given time $t_1$ and finally introduces $t_2$, the time at which the state is to be determined, in place of the variable t appearing in the solution. This state is then related to a complex of data in the same manner as was the old. Possiblity of prediction of states is a correlate of causal description, but it is not synonymous with causality, as was pointed out previously.[12] It is clear, of course, that causal description also permits a retrograde determination of states in time, so that what has here been said about prediction is also true about reconstruction.

The reader should not overlook, however, that physical causality, which we have here isolated as one of the requirements for the constructs of explanation, represents an attribute of nature only in an indirect sense; for it signifies no more than that nature is so constituted as to be simply explicable, within the framework of the present methodology, by a causal symbolism. To say that nature is causal is superficial and ambiguous.

The requirements of permanent and extensive correspondence, simplicity, and causality exhaust the conditions to be imposed upon admissible constructs of explanation. The first and last of these partly overlap.

## 5 ONTOLOGICAL STATUS OF THE CONSTRUCTS OF EXPLANATION

Do masses, electrons, atoms, magnetic field strengths, etc., *exist*? Nothing is more surprising indeed than the fact that in these days of minute quantitative analysis, of relativistic thought, most of us still expect an answer to this question in terms of yes or no. The physicist frowns upon questions of the sort: is this object green?; or what time is it on a distant star? For he knows that there are many different shades of green, and that the time depends on the state of motion of the star. Almost every term that has come under scientific scrutiny has lost its initially absolute significance and acquired a range of meaning of which even the boundaries are often variable. Apparently the verb *to be* has escaped this process.

A glance at the multifarious usage of the word exist is sufficient to convince anyone of the confusion which "exists" in connection with this point. The chair exists because I can see it; the United States exist; a definite integral exists; fear exists in the minds of people; fairies exist in my imagination. I have been using the word, I think, in a manner sanctioned by refined discourse in all these instances. Clearly, if there is no more uniformity in the standard usage of the word we had better drop the question as to the existence of the constructs of explanation.

But perhaps there is a way out. We may classify a suitable number of modes of existence and then state in which of these modes electrons and so forth exist. Without claiming more than a provisional or illustrative character for this particular classification I propose to use the five preceding examples as typifying five modes of existence. It is then much simpler to answer our question, but even so we can expect no uniform answer for all cases. Masses as well as the quantities time and space, which figure prominently among constructs, exist in a sense much like that in the first example. Electrons, atoms, etc., certainly belong to a different class for they can not, in principle, be perceived. Some may perhaps be willing to ascribe to them a mode of existence such as is possessed by the United States, others may feel, rightly, that our classification is not complete and that we should provide a separate compartment in our scheme for entities of this kind. Some of the abstract constructs employed in quantum mechanics (Cf. section 7) exist in the third mode of our classification, and there are no constructs that exist in the fourth or the fifth. Our scheme is seen to become highly artificial, but why should it not be artificial? Physical science has not developed along the lines of any of the traditional meanings of "existence," and it is perfectly natural that its constructs, when analyzed according to an extraneous principle of division, should present a bizarre appearance. The main point we are making is this: physical constructs can not be said either to exist or not to exist; their ontological status has to be fixed in accordance with a more elaborate analysis of the meaning of existence. In particular, the value of a construct bears absolutely no relation to its mode of existence.

It is apparent that the term "universe" must offer similar ambiguities. We are certainly at liberty to standardize the term in any way we please. A possible choice would be to combine all modes of existence into a universe of all things. Many prefer to include everything that exists in the manner of the chair into the universe and exclude everything else, a procedure which the critical physicist is unlikely to adopt. Logicians feel free to distinguish several universes of discourse, each containing a specifiable class of things. We are not particularly interested in any special one of these definitions, but rather in the patent fact of the freedom which we can exercise. We shall define the *physical universe* as the class of all data (nature) and of all constructs of explanation. This, we feel, is most strictly in accord with the judicious use of the term.

This whole story could be repeated in connection with *reality*, but we shall not belabor the point ad nauseam. Mention might be made, however, of an etymological consideration which could serve to regulate the usage of this word. The origin of real, the Latin res, is just as diffuse as our word thing and will not help us in setting up a standard of usage. Let us recall, therefore, that real is the very exact translation (in point of meaning) of the German *wirklich*. The origin of this word is beautifully clear: *wirklich* is anything capable of acting upon, or influencing, anything else. If we use real in this sense the word loses its force, for practically anything would then be real inasmuch as it can influence our minds; but it also loses its menacing stare, and people will no longer be mystified by a reference to reality. The world would be happier if this were true! For my part, I shall use the word in this way; or rather, I shall refrain from using it, observing that it is so weak.

## 6. CLASSIFICATION OF CONSTRUCTS

There is a point of view from which constructs can be classified more naturally than was attempted in the preceding section, a point of view that will at the same time afford interesting glimpses of the historical development of physics.

Let us consider first the field of mechanics. Here we find a great number of notions which appear to have their origin among the data of nature. On closer examination it is seen, however, that they are not truly identical with data but are derived from them by simple processes of abstraction. All large scale mechanical models are of this type. Such constructs are capable of being visualized, they are "anschaulich." By virtue of their suggestiveness they recommended themselves most directly for use in an explanatory symbolism, and it is by no means surprising that they played such an important rôle in all early theories. It is important to recognize, however, that no theory can consist of visualizable constructs alone; the models must be furnished with properties which enter the stage as definite quantities like fields of force, potential energy, etc., and these are in general not visualizable. We shall indicate the close relation which this important group of constructs bears to nature by calling them *sensible* constructs, sensible not in its appreciative but in its denotative meaning.

The next and probably more important group contains electrons, atoms and many similar entities. It is not uncommon to find people believing that these are sensible in the same manner as the former class. To them an electron is a very small charged sphere, precisely like an insulated metal shell that has been touched by a charged body. And an atom is simply a lot of such little charged spheres moving about a more concentrated group of smaller but more massive spheres, called a nucleus. But they forget that in the process of making the spheres as small as they need be, the very attribute of sensibility is lost. There is no way in which a sphere smaller than a wave length of light can be imagined, just as there is no way of assigning to it a color. The claim that an electron be an "anschauliches" model is quite as anachronistic as a revival of Anaxagoras' thesis that atoms (homeomerics) differ by their tastes and colors. Moreover, every imaginable charged sphere possesses a mass besides that which is associated with its net charge, an electron does not. For similar reasons atoms, protons, neutrons etc., in fact most elementary physical systems, are not sensible. We shall not deceive ourselves by believing them at all imaginable, and therefore call them *pseudo-sensible* constructs. They borrow, it is true, a number of aspects from the realm of nature, but possess other properties which permit us to say that they could never be the object of sensory experience.

Of course there are experiences which can inform us, by a series of plausible inferences, of the presence of an electron. This series of influences amounts to the correspondence between constructs and data of which we have spoken. Replacement of correspondence by identity in this connection would, I believe, be a logical error. Nor should we say, except metaphorically or for the sake of brevity, that the electron causes a track in a cloud chamber, a point which we have also discussed previously. It is well to emphasize that this formal attitude, so distasteful to many physicists, by no means renders the vivid association between their little particle and its visible path illusory; for there is no reason why a more abstract relation can not be as impressive and vivid as one that is concrete. The great advantage of the attitude is, beside its logical consistency, the ease with which it allows the transition from one pseudo-sensible construct (e.g. little charged sphere) to another (electron wave) when such a transition becomes necessary.

Most quantities, i.e. constructs susceptible of epistemic definition, are pseudo-sensible. Consider, for example, the notion of kinetic energy. It involves somewhat vaguely the conception of a moving body, in particular an abstraction so typically expressed as *vis viva*. But this sensible characterization alone would not be sufficient to distinguish it from momentum, hence we take refuge to mathematical description and define the one as $\frac{1}{2} mv^2$, the other as $mv$. Temperature suggests features of the sensation of hotness, yet there is no way of defining physical temperature adequately by reference to this sensation. Finally, let us take wave length. When the term is used in mechanics it represents a sensible construct, since the length of a water wave can be the immediate object of perception. In acoustics, however, the character of the term undergoes a distinct metamorphosis, and it emerges as a clearly pseudo-sensible construct in optics.

Historically this latter class of physical notions emerged nearly simultaneously with the sensible constructs, but the two types were largely confused, not so much because of carelessness in the analysis of the concepts as because of the imperfection of the early theories themselves. The fact that atoms are imperceptible, for instance, could not be known before the wave theory of light was developed. It is very essential at present, however, that we recognize the abundance, even in the more primitive systems of physical thought, of elements which make no direct appeal to nature, for it is these elements which distinguish the most recent theories. Pseudo-sensible constructs form a link between the former sensible type and the abstract ones which are now to be considered.

We have seen that an electron according to the conventional prequantum view is, crudely speaking, a piece of nature with its properties modified in an unnatural manner; it is indeed a peculiar mixture of sensible and abstract attributes. As far as its sensible portion is concerned one can not be quite sure whether it is a particle or a wave. But it is impossible to maintain that it is sometimes one and sometimes the other, since that assumption would violate most flagrantly the first of the requirements laid down in section 4.[13] Thus arises the question: Is it not possible, by suppressing altogether the sensible aspects of the construct, to arrive at a notion which is free from all conflict? To clear the way for an affirmative answer we first observe that our efforts of analyzing physical thought have exposed no requirement which would bar totally abstract constructs from admission into physical theories. To be sure, there are psychological reasons why they would be considered last, but apparently the time is ripe for their consideration. Ask anyone who presses the use of physical models to give the reason for his insistence; if he faces the issue squarely he will either say that all good theories employ models, which, of course, is not strictly true and besides irrelevant, or he will point to the difficulty and complexity of abstract theories, which simply means that he is more conversant with the others and perhaps unwilling to learn anew. We must admit the legitimacy of constructs wholly insensible, constructs which permit no more visualization than a mathematical formula.

Quantum mechanics actually replaces the electron, and any other physical system, by an *abstract* construct of this sort. Previously it was presumably round, of definite size and mass, and it carried a certain charge. Now it is described by certain mathematical symbols called Hamiltonian, momentum operator and others. These symbols are placed in correspondence with nature, much in the same way as the previous sensible and pseudo-sensible constructs, and the resulting system of explanation is just as satisfactory and proper in a critical sense as any other. The question as to the "meaning" of Hamiltonian and momentum operators is no more embarrassing than that as to the "meaning" of imperceptibly small particles; neither construct being part of nature, though both are essential parts of the physical universe. It has been said that the operators in question are inadequate to define a system because they leave a large aspect of the construct undetermined; they are consequently, it is maintained, to be supplemented by further characteristics. This attitude, however, springs from the mistaken belief that a construct must be defined as completely as a datum of nature, which is clearly unncessary if we recognize the essential distinction between these two kinds of entity emphasized previously. Let us see whether the older "definition" of an electron is complete. Can we say that the particle is truly round? Is it solid, that is, does it represent a volume distribution of charge or a surface distribution? The older theory permits an answer to none of these questions, although the tendency toward visualization will forever raise them. It is, I feel, a great advantage peculiar to abstract constructs that they do not suggest questions which presumably cannot be answered. In this sense we become more modest in our claims when we use the more abstract formulation.

Historically, Hamiltonian operators and momentum operators have developed from pseudo-sensible constructs of charge and mass, and it cannot be denied that the formal resemblance between the quantum mechanical Hamiltonian operator and the so-called classical energy expression is extremely close. In fact we construct the former from the latter in most practical cases, hence the claim that we are still operating with the old construct of a massive and charged particle. One can not deny that the quantities charge and mass are being retained in the new symbolism, which is therefore not entirely free from an admixture of sensible and pseudo-sensible constructs as indeed it should not be. We have already seen that all constructs, since they must permit of correspondence with data, possess necessarily some of the characteristics of nature, in particular they contain some allusion to space and time. The Hamiltonian operator does involve space and time co-ordinates; the fact that it also contains the constants $m$ and $e$ hardly renders it less abstract. The reader will recall that every abstract notion, if it is to be intelligible, implies some reference to sensory experience.

But as to the assertion that the abstract electron still represents a particle, our answer is different. The directions for finding the Hamiltonian operator from classical expressions have been termed a cookbook recipe, and this is a very appropriate jest. It brings out clearly the fact that, in passing from the constructs of classical physics to those of quantum mechanics, we are not *deriving* conclusions from premises, but we are making a transition which has been found successful. The starting point is a heuristic one; from a logical point of view it would be better to begin with the abstract constructs and deduce from them the others by showing how, for instance, in the limiting case of very many electrons, the abstract constructs permit a sensible interpretation. History, proceeding inductively as it always does, has inverted the logical situation. It is of course absurd to speculate about the sequence of discoveries if things had had a different start; but it is at least imaginable that the abstract formalism of quantum mechanics might have been constructed before the classical analogue was known, if mathematicians had looked more seriously for fields of application of their theories. In that event, the symbols $e$, $m$, $V$, appearing in the Hamiltonian operator $H$ would probably not have been called charge, mass, and potential, and would have derived their significance merely from the fact that, if they were inserted in $H$, and the mathematical crank was turned, the results agreed with observation. Even the functional form of $V$ could conceivably have been determined by postulating conformity of results with data. The identification of $e$, $m$, and $V$ with the classical quantities charge, mass, and potential of an electron would then perhaps have been a separate discovery. It is fortunate that we have a simpler way for ascertaining $V$, but this does not change the logical aspect of the matter. Quantum mechanics, we conclude, is not *logically* dependent upon the classical sensible notions; it is a discipline in its own right, which operates with abstract constructs. Heuristically, the most convenient way of arriving at these is via the familiar classical notions.

The abstract formalism avoids the question: what *is* an electron? To see how it manages to do this let us refer to an example which has already been mentioned, the photoelectric effect. The data may be described in a compact fashion by saying that, under the influence of light, negative charge emerges from a material surface. The explanation of this phenomenon amounts to this: We suppose a region of space (the interior of the material) to contain electrons, a term which is a label for a set of directions. These directions include a method for finding the charge at any point of space, a method which depends on the mathematical form of the Hamiltonian operator. If we calculate this charge at a point outside the material the result is zero. Next we wish to take account of the datum: presence of light. This changes the form of the Hamiltonian operator in a well known manner, and if we apply again the method for finding the charge after this change has been incorporated, the result at a point outside the material is no longer zero. Nothing has been said in this procedure about what an electron *is*, yet the result is entirely satisfactory. We are not, therefore, curtailing the physical universe by using this abstract mode of explanation; we are omitting constructs which, at least in the present state of knowledge, have no counterpart among data. We are subordinating our choice of constructs more effectively to the requirement of simplicity.

If the physicist were to restrict himself exclusively to the use of sensible and pseudo-sensible constructs his dream of an ultimate system of explanation would never come true. This point argues against the view that physical reasoning should employ only mechanical models. We have seen in section 4 that the correspondence between constructs and data must be an extensive one. The ideal system of explanation is one in which every quantity appearing among the constructs has its counterpart on the side of nature, thus rendering the initial correspondence universal and avoiding all breaks. Now it is a fact that every set of sensible or pseudo-sensible constructs, and hence every model, necessarily implies more than the data which it is designed to explain: it implies something which should correspond to its internal structure. Often this something is found, and the model, originally a unit, is divided into smaller mechanisms. The question as to the significance of the internal structure of the smaller mechanisms arises again; the difficulty has been pushed in the background but not eliminated. Continuity of space acquires a parellel in the never ending conquest of the physical universe.

This is of course precisely what has gone on in physics before the advent of recent theories. The original elements of Empedokles revealed themselves as a collection of atoms. For a very long time it occurred to nobody to worry about the structure of an atom, but finally it gave birth to electrons and nuclei. At present we are splitting up the former and wondering about the constitution of the latter. The fact that on the one hand the parts of which a nucleus consists are, so far as we can speak of size, about as large as the entire nucleus, and the amazing failure of finding any empirical clue to justify the assumption of a coherent electron on the other, adds material evidence to the formal difficulty of never ending subdivision. The introduction of abstract constructs, giving rise to spatio-temporal indeterminacies as we shall see, terminates this process of dissection. This argument for abstract notions is by no means conclusive; yet it confers upon them a quality which makes us regard them with favor at present, and gives us the stimulus to try them out. No one can say that in the future their use may not be abandoned.

## 7. ELEMENTARY DIFFUSENESS AS A RESULT OF ABSTRACT PHYSICAL CONSTRUCTS

At the end of section 3 we suggested a division of constructs into two groups: physical systems and quantities characterizing physical systems. In that connection attention was called to certain combinations of systems and quantities, called states, but the exact manner in which these combinations are made was not examined. In the preceding section we have introduced a different classification of constructs from a more qualitative point of view. This classification will aid us, however, in obtaining a deeper insight into the meaning of physical states.

Sensible and pseudo-sensible constructs have their origin, at least partly, in the domain of the physical universe here called nature. It is not surprising, therefore, that those physical theories which employ, or claim to employ, only these two classes of constructs imitate in their methodological formalism the conventional analysis of data. Now data are usually conceived of as *things* with definite *properties*, the relation between properties and things being one of possession. In the field of symbolic construction this relation has found its parallel in the direct assignment of quantities to systems, in a possessive sense. Thus it comes about that theories which abstain from the use of abstract constructs formulate a state in just this manner, namely by ascribing to a physical system a sufficient number of quantities, each of definite amount. In mechanics, the state of a particle is described by specifying its position and its momentum.

All quantities, through their epistemic definition, are measurable, and the result of a measurement is an arithmetical number, or a range of numbers. Hence a state formed in the manner under discussion can always be symbolized as a group of numbers, and as the state changes, the numbers in this group vary. Calculations made on the basis of this scheme involve these numbers; they contain none of the more abstract mathematical apparatus except in so far as they facilitate the treatment of numbers. The correspondence between constructs and data is of a character that assigns to a number or group of numbers forming the result of a calculation, a definite number defining the result of an experiment.

For the sake of completeness and precision these remarks should be qualified somewhat. Observations are not exactly repeatable, hence we can not speak significantly of a single number as defining the result of an experiment. But there is a method, developed by means of the theory of errors, which permits the assignment of a single number to a series of slightly divergent observations. It is this number to which our statement referred.

When no effort is made to limit constructs to the pseudo-sensible type one meets with an altogether different situation. The systems themselves, as we have seen, no longer partake of the attributes of data, and quantities can not reasonably be assigned to them possessively, as before. To be sure, physical quantities, the only measurable kind of construct, must in some way be brought into the picture because they form the last link in the correspondence between constructs and data. But since they need not be assigned to systems in as direct a manner as before we can proceed more freely in the composition of states. In particular, any reason for expressing a state in terms of a group of arithmetical numbers has ceased to exist. We may, if we wish, bring in the entire machinery of abstract mathematical construction, provided we make sure that a transition from the selected states to physical quantities, and thence to data, is possible. Since data are dominated by arithmetical numbers, this transition must ultimately lead to numbers, but not of necessity to a single number, for it may well be that our theories determine data in terms of aggregates of numbers, that is, statistically.

The preceding paragraph sketches a general outline into which such theories as quantum mechanics fit quite accurately. States are here defined by means of *state functions*, functions of space and time co-ordinates which make no explicit reference to quantities. Only in their choice of space cöordinates do they reflect the abstract properties of the system whose state they represent. Yet while they do not suggest quantities directly they are related to them through well known rules of operation. Suppose, for instance, that an atom is in a state symbolized by the state function $\psi(x_1, \ldots x_{3n}, t)$, $n$ being its number of degrees of freedom.[14] Then, if we wish to find the value of the quantity: total energy, we perform the operation $\int \psi^* H\psi dx_1 \ldots dx_{3n}$, where $H$ stands for the Hamiltonian operator appropriate to the atom. But it is found that the number which results upon integration, while it refers to the total energy, does not in general correspond to any single number among the data obtained by measuring the energy of the atom; instead, it turns out to be their *weighted mean*. This was meant when we said that the relation between states and quantities is less direct. It is possible to get even more information from the state function, for we can, by simple expansion of $\psi$ in a series of well defined functions known as "eigenfunctions of the energy," or "pure states with regard to energy," find the probability that, in a given measurement, the result will be a preassigned number. In this sense a state function implies, not specific values of a set of quantities, but statistical aggregates of such values. The example here chosen is typical of all other quantities. I have given fuller details of the procedure in an earlier paper.[15]

Although the constructs employed in quantum mechanics appear far removed from the realm of nature, the results to which it leads are in better accord with data. For physical data, at least in their cruder state (i.e. before the divergence of the individual observations has been removed by special procedures) do form statistical aggregates and not single numbers.

It should be observed above all that the requirement of causal description, discussed in section 4 c, has not been dropped from our methodology. The physical states, to which alone this requirement refers, are still governed by unique laws of temporal variation. In the present case, the state functions obey a first order differential equation in the time, and if $\psi$ is given at any instant of time its later or earlier form can in principle be calculated with precision. Data, however, are no longer exactly predictable. Another point worthy of special emphasis is this. The recent theories, exemplified in the preceding paragraphs, do not fall outside the frame of the methodology which has here been developed. It has not been necessary to recast the fundamental principles of physical reasoning in order to bring them into line. By placing into prominence certain abstract features which have always been inherent in physical explanation it is possible to reconcile apparently divergent tendencies among the various theories of physics, and to restore to the physical world the uniformity which it threatened to lose.

Let us not forget, however, the change in the manner of predicting elementary data which is consequent upon the use of abstract constructs. We have shown how the greater degree of freedom in connection with the definition of states permitted by them elevates us above the necessity of numerical description, and how this, in turn, makes possible a peculiar diffuseness in the data to be represented by the theory. To put it another way, if we were to design a theory after the classical fashion, i.e. one whose constructs are all pseudo-sensible, which would predict similarly diffuse data, then *such a theory would be acausal*. This is the precise meaning of the famous complementarity principle formulated by Bohr and advocated so strongly by Heisenberg. Two special instances of what has here been called elementary diffuseness are of great importance. One is known as Heisenberg's indeterminacy principle; it has been discussed extensively both in the physical and in the philosophical literature—though often very superficially—hence we may here be justified in dismissing it without further comment. The other has to do with an interesting modification in the concept of a "pure case." This we wish to describe briefly.[15a]

The customary notion of a pure case as that of a (constructional) situation in which a multitude of systems, while differing in the assignment of the amounts of certain quantities, all possess the same amount of one particular quantity. Thus a group of small magnets (for instance atoms) flying through space with different velocities, different energies, each being, of course, at different points of space, might have their magnetic moments all oriented along a certain axis. This group is then said to represent a pure case with respect to orientation of magnetic moments. (The example here chosen is of concrete importance, since the corresponding experiment can actually be performed. Cf. any textbook on the Stern-Gerlach Effect.) The test as to whether or not such a pure case is present would be to pass the stream of atoms through a suitable magnetic field devised to stop all atoms which do not have their magnetic moment along the direction in question. (We need not remind the reader that we are here using the customary technical lingo which obscures distinctions between constructs and data.) Now the interesting point is this: No matter how hard we try to bring about the state of perfect homogeneity regarding the direction of magnetic moments, some atoms will always be stopped. Evidently, the attainable degree of homogeneity is not the maximum imaginable, it is distinctly smaller. There is an inevitable diffuseness in the orientation of moments, which follows clearly from the abstract definition of states. This example may in fact be regarded as partial empirical evidence for the superiority of the use of abstract constructs.

## 8. SCIENTIFIC PROCEDURE

We have concluded the search for the object of physical inquiry and thus disposed of the major and more problematic part of our task. Though the working physicist is not in general fully conscious of the rules of the game which we have outlined, he allows himself to be guided by them instinctively through a sense of fitness he has acquired in his training. This background of method is the birthplace of his problems; failures of data and theories to conform, in the manner of this scheme, provide the principal urge for his researches. And while it is true that valuable achievements are often inspired by other motives, whether or not they are to be incorporated into physics is always decided by reference to this methodology. The results of electrical engineering are not counted as physical issues unless or until they enmesh themselves in the relations and requirements here set forth.

Yet these rules are not to be regarded as a program to which physical activities must always conform; for then our methodology would be final, which as we have pointed out, it can not be. Rather, the rules are the results of our attempt at crystallizing what the physicist is doing. Hence their normative value is not absolute or enduring. Their exposition is important, however, because the development of physical science is not likely to be shifted suddenly from its accustomed course, so that the older rules of guidance may well be applied to new unruly theories. One should certainly not abandon time-tested principles without compulsion.

To complete our task let us consider briefly *how* the physicist solves his problems, that is, how he fits his details into the abstract methodological scheme. The latter involves a formal bifurcation of the physical universe into nature and the domain of constructs; to this there corresponds the usual division of physical activity into experimental and theoretical research. In so far as we have, in some of our examples, drawn a sharp line of demarcation between nature and constructs, our considerations may have appeared forced from a practical standpoint, for such a line does not conventionally exist. In principle, however, it can and should be introduced. In actual investigations it is often impossible to maintain a sharp boundary between practical and theoretical procedures, because the manner in which we explore nature is already dependent upon available constructs. Psychologically, but not logically, the sharpness of the division is an illusion. Thus it would be difficult to point out a simple experiment that does not utilize theory in some form.

Experiments can be performed with two different ends in view corresponding to the fact that nature is both start and terminus of every complete cycle of physical inquiry. Those experiments which establish a start are usually excursions into a new range of phenomena; in the majority of cases they are conducted by accident. The other, more common type, is designed to test a theory, that is, to confirm or disprove the extensibility of a set of constructs of explanation. The results, if confirmatory, are then said to have been predicted by the theory. To this category also belong the experiments which are performed to decide between several conflicting modes of explanation.

Turning now to theoretical physics we may ask: How do we go about the business of constructing theories? This question could be dealt with at considerable length, but I do not wish to burden this already comprehensive paper with incidental details. There are three main procedures, and each could be illustrated by numerous examples. The first is to begin with an internally coherent but detached system of constructs and work out the data which it should entail; but this is rarely if ever done at present since we are already in possession of too large a collection of constructs. The second is the reverse of this, namely to allow data to suggest at once a formalism of explanation closely in line with observations. This is possible and indeed common in fields where the prominent constructs are sensible or pseudo-sensible, but of little avail in others. Large scale mechanics has been developed by the use of this procedure. Finally we come to the method most widely employed. An existing set of constructs which fails to correspond correctly to an enlarged domain of nature is modified by generalization, that is by removing qualities of constructs which, though they slipped inadvertently into their original definition, are not required in accounting for any phenomenon observed. As a result of this progressive removal of obnoxious hidden properties modern physics has a tendency toward abstractness and assumes increasingly the character of mathematical analysis. The revision of the constructs of time, space, and mass in the special theory of relativity illustrate the point. Independence of these entities had unwittingly been included by Newton in their definition, chiefly, I suppose, to render them sensible. When Einstein replaced this independence by suitable relations more general than the assumption of absolute constancy, he was able to make these modified constructs represent a far larger domain of physical data.

## 9. REFLECTIONS UPON CONTEMPORARY THEORIES

As a supplement to the main theme of this paper I wish to add a brief and non-analytic exposition of two very recent theories, which, by their strangeness, have caused considerable consternation; our aim being to determine more exactly those features in them which stand in contrast to present methodical requirements. In the account that follows we shall exercise little care in distinguishing constructs from data; for if a physical situation were described with all the fullness of its logical and metaphysical implications, the result would be as tedious as a story in which each word is paraphrased by interpolations calling attention to the particular shade of meaning to be conveyed.

The first is Dirac's theory of the positron, the newly discovered "particle" having a mass equal (within an experimental uncertainty at present quite large) to that of the electron, and a charge equal in magnitude but opposite in sign to that of the electron. Our story starts with Schrödinger's quantum theory, in particular its application to an electron. The reader probably knows that this theory explains in a striking manner the behavior of the negative element of charge both when it is nearly free, as in the experiments on electron diffraction (Davison and Germer), and when it is bound in atoms or molecules. Most spectroscopic data are beautifully accounted for on the basis of its formalism. Nevertheless there were peculiar difficulties when observations were carried to sufficient detail. Spectral lines which according to the theory should be single turned out to be doublets on closer examination; and when lines were emitted by atoms subjected to a magnetic field (Zeeman-effect) they exhibited a complexity of which the theory could give no account.

In fact these difficulties were known before the development of Schrödinger's theory, for Bohr's theory, the predecessor of present quantum mechanics, had similarly failed to explain these anomalies. But in connection with the latter the discrepancies had been removed by the brilliant hypothesis of Uhlenbeck and Goudsmit, who assigned to the electron—in Bohr's theory still a pseudosensible construct—a motion of rotation, now called a spin. The literal transcription of such an hypothesis into Schrödinger's theory was impossible, for there the electron had lost its sensible character. But something similar could be done: an additional degree of freedom could be assigned, in a formal manner, to the electron, and by revamping somewhat artificially the Schrödinger theory it could be forced to give correct results.

Pauli, in his theory of the spin, succeeded in achieving this end. In a fundamental sense, however, the difficulty still loomed large.

It occurred to a number of investigators that Schrödinger's theory might be at fault because it does not satisfy the requirements of the relativity theory. Various attempts were made to remedy this defect, mainly by adding terms whose effect was inappreciable when the velocity of light was considered large, so that, in the limit of an infinite velocity of light the Schrödinger equation was restored. This procedure yielded some interesting results but did not obviate the difficulties for whose removal the spin had been invented. The spin still had to be tacked onto the theory in the manner of an afterthought in order to make its predictions correct.

This unsatisfactory situation was greatly relieved when Dirac succeeded in his attempt of reconciling quantum theory and relativity. Instead of producing relativisitic invariance of Schrödinger's equation by superficial means he recast the entire formalism on the basis of novel considerations that are logically very simple indeed. His modification amounts in fact to *two* improvements: 1) The insurance of the correct relativisitic behavior of the equations; 2) the establishment of a correct statistical interpretation.[16] When this was done the spin was no longer needed; the new theory produced automatically all the data which Schrödinger's equation and the spin theory jointly had predicted. Physicists had every reason to rejoice over this unification which represented a triumph of the same order as the consolidation of constructs attained in the relativity theory itself. The implausible spin, as we pointed out, had vanished from the picture. The term, to be sure, has been retained. Even now physicists make calculations in the basis of the older theories in which the spin figures as a separate entity, but this is done merely for the sake of convenience. The mathematics of Dirac's theory is sometimes difficult—a discipline of great logical simplicity is often forced to employ unfamiliar kinds of analysis—but it can be shown that it reduces to the form of Pauli's spin theory in certain limiting cases. Hence, when we are dealing with these limiting cases, we avail ourselves of this opportunity to simplify the computations.

Unfortunately, the happy state of affairs was soon disturbed. Upon closer inspection it was seen that Dirac's theory made a perfectly grotesque prediction. It implies the occurrence of electrons with negative kinetic energy, or negative mass. Translated into terms of data, there should be ionizing agents (as manifested in cloud chamber photographs) which grow more effective as they are opposed by matter, loosely speaking, there should be particles which go the faster the more energy they lose; under an attractive force they should be repelled. Not only are such data unobserved, they would also disqualify the entire theory of Newtonian mechanics by invalidating the law of action and reaction. We are confronted with a case where the correspondence between a set of constructs and certain parts of nature can not be extended to other parts, a situation with which we have already dealt in a general way. This limitation of theories, as we saw, is frequently encountered, and a theory is not always rejected if it appears. Hence, one might say, let us set an arbitrary limit to its validity by saying that Dirac's theory cannot be applied to make such monstrous predictions in the same sense that Hooke's Law, for example, must not be applied to inelastic bodies.

This escape is blocked, however. For on striking out the negative kinetic energy states the Dirac theory collapses. Its mathematical structure is founded essentially upon these states (state functions do not form a "complete set" without them); furthermore, their existence is already presupposed in some of the successful applications of the theory (Klein-Nishina formula). Hence, what are we to do?

At this point Dirac made a sweeping suggestion. Instead of denying the occurrence of these obnoxious electrons he postulated their universal presence. According to the Pauli exculsion principle every energy state can "contain" at most two electrons. Let us now suppose that almost all negative kinetic energy states be filled and that the incumbents of these states be *unobservable*. The vacancies in this distribution of negative electricity are then observable and exhibit the normal properties of an electron with positive charge, as can be shown analytically. If these "holes" are rare the stability of ordinary, i.e. negative electrons with positive mass is insured for they can not convert themselves without limit into the unobservable paradoxical kind because of the lack of vacancies in their distribution. Thus Dirac was led to the prediction of the positron even before it was discovered.

But the story is not yet complete. If there are unoccupied energy states among the undesirable electrons, an ordinary one may occasionally pass into one of these states by changing its energy from a positive to a negative value. Since the unoccupied state represents a positron, this process is equivalent to the simultaneous annihilation of an electron and a positron and the emergence of energy in the form of radiation. A mechanism of this sort has often been postulated in connection with cosmic processes and is not unlikely to be operative. But the reverse process, i.e. the removal of an electron from a negative kinetic energy state to one of positive energy, which by the same interpretation should correspond to the creation of a pair (positron + electron) with accompanying absorption of energy, is manifest in numerous cloud chamber photographs and agrees with observations even in many quantitative respects. These distinct successes make it difficult to look upon Dirac's proposal as an idle speculation.

In what way does the theory fail to satisfy the requirements outlined in this paper? Its constructs are abstract, which, as we now know, is no comment on its value. The correspondence between its symbolism and nature is very widely extensible as the foregoing remarks suggest. But it falls down in the point of simplicity because of its introduction of constructs to which, *by definition*, no data shall correspond. In this respect the theory recalls somewhat the abortive attempt of explaining the necessary mechanical properties of a luminiferous ether by installing in it a multitude of invisible gyrostats. One may well feel that Dirac's theory is incomplete in its interpretation and use of the negative kinetic energy states, but it seems unreasonable to suppose that it is altogether misleading.

Dirac's hypothesis is not the only recent proposal for eliminating paradoxes by inventing constructs without counterpart on the side of nature. Fermi's suggestion regarding the treatment of the continuous beta-ray spectrum is another noteworthy example. The difficulty is this. Some radioactive elements are known to emit electrons, which appear to have their origin in that very limited region inside an atom called the nucleus. It is quite difficult, in the first place, to think of an electron as being enclosed in the nucleus, but if we must place it there, as the observations suggest, then its energy must be "quantized," i.e. it can not have any energy it pleases, only one or another of a discrete set of values. This, at least, is the customary argument (which could be criticised). Now the emitted electrons have a continuous distribution of energies, and this is not because they have lost a portion of their original energy by encounters with extra-nuclear electrons as has been shown experimentally. Thus there is evidently a certain amount of energy which slips through the net, for a discrete energy distribution can become continuous only be subtraction or addition of energy, and the latter possibility is in this case excluded.

A trivial way out of the difficulty would be to abandon the principle of conservation of energy, a very radical step which has often been discussed. It is not necessary, of course, to question the constancy of the total amount of energy in the universe for the sake of these data; they only suggest that the principle may not be applicable in detail to single elementary phenomena, whilst in the aggregate processes manage to balance energetically. In my opinion such a proposal has very little force indeed. In the first place, if it is taken literally, it means simply that the elementary processes take place in systems which are not closed, in other words, that there are interactions between the nuclei, or different parts of one nucleus, which the theory is neglecting. Thus the problem is not solved, it has only shifted its ground in so far as we must then look for these unknown—and implausible—energy exchanges. But the suggestion can not be taken literally, and this is the second, more important point we are making. The suggestion has its origin in a confusion of concepts which is very common. We know that atoms and electrons, from the quantum mechanical point of view, are abstract constructs to which quantities, like energy, can not be directly assigned. To be sure, we continue to speak of them as though they were pseudo-sensible, principally to avoid cumbrous discourse. Unfortunately, this ambiguity of speech has entailed an ambiguity of thought in as much as physicists are tempted to cling to the picture of particles. The suggestion that energy be not conserved in detail can only be made by one who succumbs to this temptation; for quantum mechanically it is perfectly clear that energy need not be conserved in a single observation. If the system, in this case the electron, happens to be in a state which is not an *eigenstate* of the energy, this quantity is only statisically determined anyway, and one fails to see the meaning of the proposal.

An alternative explanation of the continuous beta-ray spectrum involves the assumption of unobserved entities, called neutrinos. Crudely, one thinks of them as particles capable of carrying away energy. If one lets them fly off simultaneously with the observable electrons the energy balance can, of course, be established. Fermi has recently given quantitative precision to this thought. One might suppose at first sight that such a theory must be highly indefinite and could prove nothing. Yet it is surprising how, in Fermi's able analysis, the results become well shaped and definite. It turns out that, if negligible (rest) mass and no charge are assigned to the neutrino and the latter is described by Dirac's theory (which is known to be applicable only to electrons), we can account not only for the existence of continuous beta-rays, but also for their distribution-in-energy.

Methodologically, this hypothesis sins against the requirement of simplicity in the same way as did the former, the only difference being that the neutrino is not in principle unobservable, but merely unobserved. Hence the proposal is capable of empirical proof, theoretically. However, the assumed properties of the neutrino place it in a position practically beyond refutation, for it must be difficult to detect a particle without mass and charge. In one sense, the hypothesis of Fermi is much more violent than Dirac's. While Dirac postulates only *one* kind of unobservable entity, Fermi implicitly assumes *three*. If the neutrino is to be governed by Dirac's equation, negative kinetic energy neutrinos are automatically called into being, and if one wishes to dispose of them by Dirac's famous trick of filling all possible energy states with them and then invoking the exclusion principle, he must account for eventual vacancies by admitting the existence of highly paradoxical entities which could be called "anti-neutrinos."—Fortunately there is little reason to be alarmed at these monstrosities, for we have already indications of a more technical kind that Fermi's hypothesis is at odds with other physical data.

In concluding I desire to make two points. First I wish to answer the critic who maintains that the division of the physical universe into constructs and data, which underlies the entire methodological scheme here presented, is artificial, unnecessary, or misleading. I suggest to him that he analyze logically, and without metaphysical bias, a simple proposition of the form: Light is an electromagnetic disturbance, and see if he can get along with less than two classes of things. The position here outlined is not founded upon philosophical convictions; it results from an attempt to classify physical concepts in the simplest possible and at the same time most inclusive manner. I have tried to analyze matters from a more definitely positivistic point of view but have not succeeded.

Secondly, I can not admit that the stand here taken with regard to the abstractness of physical constructs is too radical, nor that abstract notions could not be ultimate. If anything, we have erred in the other direction, when we claimed that all constructs, including the abstract ones, must permit some reference to space and time, these forming the substratum of nature. Laue[17] and Schrödinger[18] have recently advocated that space and time may lose their significance in connection with ultimate constructs, thus opening an avenue to even greater freedom than we have here permitted.

## NOTES

[1] As presented in the work of Mill, Jevons, Russell, Keynes.
[2] E.g., by A. S. Eddington, *The Nature of the Physical World*.
[3] Matter is a datum, but mass is a construct in the sense here advocated.
[4] See for instance E. Mach, *Science of Mechanics*; R. B. Lindsay, *Physical Mechanics*.
[5] This is contrary to the view expressed by P. W. Bridgman in *The Logic of Modern Physics*. See, however, H. Dotterer, *Monist*, 44, 231, 1934.
[6] N.B. We do not question the possibility of measuring its charge or its mass, etc.!
[7] This happens to be true; for if we allow the states of the constructs to develop under the influence of a light wave the distribution of frequencies in the spectral line changes in accordance with the initial rules of association between constructs and data. For instance, if the exposure to the light wave is short, the absorption line is broad; if it is long, the line is narrow. But it is important to observe that in this instance a constructional state does not determine a datum directly, but a probability distribution among data. This, however, is not contrary to the principles outlined above, which do not require the rules of association to be point for point direct.

Professor F.S.C. Northrop, *Jour. Unified Science* IX 125, 1939 has given an excellent analysis of what I have called rules of correspondence. He denotes them "epistemic correlations."

[8] It is unfortunate, perhaps, that one must use such stilted language to convey ideas clearly. In simply saying: "Caloric is a form of matter," we are guilty of logical confusion. The construct caloric is not the datum matter, nor a form of it. The trouble is that the word matter has a double meaning: 1) that of a datum, 2) that of a construct (collection of massive particles, fluid). Common speech, being developed on practical concerns, rarely realizes epistemological distinctions.

[9] Whether we regard states as a third class of constructs (beside systems and quantities) or whether we look upon them as associations of systems and quantities is a matter of practical indifference.

[10] Cf. H. Margenau, *Phil. of Science*, 1, 133 (1934).

[11] Cf. for instance Poincaré's example of astronomers who cannot observe the longitude of a planet. *The Foundations of Science*, p. 111 et seq.

[12] Ref. 10.

[13] Occasionally one finds in the literature the assertion that physicists apply the wave theory for certain purposes of calculation, the particle theory for others, and that they must know in advance which theory to apply in order to get the correct result. If this is taken to reflect the physicist's basic attitude, then the statement is grossly misleading. The point is that we have a reasonably satisfactory theory of the electron which assigns to it neither the character of a particle nor that of a wave. In principle, this theory can be applied to all specific cases; but its rigorous application is often very cumbersome. Fortunately we know, however, that in certain limiting cases this very theory leads to the same results as the wave theory, and in others it leads to the results of the particle theory. On the basis of this circumstance we often permit ourselves to use either of these latter, simpler theories in the solution of specific problems, or even to illuminate our discourse by reference to particles or waves. But there is absolutely no conflict between these notions.

[14] $n$ need not be thought of as number of "particles" multiplied by 3; in fact the number of degrees of freedom, if we desire to retain the term in this connection, should be regarded as a suitable number ascertained by trial.

[15] H. Margenau, The Monist, 42, 161 (1932).

[15a] Our use of the term "pure case" differs from that of von Neumann.

[16] To this end the differential equations were written in first order form, which forces statistical weights to be always positive. The second order form would admit such absurdities as negative probabilities.

[17] M. v. Laue, *Naturwissenschaften*, 20, 915 (1932); 22, 439 (1934).

[18] E. Schrödinger, *Die Naturwissenschaften*, 22, 518 (1934).

CHAPTER 5

# METAPHYSICAL ELEMENTS IN PHYSICS

## 1. WHAT IS METAPHYSICS?

Our time appears to be distinguished by its *taboos*, among which there is to be found the broad convention that the word *metaphysics*[1] must never be used in polite scientific society. When I infringe upon this custom, and the title of this paper leaves me without defense to this accusation, the reaction is likely to take one of two standard forms: the so-called "working" physicist will say (after muttering one of his milder oaths), "Why bother about metaphysics when there is enough physics to occupy every one's attention, and even more? A person who talks metaphysics nowadays gives evidence of an outmoded taste and of insufficient knowledge." Well-meaning friends and kinder souls will tolerate the infraction of etiquette, expecting that I shall redeem myself by digging up some unfortunate bits of metaphysics which have still been hidden in modern physics, by exhibiting them with scorn and disapproval and finally disposing of them in a fitting manner.

I am going to shock both groups, the first without apology or regret, the second with some solicitude and in the hope of winning their sympathy for my point of view. For I shall confess to the doctrine that there not only are, but ought to be metaphysical elements in physical science. The sense of this assertion will be clearer after a brief discussion of the historical meaning of metaphysics.

It needs to be pointed out with emphasis that the Greek preposition *meta* is not synonymous with *anti*; moreover, that the use of the word *meta* has, in the whole of the history of philosophy, never been intended to signify the relation of the *subject matter* of the field of endeavor called metaphysics to that of physics. As is well known, the preposition had a purely local sense, i.e., it referred to the particular one of Aristotle's books which the Roman editor happened to place after the books on physics. By a peculiar shift of meaning, for which it is hard to find a reason other than thoughtlessness, the label metaphysics has at present come to be attached to the field of speculation, vast and vague, which lies beyond verified or even verifiable physical theory.

There is, perhaps, a contributory cause for this misconception aside from etymological confusion, namely the identification of metaphysics with ontology. Writers on traditional philosophy have divided the subject matter of metaphysics into two great branches: ontology and epistemology. The first of these deals with questions of reality and being, the second, usually translated as theory of knowledge, concerns itself with why and how knowledge can be acquired. Ontology, which flourished as a scientific discipline in the days of Greece, continued to vegetate in scholasticism and produced its greatest flowers in Spinoza, Leibnitz and Berkeley, was utterly destroyed by Kant's criticism. The noteworthy fact is that Kant used the theory of knowledge in eliminating ontology. It was Hume who attempted to destroy both ontology and epistemology simultaneously; he succeeded, but only at the expense of destroying science, too. Fundamental hostility to metaphysics is *not* characteristic of the modern era, it has been alive at all times. And if the history of philosophy teaches anything at all it is this: Ontology is unnecessary as a basis of scientific inquiry, and this can be shown with Kant's skillful use of the theory of knowledge; complete skepticism is capable of eliminating *all* of metaphysics but causes science also to collapse, as Hume has shown. As physicists, we do want science, and we must have metaphysics too. But we reject ontology.

These remarks are not intended to provide positive evidence for the present point of view, which historical considerations could never do. They are here included to fix the place of our ideas within the framework of larger philosophic contexts. Nor do we wish to accept the details of Kant's teachings *in toto*, for Kant himself was unwilling to use the weapon of his powerful critique to its full capacity. But his philosophy stands as a memento to those who endeavor misguidedly to construct, from the results of physical investigations, a so-called objective world which is independent of its architect, and to the realist who professes to build his world from perceived colors, sounds, aromas, and tactual impressions.

Having disposed of ontology,[2] and having suggested (in far too sketchy a manner to be convincing at this point) that epistemology's fate is somehow linked with science, let us look at epistemology more closely. Inspection will show that this subject, too, can be shorn of some traditional superfluities without destroying science; for the question as to the *why* of knowledge is no longer of importance. In pre-critical times it formed the link with ontology: The mind had to come in contact with absolute being; essence had to communicate itself to the knowing subject. At present we may say, idealistically, that knowledge ennobles, or, pragmatically, that knowledge is power, depending on our individual inclinations. Or we may decline to answer the question altogether. But a similar variety of attitudes cannot be evoked by the question as to the *how* of knowledge, provided we refer to scientific knowledge. *For it is precisely the manner of attainment which distinguishes scientific from other knowledge.* Only if the how contains a transcendental reference to being which is not knowledge, as it does when concern arises about the relation of experience to truth, do we reject the question.

To sum up, then, metaphysics has been reduced to methodology of science. Thus far we follow the positivistic process of disintegration dominating Western philosophy at the present time. But further we must refuse to go. It is methodology which constitutes a branch of knowledge as a science, and if scientists themselves relax their vigilance in regard to methods of inquiry and of discovery, if their sense of order and fitness with respect to organization of knowledge becomes dull, then it may well happen that obnoxious ontological elements find their way back into science, depriving it of its foremost distinctions. The danger of such a contingency, I fear, is greater than we commonly believe.

Nor has physics been completely purged of ontological impurities even at present. There are many competent physicists, some of them brilliant leaders in research, who while regarding the electron as an ultimate constituent of some unchangeable reality, look upon the $\psi$-function of quantum mechanics as a formal and useful artifice having no "real" existence. Such inconsistencies, which spring from lack of attention to methodological considerations, have done as much to estrange philosophy and physics as the failure on the part of philosophers, so keenly felt by some physicists, to be conversant with the latest physical theories.

But if methodology is all we are going to talk about in this paper, why call it metaphysics? In the first place, because it has been treated under that heading for centuries and by the most meticulous writers. To yield the terms at present would almost certainly be misinterpreted by philosophers as capitulation to one of the standard brands of positivism. In the second place, methodology is not considered by physicists, at least in this country, as an integral part of their science; publications dealing with it still appear in journals customarily devoted to problems of metaphysics. Finally, there is an important sense in which methodology is different from the typical attitude of the physicist at work, a sense which Whitehead[3] has expressed in his distinction between "homogeneous and heterogeneous thought about nature." Were we to deny to methodology the attribute of metaphysics, we should run the risk of failing to draw a fundamental distinction. For instance, we shall be compelled, as a result of further considerations, to regard causality not as a regularity describing sense impressions or inherent in them, but as a mode of deliberation imposed upon them (though not in so fundamental a sense as to make causality a "category" à la Kant). This state of affairs and many others would, in our opinion, be inaptly described unless it were called metaphysical.

Terminology is always troublesome. I know from past experience that statements meant in all innocence are often suspected of hidden meanings. Thus the foregoing phrase, mode of deliberation imposed on sense impressions, simply swarms with possibilities of interpretation which are not intended. There is no need of defining thought or deliberation, nor sense impression; these forms I regard as primary and understood by all. But if you are tempted to ask: imposed by whom? maliciously hoping to have me answer "the human mind" or "God," I refuse to enter the trap. The answer is quite unnecessary: you can have a wave without an ether; you can have a datum without a giver; you can have a fact without an ulterior agency. This point is worth mentioning becuase the illicit inference from object to existing subject has done much harm in philosophy.

Thus far my aim has been to separate out and to describe that portion of metaphysics which the physicist must regard as possessing relevance to his science, and this was done by eliminating certain branches of traditional philosophy which have become dispensable. Another approach, which will now be sketched, leads to precisely the same irreducible metaphysical residue. Modern positivistic criticism has emphasized forcefully the precariousness of propositions which can in principle not be verified. In the parlance of empiricism such propositions are meaningless. It is my conviction that this recognition should be one of the premises of the philosophy of science; it enjoys, indeed, almost universal assent. To go further than this, and to prescribe the exact manner in which verification must be conducted appears to be unnecessary because the fundamental limitation just stated is sufficient. It is for this reason that I do not regard operationalism as a necessary scientific doctrine. For operationalism in the extreme form which requires that all verification be of the manipulatory type—a form which is disavowed even by so ardent and distinguished a protagonist of the operational point of view as Bridgman[4]—certainly does injustice to theoretical physics, while in its modified form which permits inclusion of mental operations it amounts to no more than the axiom of verifiability.

Let us apply this axiom to propositions involving various kinds of experience, first the simplest: scientific data in the form of *sense perceptions*. In all that follows these will be taken as primary facts; more will be said about them later. Perceptions stand in very simple relations, e.g. spatial or temporal, to other sense perceptions, giving rise to propositions which can be verified by immediate inspection.

But they also bear relations to other elements of scientific interest, elements not given in apprehension and generated in more indirect processes of reflection; they will here be called physical *constructs*, or symbolic constructs (although the term "symbolic" which I have used formerly may easily cause misunderstanding). To exemplify the distinction: the perception "blue" is related to a certain range of wave-lengths (construct) of the electromagnetic spectrum. Propositions involving such relations are or are not verifiable by methods familiar to the physicist. In the present example this is clearly the case. Let us now see how verifiability may fail.

The most obvious failure occurs when scientists, perhaps by a lapse of caution, proceed to extrapolate relations of the type in question beyond consciousness, as in the statement: the blue color is caused by an external object other than that whose attribute the color is in consciousness. Propositions of this sort, though very common, are unverifiable and hence meaningless. They define the domain of metaphysics which should be abandoned in science, for they arise from an altogether erroneous notion of causality and have as such been strongly criticized by Kant and his followers. Objectionable constructs of similar nature appear most frequently when relations valid between perceptions, or between perceptions and verifiable constructs, are projected in some vague way beyond experience; we shall call them *ultra-perceptory*. To this class belong, besides the external world, the *Ding an sich*, such physical notions as the luminiferous ether, simultaneity in different Lorentz frames, the phase difference between the electrons in an atom, and perhaps the neutrino. Physical constructs which, through their relations with perceptible facts, are verifiable, will be called *infra-perceptory*.

Was it necessary to introduce such terms as ultra- and infra-perceptory? In the minds of many, reasoning of this sort is loose, designed to smuggle across the threshold of uncritical acceptance larger chunks of metaphysics than the reader would, without befuddlement, be willing to admit. Hence let us look at the situation carefully and without preconception. There can be no question about our having perceptions (sense impressions). There can be no question about our having thoughts, or ideas, which involve elements (often called concepts) not identical with immediate perceptions. The variety of such elements is so great as to be confusing and therefore requires careful study. It would be perfect folly to deny either their prevalence or their scientific usefulness. Whether we could get along without them is by no means the question: amebas probably do. The point is that their complete destruction would destroy science. Let us fix our attention on two such elements: the electron and the transcendental God. Both are non-perceptory. But the electron is distinguished by its entering into discourse in a meaningful manner; the statement: the electron has a charge equal to $4.70 \times 10^{-10}$

e.s.u. may be related to sense impressions and may thus be verified. The attributes of God, however, while they can be stated for emotional effects, cannot be thus verified. The latter idea arises, figuratively, if we point our inquiry into the void beyond perception and is therefore called ultra-perceptory; the former is found, again in a rough figurative way, "on this side" of perception and is called infra-perceptory. Thus it would seem that we proceed with a minimum of impediment if we recognize in our activities no more than (a) the difference between the perceptory and non-perceptory elements, (b) the difference between ultra- and infra-perceptory elements among the latter.

I apologize for the use of new terms which I coined with very great regret. But I could find no words that are not preempted. It seems that in philosophy terms once used in a philosophic system acquire meanings and flavors which spoil them for later use. For instance, on careful consideration I had to conclude that Carnap's criterion of reducibility, which effects a distinction similar to that between infra- and ultra-perceptory concepts, is not applicable because of its emphasis on discourse rather than scientific experience (among other minor things), while Kant's distinction between immanent and transcendental concepts is not the same in a technical sense as that here contemplated.

To study the role of, and the relations between perceptions, infra-sensible elements of our experience, and between both, is the legitimate object of metaphysics. This subject matter is always at hand whether the scientist sees it clearly, dimly, or not at all. In many positivistic systems of science it is simply disavowed, with the result that it drifts like a mushy residue through the entire body of organized scientific fact, uncontrolled. The scientist attentive to methodology exposes it, analyzes it, and shows its structure, for it has structure indeed.

## 2. GENERAL FRAMEWORK OF PHYSICAL METHODOLOGY

To go from here to a more detailed inspection of the methodological structure of physics, it is necessary to agree on some very general conventions, particularly respecting terminology. The choice of convention, in fact the very choice of terms, is likely to add some philosophic flavor to the scheme. Fearful of giving occasion to purely verbal arguments in this connection, I should like to emphasize that the account of the general methodological framework to be presented contains much which is arbitrary. I do believe, however, that it is more closely integrated with modern physical research than many other more definitely traditional or more definitely operational formulations current today.

In the first place, it is well to have a name for the type of experience called perceptions, or sense impressions. I mean by them the class of immediately intuited awareness, such as the color blue, the perceived noise, or a greater complex of awareness as the seen chair or sunset. The totality of these I propose to call "sensed nature." In fact I shall often omit the word sensed, because reference will not be made in this discussion to any other kind of nature. This limits the meaning of nature drastically and perhaps artificially. It leads to the apparent absurdity that sensed nature possesses no continued identity: the chair is part of nature only while it is being seen. All statements about sensed nature are of a very primitive type; the slightest generalization impairs reference to nature except in an indirect way. A proposition about dogs does not deal with sensed nature unless it is a demonstrative proposition about a particular dog. Statements about past perceptions no longer deal with nature. Atoms, electrons, light waves are not part of sensed nature. There are many other infelicities inherent in the use of the term, but I feel that *they should be incurred* in order to endow the word with the precision to which the physicist is accustomed. For his *data* are the elements which constitute nature in this sense; when he has recourse to observation in order to verify his theories he refers to sense impressions of this narrow kind.

It is perhaps pedantic to exclude from sensed nature all past perceptions. Whether this restriction is upheld rigorously makes very little difference, for one can always take the stand that a seen record, a mark on paper, which may belong to nature, is equivalent to past perceptions. At any rate it is the sense of the present terminology that recourse to facts reconstructed in memory does not constitute recourse to nature.

Aside from these minor advantages, what is gained by this arbitrary definition? One answer is this: sense impression possesses a degree of immediacy, of singleness and of freedom from choice which is common to no other element of experience. This is precisely the attribute we wish to impart to nature. The very fact that the *same* set of perceptions can occasionally be explained or symbolized by *different* theories in a more or less abstract manner confirms this point. A spectral line seen on a plate has a singular vividness which the wave-length associated with it has not, hence the former is here said to be part of nature while the latter is not. It is true, of course, that sense impressions are not always above scientific suspicion; they may be what is commonly called "subjective" (color blindness). If this is felt to be an imperfection then nothing can be done about it; for we know of nothing that is more surely given than sense impression. To repeat: immediate perceptions, and they alone, are said to constitute sensed nature because of the involuntary, immediate spontaneity with which they occur.

How are we to widen this definition in order to make it conform more nearly to common usage, but without opening the flood gates so far as to admit into nature every item of the physicist's repertory? Do we wish to count as part of nature purely abstract entities which have no direct observational counterpart, such as vector potentials, quantum-mechanical state functions? If we do there appears to be very little gain in using so vague a term. If we do not, I fear one has to limit the meaning of nature to that class of experience to which in fact the physicist has recourse in his observations.

Other definitions are possible, but only on the basis of special, avoidable metaphysical assumptions. Indeed the unsophisticated person who sees no reason to exclude from nature the tree in front of his house while he himself is away makes tacit metaphysical assumptions which could not stand in the face of Hume's or Berkeley's criticism. Now my private opinion is that these assumptions are entirely valid and that, therefore, I may just as well include the tree in nature. But I think it important to point out that *physics can get along with an interpretation of nature which is not tied to these metaphysical assumptions*.

It is clear that the terminology here employed differs from that in Whitehead's *Concept of Nature*. Unfortunate as this may be, one should not forget this fact: during the 20 years since Whitehead's book was written physical theory has undergone radical changes. It has acquired an abstractness which makes a revision of older ideas worth while. In particular, while there was sense in populating nature with all the mechanical entities known to classical physics, that sense is certainly lost at present. Closer agreement with common usage could perhaps be achieved by introducing, in addition to sensed nature, a *postulated* nature which is the sum total of all physical constructs. But in doing this we should be guilty of the sin of "bifurcating" nature, to which Whitehead so vehemently objects. At any rate, nature in this composite sense is nothing other than what we shall later call the physical universe.

In belaboring this point I confess to an endeavor of justifying past actions. For I have formerly used the term nature in this monstrous way[4] and have seen eyebrows raised in print. Nevertheless the discussion here was not altogether a digression because the idea described is needed later on. Conscientious objectors may call it the class of immediate perceptions or anything they like.

Sensed nature subsumes but a small fraction of the features which constitute physical experience. Of the remainder, we now single out all infraperceptible elements of interest to the physicist. To name a few: *mass, force* (except when they are taken in an immediate and particular sense; in physics, as everywhere else, a word often signifies a perception as well as the generalized, abstracted features of a class of similar perceptions), *energy, charge, wave-length of light, field strength, potential, probability, amplitude; —voltmeter, crystal* (last parenthesis repeated), *magnetic field, atom, photon, electron, meson*. These typical examples have been listed in a special order on which we shall comment later. All such elements, which have the attributes of being infra-perceptible and of occuring in some physical theory, are here called "*constructs of physical explanation*." This term was chosen[5] to emphasize the following facts.

These elements are not and can never be part of sensed nature; they serve in fact the purpose of *explaining* nature. By physical explanation the physicist means nothing more than the establishment of organized relations between constituents of nature and these elements. Every other interpretation, as the search for the Causes of Things, draws in more metaphysics than the minimum we are admitting. Moreover, these constructs do not partake of the spontaneity and immediacy that distinguishes the elements of nature; they can, in fact, be freely chosen, can be generated *ad libitum*. There is no criterion able to distinguish their suitability for physics *at birth*. It is to expose clearly this freedom in the manner of generating them that they are here called constructs. We wish to avoid *ab initio* the misconception that a good construct must be one that is visualizable, or mechanical or any other such thing.

But certainly, not all constructs of explanation are of equal interest to the physicist. Some he rejects, others he retains in his theories. It is the central problem of metaphysics, or methodology of science, to establish the conditions under which constructs of explanation are to be rejected or retained. And this is not a matter for casual notice, but worthy of the physicist's careful attention. The problem is of course not solved, but an attack will be made upon it in Section V.

There are several ways in which constructs can be classified. It was noted that some recall elements of nature more directly than others (the first two in each of the two groups of examples separated by a dash). The last examples in each group are highly abstract. In a very rough sense, some are closer to nature than others. In view of this circumstance I have in a former publication[5] divided all constructs into three not sharply separated groups: sensible, pseudo-sensible, and abstract constructs. The emphasis entailed by this distinction is, I think, well placed; for it exposes the fact of great diversity among them and eliminates the widespread preconception that all constructs must in some manner be sensible.

Another, perhaps more important principle of division is this. The examples preceding the semicolon (first group) are all capable of being measured quantitatively and can thus be specified by numbers; the second group cannot. I am perfectly aware, of course, that the *density* of a crystal, or the *mass* of an electron, etc., can be measured. Each entity of the second group comprises, as it were, a number of those in the first. I call constructs in the first group *quantities*, constructs in the second, *physical systems*. The former are measurable and specifiable by numbers; the latter are not.

Formerly[5] I have examined quantities and systems from the point of view of how they can be defined, concluding that quantities can be defined in two ways (constitutively and by operations), systems in only one (constitutively). There is nothing I now wish to add to this discussion, except to say that I do not regard the manner of definition as of equal importance with the foregoing distinctions.

What formal role do constructs play in physical explanation? Stripped to its essentials the role is this: Sensed nature as such, though spontaneous and vivid, is felt to be contingent, incoherent, unpredictable. But it is found that by mapping, so to speak, the elements of nature upon constructs, a constructional ("symbolic" if you please) system is obtained which is more co-herent and more tractable than nature. Arbitrary rules of combination can be imposed upon these constructs in such a way that one can reason about them. After such purely rational transformation of the constructs, the results are mapped back upon nature, where they now "predict a phenomenon." If nature verifies the prediction, the particular act of explanation was successful. Complexes of constructs, together with their rules of combination *and* the rules for mapping, which are—within limits—generally successful are said to be verified theories.

Now an analysis of verified theories shows that the rules of combination to which constructs are subjected, have specifiable features which are isolated and studied in metaphysics.

The preceding exposition concerning the role of constructs is admittedly vague. Some detail will follow, but to exhaust the matter would require too much space. It will suffice here to fix in mind the following simple outline: Every physical explanation involves a circuit which starts and ends in two different elements of nature. It is never closed. It swings through the field of infra-perceptory entities, sometimes remaining in the neighborhood of nature (mechanics), sometimes going far out into the range of pseudo-sensible and abstract constructs (electrodynamics and quantum mechanics). Methodological considerations guide its way while in the field of constructs.

### 3. REALITY

Before going farther it seems well to settle a question which may have troubled the reader in connection with these formal speculations. Which of the numerous entities appearing on the scene of the present analysis are *real*, and which are not? As a physicist, this question holds no concern for me. In fact, I believe I should have very little interest in it if I were a philosopher. For there is absolutely nothing to be found in the part of epistemology needed as a basis for science, which would provide a criterion for reality. To assign reality only to what we have here called nature would be very arbitrary indeed and would, I think, offend the sensibilities of scientists severely. If, therefore, I were to define the realm of physical reality I would let it include nature plus all constructs which occur in physical theories held valid at present. To this realm I would also apply the name: physical universe.

In the eyes of many, such a use of terms would appear to have two defects: First, reality would change as new discoveries are made and as new theories are developed. To deplore this would imply a desire to find *absolute* reality. But absolute reality is ultra-perceptory and hence is of no interest in science. For no amount of observation could ever verify absoluteness. The view here taken might be described as affirming *dynamic* or *constructive* reality, and I fail to see that it is even esthetically less satisfying than the postulate of an absolute, static reality. The physicist does not disover, he creates his universe.

The other apparent defect of the present view lies in its inability to specify exactly whether something is real or not, even at present. There are theories whose status is in doubt, hence it is questionable whether constructs which they alone contain (cf. neutrino) are real. But this penumbra reflects precisely the attitude of physicists toward these constructs; its presence is not a fault but a virtue. The answer to such verbal questions (I would not call them meaningless because they can be answered on the basis of our definition) as: can a thing be real without being known, is obvious from our point of view.

At this juncture, permit me to digress; permit me to pay homage to a construct which is definitely ultra-perceptory, but which is nevertheless held dear, in a religious sense, by a great and distinguished number of living physicists. It is the idea of an *ultimate* reality toward which dynamic reality slowly strives. One sees dimly the figures of Plato, of Kant, and of Goethe behind this misty hope, for only hope it it. No one can say whether, as physical theories become more refined and general, they will exhibit a tendency toward unifcation and not toward diversification. At present, opinion on this point is sharply divided. If, however, a contact between science and art, or science and religion *must* be made, I would rather have it occur at this place than anywhere else.

## 4. THE LOGICAL POSITIVISTS

There are some similarities between the present point of view and the tenets of logical positivism.[6] Aside from displaying a high-strung disaffection for the verbalism of metaphysics, which we share but mildly and without missionary feeling, the logical positivist also endeavors to tie the concepts of science to the solid ground of immediate observation. He recognizes, as we do, the *ultimate* qualities of perceptual experience and its unique function in verification. His criteria for discarding concepts (or "terms") from intelligible discourse are in essence not unlike the rules which permitted us to ban ultra-perceptory constructs from physics. He also emphasizes the need for methodology as distinct from the more normal activities of the scientist, although he calls it logic of science, reserving, it appears, the term methodology for the description of the external behavior of working scientists. His "observable thing predicates" function very much like our "sensed nature," and his "thing language" might be said to contain all propositions involving elements of nature plus sensible constructs. His failure to draw as sharp a distinction as we advocate between terms like hot, red, etc. ("nature") on one hand and generalizations like temperature ("construct") on the other is not perhaps of great final importance, and is probably due to the fact that he has a greater desire to break, once and for all, with traditional philosophy than we profess. However, the differences between the points of view are more noteworthy than their similarities.

For the claims of the logical positivist are much wider than those here made. In disposing of elements such as ultra-perceptory constructs, I have tried to be careful to say that they are not needed in physics or in that part of metaphysics which is rightfully associated with physics. But I have left them otherwise intact. We have thus reserved space for non-scientific pursuits whose legitimacy the physicist has no right to question. The present point of view can at best be called *restricted* positivism. The success of physical procedures does perhaps distinguish physical methodology above all others, but success alone cannot justify dogmatism. The logical positivist claims, or at least infers, that terms not reducible to the thing language are devoid of *meaning*, when we should prefer to say that they cease to be the physicist's professional business. Thus, when logical positivism turns towards non-scientific fields of endeavor it tends to become dogmatic yet without dogma: the present analysis simply becomes inapplicable.

A very definite departure from the precepts of logical positivism should be seen in our willingness to consider seriously the primary elements of experience (perception, thought), instead of starting the analysis on the secondary plane of *language*. The thesis that cognition which cannot be expressed at all is uninteresting, is true but irrelevant. For if it is inexpressible it is also presumably unverifiable and as such would not constitute part of the subject matter of physics. On the other hand, if cognition is simply unexpressed but verifiable there is no reason for its exclusion. It has always seemed to me that recent positivistic doctrine comes very near the assertion that there is no difference between an idea and its expression. Once this is granted philosophy does become identical with syntax, and language may profitably be made the starting point of methodological analysis. I doubt, however, whether many physicists or philosophers would subscribe to this radical assertion. There is certainly not even a one-to-one correspondence between ideas and their expression in verbal form, since an idea can be expressed in a multitude of ways. Hence analysis of language, if carried through as such, would of necessity be vague, would maintain itself continually on the periphery of meaning. The reason why logical positivism is not entirely vague is that it does not conform to its own maxim, but operates with *concepts* under the guise of "terms." From a wholly unsophisticated point of view, it is hard to see how it could proceed otherwise. Aside from the fundamental error of confusing primary scientific experience with language, the doctrine in question accentuates a phase of activity in which the physicist, of all scientists, is least interested. To him the grasping of an idea, without the encumbrance of lingual context, is often far more important than its verbal exposition. I maintain, in fact, that there are physical constructs and physical theories which no one can comprehend by merely exposing himself to lingo; one sees their light only after one has actively worked with them. Nobody understands the exclusion principle unless he has applied it, himself, to a specific case. Emphasis on language, or even more general symbols, is simply out of place.

It is a curious fact that the positivist, with all his emphasis on observation, should fall back upon language as his medium of verification. Language, even the thing language, is not part of sensed nature except in the immediate act of perceiving the words. Otherwise words are symbols, constructs of a special kind, and possess none of the immediateness, spontaneity, and contingency we have ascribed to sensed nature. It thus appears that the positivistic element, for the sake of which metaphysics was first sacrificed, has now been liquidated also by this movement in its desperate efforts to defend itself against the inroads of abstract science. Mach saw this point quite clearly; the logical positivists appear to have overlooked it.

According to Carnap an abstract scientific term gains meaning through reduction relations which connect it with terms of the thing language. There can be no quarrel with the truth of this statement, but certainly with its significance. For it is nothing but a weak reflection of the recognition that the physicist *understands* his constructs, that there exist rules for correlating them with the elements of sensed nature. These rules are sometimes intuitive and unstated, more often operational procedures, occasionally of a more intricate logical character. If necessary, they *can* always be stated, and alas, they can be stated in a vast variety of ways! This is the price one has to pay for indulgence in verbalism: as soon as one chooses a particular reduction relation to "define" a term, there arise a million others, also clamoring for acceptance. Simplicity and coherence cannot thus be achieved.

## 5. THE METAPHYSICS OF SYMBOLIC CONSTRUCTION

Constructs possess no generic properties which make them good or bad, acceptable or unacceptable; they derive their validity entirely from the success they achieve in providing satisfactory physical explanations. The word satisfactory should not give pause, for it will be elucidated presently. In the sense, then, that validity and hence reality depends on *function*, my view is definitely *pragmatic*.

But clearly, a very great number, if not an infinite number, of constructs will provide a physical explanation of a given set of data. How are we to decide which of these explanations is satisfactory? What restrictions must the constructs satisfy in order that their scheme be acceptable? The answer to these questions is important, since it fixes that elusive quality which bestows reality on constructs. We give it, not as a result of reflection on what *ought* to be, but simply as the result of an empirical analysis of present, accepted physical theories. The rules to be discussed should not be taken as categories in the Kantian sense, nor as having any intrinsic evidence *a priori*. They transcend empirical observation only in that they regulate and rise beyond it, for which reason they might be termed empirical in a secondary way; they have normative power in our thinking only because they command almost universal assent. When they change ever so slightly, as they have recently, the structure of physical science shakes.

In other papers[4] I have attempted to give a more complete analysis of the rules concerning constructs than is called for here. Let me, then, give a brief summary with a few clarifying remarks about matters of interest to the physicist. Two distinct sets of rules are to be distinguished.

(1)  Rules of correspondence between the elements of sensed nature, and constructs.
(2)  Rules governing the use of constructs.

We first consider the former class. Its purpose is to make possible what was previously called the process of "mapping" between sensed nature and constructs. The rules of this class are similar to Professor Northrop's[7] epistemic correlations. Their complete analysis and classification are still outstanding; for the present purposes a few descriptive remarks will suffice. By far the greatest number of the rules of correspondence are intuitive and so ingrained in our normal behavior that it is difficult to recognize them as not belonging to sensed nature. Thus the various patches of light now perceived at once transform themselves into the construct: physical object (desk). Every physical object is a construct and is related to certain groups of sense perception through such intuitive rules of correspondence. It seems to be a characteristic trait of the occidental mind that it performs this correspondence automatically and with an awareness of its inevitability. Oriental art, according to keen observers. cannot be fully grasped without deletion of the sensed —external object correspondence.

A more definitely physical kind of correspondence is the operational one. It is usually more indirect, proceeding in most cases by way of the external object. Thus patches of color, or a group of tactual impressions, after having been identified as a physical object, may then be said to have a certain mass. Mass is a construct which is set in final correspondence with sense impressions, although no precision is lost if we say, as is customary, that it refers to, or is a property of, the object in question. Operational correspondence implies the existence and the specification of standard manipulations by means of which the construct is established. Whenever such operational correspondence is obtained the construct takes on, in the view of most physicists, a peculiar and desirable clarity, a clarity which has often begotten discovery. It is therefore a good rule to encourage the use of the operational method. But in singing its praise I would not declare it Lord of All.—As was already noted, every *quantity* in physics stands in operational relation with other constructs and finally with sensed nature.

More abstract rules of correspondence form the bridge between, say, "wave functions" and sense experience. They may, though perhaps not without strain, be phrased operationally. We wish here to emphasize the lack of directness of the correspondence between abstract constructs and nature. Suppose we map a wave function on nature—the opposite process is a little more difficult to describe. The first steps, the calculation of probability amplitudes for the values of some observable and then the probabilities, are purely mathematical. Next, the interpretation of the probabilities, is more definitely operational but it lands us, not with a single predicted sense impression, but with *many*, as the nature of probability statements requires.

This is by no means a dilemma; it is simply a reminder that the correspondence between constructs and sensed nature need not be point for point direct. However, if a construct were ever invented which did not permit reduction to the elements of sensed nature at all, it would have to be abandoned as being ultra-perceptory.

We now consider the rules of the second class, those governing the use of constructs and their relations amongst one another. Here we encounter first the rules of formal implication which are employed in physics as everywhere else. To them is devoted the study of formal logic. Being an integral part of physical methodology, formal logic has of course an important bearing upon physics. Suffice it to say here that thus far physical theory has moved entirely within the compass of two-valued logics, that there has been no significant departure from the law of the excluded middle, notwithstanding considerable discussion of this point. Some day, no doubt, such departures will be tried; and I believe they will be more incisive than the present fashion of reinterpreting the meaning of probability as involving a scale of truth values, and subsequently claiming that the physicist, because he uses probabilities, uses many-valued logics.[8]

Aside from the rules of logic, the use of constructs involves four large principles: permanence, extensibility, causality, and simplicity. This list may not be exhaustive. Moreover the division here given may not be unique, and certainly the terms may be ill chosen. I am using the scheme because in my own thinking it has attained increasing clarity and usefulness.

*Permanence* of the constructs of explanation is a very basic requirement. I mean by it several closely related things: first, that constructs do not change their meaning; second, that the rules of correlation implied in a given construct are stable; and third, that constructs persevere in time. Illustrations of these simple properties are hardly necessary, for it is difficult to name even a discarded physical theory which violates these rules. The mass or length of an object does not change from instant to instant, and when the physicist has established the existence of a certain potential at a point he assumes it to be invariable except when significant changes occur in sensed nature. But there is one aspect, arising on the very lowest plane of physical inquiry, in which permanence *is* of interest. As was shown, the elements of sensed nature are spontaneous and fleeting. The patch of brown disappears when I look away, tactual impression ceases when I remove my hand; yet the table, as a physical object, persists. The table is a very simple kind of construct, but is already endowed with the property of permanence. The present consideration shows that the (naive) realism of the unsophisticated is indeed a simple instance of physical explanation: it constitutes a circuit which starts at a given set of sense perceptions, swings by intuitive rules of correspondence to the construct: physical object possessing permanence, and returns to a set of perceptions similar to though not identical with the original group. More advanced physical explanations are not different, in a fundamental sense, from this elementary circuit; they differ only in complexity and in the use of more abstract constructs.

Next, I illustrate the meaning of *extensibility*. It is said that Newton conceived of the construct: gravitational mass (which derives its full significance from, and implies, the law of, universal gravitation) when he saw an apple fall. This concept of mass, we may safely assume, would not have been retained in physics had it only accounted for the falling apple, or even had it been applicable to all falling apples. The point at issue is that its reference could be *extended* to all known bodies. Indeed Newton looked for "confirmation" of his theory to astronomy and found it applicable to the moon. This extensibility is generally expected of physical constructs; it is the rule we are here discussing. There is no shortage of other examples. A question, however, arises here. Is it necessary for a construct, such as mass, to be extensible to *all* phenomena? The answer is very clearly: no; but the question is worth further analysis. To make it brief we recall that mass can correspond to nature ("phenomena") only via the construct: physical object, and there are many elements of nature which cannot be subsumed under this class. Perhaps the original question should have been: Can the construct "mass" be applied to every construct "physical object"? In that sense the answer is probably yes.

The *degree* of extensibility is of very great concern to physicists, and it is part of their science to outline ranges of extensibility for their various constructs. Whenever one construct is more extensible than another and otherwise equally acceptable, the latter is abandoned in favor of the former. As a case in point we cite the renowned "conflict" between the undulatory and the corpuscular theory of light. The situation was merely this: the idea of a wave, together with other attendant constructs (wave-length, frequency, etc.) was found to be widely extensible in the field of optics, but not completely so. The photon construct proved valid in connection with a large part of experience. There was no fundamental reason why these two complexes could not have lived happily side by side. The uneasiness felt about the situation was an expression of surprise at the unexpected lack of extensibility of the two constructs, not of a basic dilemma. The feeling was proper, for it gave theorists the impetus to look for a new, more extensible set of constructs which we now possess in the quantum theory of radiation.

I proceed to *causality*, discussed at much greater length elsewhere.[9] To understand its meaning properly, a few of the details of Section II must be recorded and amplified. There a distinction was made between quantities and systems. Now it happens that physical theory always involves the representation of systems in time. With every physical system there is correlated a group of quantities, and these are said to define a *state*. The complete knowledge of its state is tantamount to a representation of the system. I hasten to make these generalizations more concrete.

A small sphere is a physical construct of the type described as a system. To it are assigned numerous quantities (of which the physicist conveniently thinks as properties, in a possessive sense, though this is not always true) like volume, color, odor, temperature, mass, velocity, position. Of these a certain set is selected, velocity and position, and this is said to define its *mechanical state*. Why these two are selected will become clear presently.

A body of gas is a system having numerous properties including all those mentioned in connection with the sphere. Of these, a certain set, volume and temperature, is selected, and it is said to define the thermodynamic state of the gas.

A distribution of electric charge together with an electric and a magnetic field constitutes a system to which can be assigned quantities like field strengths, potentials, colors, odors, charge density together with all its time rates of change, and many others. Under certain conditions which are not of interest in this connection, $E$, $H$, and $\rho$ can be selected to define the electromagnetic state of the system.

In these three instances the selection of quantities constituting a state has been made in such a way that specification of the values of these quantities at any one instant of time allows their determination at all other times. The determination is made through *laws of physics* which are specific rules for combining constructs. *Causality implies that this procedure be possible.* A theory is causal if a set of quantities can be found whose propagation is self-regulatory through laws which remain unchanged in time.[10] A theory can fail to be causal because no set of quantities can be found which is governed by timeless laws, but also because the wrong choice of constructs has been made in defining a state. Ordinary mechanics would not be a causal theory if states were defined in terms of colors and sizes.

Causality therefore has meaning only when applied to constructs of explanation, i.e. to theories. To speak of sensed nature as being causal or noncausal has no significance whatever, and arguments about this point are purely verbal. Causality is a methodological rule to be imposed on constructs, and *there is so far no example of a valid physical theory which ignores it.* Hence I include it in this list.

But is not this claim in flagrant contradiction to the pronouncements of the very men who invented quantum mechanics, and whose authoritative voice has repeatedly proclaimed it to be a non-causal discipline? It is in contradiction, and I note it with utmost deference. But the divergence of views is not one of physical, but of methodological interpretation. Let us first examine quantum mechanics, and then the thesis that it is non-causal.

In its original (Heisenberg) form quantum mechanics operated with matrices which took the place of continuously varying classical observables like position, momentum, energy. States were usually defined in terms of these matrices and it was found that the laws regulating these states did not permit prediction of specific values of all observable quantities. During this early period, a very clear convention regarding the meaning of state did not prevail, hence it is not very fruitful to speculate about the causal status of the early theory. Very soon, however, chiefly through the work of Schrödinger and Dirac, functions were introduced into the formalism whose initial purpose was to facilitate the calculation of matrices. They were called wave functions because in the simplest instances (single particle case) their mathematical properties were precisely those of standing waves in non-homogeneous media. But this usage is fading because of the recognition of a larger meaning of these functions. They are now generally and more properly termed state functions. These state functions have as arguments some classical variables; in the most familiar case (Schrödinger representation) these are space coordinates and time. When the state function describing the state of a system (e.g., the hydrogen atom) is given, the *distributions of all observable probability aggregates*[11] associated with the system is fixed, though it is of course not possible to fix the outcome of every single measurement (uncertainty relation). Furthermore, the state function satisfies a first-order differential equation which allows its determination at every instant when its functional form is specified at any given time. *If states are defined in terms of this function then quantum mechanics is a causal theory.*[12]

The feeling that $\psi$-functions define states gains ground slowly, partly because in its matrix form quantum mechanics did get along without them, partly because $\psi$-functions are unobservable and abstract. A state, the contention goes, ought to be defined in terms of positions and momenta of particles, as it is defined in classical physics. When that is done no laws can be found which bestow causality upon the description. Hence the claim that causality has died at the hands of the physicist. It is a valid claim for anyone unwilling to modify the classical mechanical definition of state when he turns to quantum mechanics.

But electrodynamics would also be a non-causal discipline if we insisted on defining states in terms of positions and momenta of charges; thermodynamics would be in the same condition if we insisted on writing an equation of state involving only the position and velocity of the center of mass of the macroscopic body. In fact every physical theory can be made acausal by the wrong choice of state. Thus I should say that the acausality entailed by the use of classical mechanical states in quantum mechanics shows this description to be unsuited. Modern physics threatened to become acausal, particularly when the failures of the Bohr theory became known, but *quantum mechanics in its present form has restored causality*.

For, after all, what is sacred about positions and momenta of elementary particles? They are certainly not parts of sensed nature; they are constructs of explanation like all the other quantities which enter into the definition of states. I believe that the extreme and persistent emphasis with which physicists often pronounce the ultimate character of these cherished quantities reflects the attitude of enthusiastic positivism wildly overshooting its mark. That a great change has been accomplished in the perfection of a theory whose predictions as to definite elements of nature are essentially statistical, no one could possibly deny. But this change is badly described by saying that causality has been abandoned. In the next section we shall examine what has happened.

Physicists sometimes incline to the view that state functions are merely a useful artifice, introduced for the sake of convenience and hence not to be classed with other more valid constructs like the matrices representing observables. Years ago there was some justification for this contention. But when a powerful axiom like Pauli's exclusion principle refers so specifically to state functions as to prescribe their symmetry, one can no longer deny them a rank equal to that of other constructs.

I repeat: causality, in the sense defined above, is one of the rules governing the choice of those constructs which are the building stones of the physical universe, or of reality.

Lastly, *simplicity* guides that choice. What is meant by the principle of simplicity is perhaps roughly evident. To discuss it fully would require a great deal of space.[13] Since its inclusion is by no means unique to the methodology here proposed I shall omit its further consideration. In physics there is usually no question as to which of two theories is the simpler and therefore the one to be retained.

It will be noted that none of the rules here recorded is absolute. We have not required that a theory, or a set of constructs, be universally extensible, nor that a construct be rigidly permanent. Where then, you may ask, is the force of these proposals? Do they not amount to the statement: Physical constructs are permanent, extensible, causal, simple, *unless they aren't*? I answer no. For the power of the metaphysical rules becomes obvious whenever two rival theories compete for acceptance; it is with their help that physicists decide, often unwittingly, which theory to retain. The rules determine reality as decisively as anything one could name, with the possible exception of the contingencies called sensed nature.

Before concluding this lengthy section it might be well to append a brief tabular summary of the metaphysics of symbolic construction. We distinguish two large classes of "rules," with a number of principles subsumed under the second:

(1) Rules of correspondence (establishing connections between sensed nature and constructs).

(2) Rules governing the use of constructs.

        (a) Formal implication (logic)
        (b) Permanence
        (c) Extensibility
        (d) Causality
        (e) Simplicity.

## 6. METHODOLOGICAL CHANGES ENTAILED BY QUANTUM MECHANICS

There is no need here to stress the peculiar importance of the *number* concept in physics. It makes its entrance through the type of construct previously called a quantity, which by virtue of its operational definition makes contact with nature via numbers. Since states always relate in the final analysis to quantities, states, too, have an important numerical aspect. But let us consider quantity more closely. We have agreed that it must, at every instant, bear a unique relation to numbers, because the operational rules of correspondence enforce this situation. More specifically, knowledge of a state must in some way assign numbers to quantities. I shall say briefly that physical quantities (like energy, momentum, etc.) must generate numbers. All this follows from foregoing considerations.

There is no rule, however, which requires that quantities *be* numbers. It is of course the simplest possible assumption to hold that momentum is the exact equivalent of the mathematical function $mv$, i.e., that it *is* the number, in some system of units, which the function takes on at specifiable instants of time. This functional representation of physical quantities is the simplest possible one to do justice to the requirement that quantities must generate numbers. It is indeed the most direct, the most intuitively obvious and, as long as it satisfies the metaphysical rules, the most desirable.

However, we should not lose sight of the specific character of this representation. Mathematics now abounds with concepts which are not functions yet allow the assignment of numbers under certain conditions. One need only mention matrices, differential operators of various forms, integral operators, groups and tensors. In classical physics they were known, of course, and used in a half-hearted way. When vectors and tensors were first employed they were regarded as useful figments, as a shorthand way of writing numerous functions. The primitive representation indicated above was in fact never abandoned.

Quantum mechanics now asks us to take the other possibilities more seriously. It says in effect that the (constitutive[4]) definition of momentum as $mv$ was wrong and that $ih(\partial/\partial q)$ should take its place, and it says this in a perfectly serious way. The statement is not to be regarded as "symbolic" or "metaphoric" any more than was the classical proposition momentum = $mv$. Other quantities may still be numbers (like coordinates in Schrödinger's scheme; mass, charge in general), others again may be matrix operators (Pauli's spin). It is this fundamental shift, this widening of the range of choice in the meaning of quantity, which distinguishes quantum mechanics methodologically.

When this shift is once accepted, other changes follow as a matter of course. Differential operators of the type most frequently occurring in physics give rise to numbers (characteristic values) through equations which automatically introduce characteristic functions, functions which play an important rôle in the definition of states. But a given operator yields in general not one, but a large set of numbers. These are the possible values which the quantity may take on when measured. Now it happens that a *state*, in the sense here defined, will not cause a given operator to generate (according to simple though abstract rules) one, but many numbers. This situation could never arise in classical physics because $mv$ generates but *one* number for given $m$ and $v$. How, then, are these numbers to be interpreted? The successful answer has proved to be: as probable. And thus it has come to pass that quantum mechanics is a basically statistical discipline.

Viewed from a methodological point of view the change in question is hardly as profound as many philosophers of science appear to believe. It has not touched the major metaphysical principles; but it has done two things: The first was to pass from pseudo-sensible to abstract constructs in the choice of states and quantities; the second, necessitated by the first, was to generalize the rules of correspondence between constructs and sensed nature in such a way that to *one* quantity there may correspond *many* elements of nature.

To me, it is a source of amazement and satisfaction to see how enormous and brilliant an improvement in the theoretical structure of physics can be wrought by relatively minor shifts in its methodology.

## NOTES

[1] Retrospectively (1976), Bunge's term "metascientific" seems to be a better word than metaphysical. See his *Metascientific Queries* (C.C. Thomas, Springfield, Ill. 1959), where he presents a detailed inventory of the main features of science including the present guiding principles or rules governing the use of constructs. He also discusses the problem of reducibility. For a similar, perhaps more specific view on this issue see *Integrative Principles of Modern Thought*, p. 41 *et seq.* (H. Margenau, Ed., Gordon and Breach, N.Y. 1972.

[2] As having any relevance to scientific discourse! We do not deny the existence of, and a great body of interest in, ontological problems. The present analysis does not touch upon other fields of philosophy such as esthetics, ethics or value theory, except in the claim that physics is possible without taking cognizance of them.

[3] A. N. Whitehead, *The Concept of Nature*.

[4] P. W. Bridgman, *J. Phil. Sci.* 5, 114 (1938).

[5] H. Margenau, *J. Phil. Soc.* 2, 48 (1935); 2, 164 (1935).

[6] For a brief exposition of logical positivism the reader is referred to R. Carnap, *Logical Foundations for the Unity of Science*, International Encyclopedia of Unified Science, Vol. 1, No. 1.

[7] F. S. C. Northrop, *J. Unified Sci.* 9, 125–128 (1939). The significance of epistemic correlations in scientific method.

[8] See in this connection H. Margenau, *J. Phil. Sci.* 6, 65 (1939).

[9] H. Margenau, *J. Phil. Sci.* 1, 133 (1934).

[10] For a more detailed exposition see reference 8.

[11] For definitions and further discussion, cf. R. B. Lindsay and H. Margenau, *Foundations of Physics* (John Wiley and Sons, New York, 1936).

[12] For a more detailed discussion of the "acausal-jump theory" cf. *J. Phil. Sci.* 4, 337 (1937).

[13] In a somewhat wider context, simplicity in physics was analyzed by R. B. Lindsay, *J. Phil. Sci.* 4, 151 (1937).

He provides both an historical survey of the development of the idea of simplicity and an analysis of its various interpretations. He concludes with a pragmatic "definition" which describes its current function and expresses the view of most physicists: "Of two theories descriptive of the same range of physical experience that is the simpler which demands the shorter time for the normally intelligent person to become sufficiently familiar with it to such an extent as to obtain correct and useful results." The logician may not like this "definition", yet is the principle that guides the working researcher.

An exhaustive study of the various meanings of the term simplicity and a careful evaluation of their relevance for science is given in Bunge's *Myth of Simplicity* (Prentice Hall, Englewood Cliffs, 1963). The reader will find that the term "myth" is not meant in an entirely derogatory sense but alludes to the usually unanalyzed variety of meanings of the word simplicity. My own later writings (e.g., *The Nature of Physical Reality*, McGraw

Hill, N.Y. 1950), while presenting no single definition of simplicity that would satisfy the logician, use that term as part of a continuous spectrum of "metaphysical requirements" without sharp boundaries. Its meaning is close to elegance, extensibility, creativity. All of these the working scientist somehow understands intuitively. I note furthermore that simplicity, to be useful in characterizing a physical theory, must be joined with other metascientific requirements. In that way our approach avoids "simplicism," a true myth which Bunge identifies with the attempt to define scientific validity in terms of some chosen form of simplicity exclusively.

CHAPTER 6

# IS THE MATHEMATICAL EXPLANATION OF PHYSICAL DATA UNIQUE?

## 1. THE FUNCTION OF MATHEMATICAL EXPLANATION IN PHYSICAL SCIENCE

Within cognitive experience, traditional philosophy and common sense distinguish two polar epistemological components: those called *data* and the rational elements called *concepts*. The former are used as protocol (*P*) experiences and are taken as authoritative, as the last instances of appeal of any theory; the latter owe their genesis to reason rather than observation, we feel a rational responsibility concerning them and modify them with some freedom. They are *constructed* as counterparts to data in order to provide a larger measure of coherence and intelligibility than *P*-experiences alone contain. Whether concepts, originally constructed to satisfy inductive suggestions coming from observations, are ultimately accepted and retained as valid explanations of datal experience depends on the way in which they meet a set of complicated requirements. Primary among these requirements is empirical confirmation, the need for conclusions drawn from theory (which is itself a set of constructs joined by logical and mathematical relations) to be confirmed within a certain tolerance by observations. Other requirements, more vague in their formulation, demand simplicity, mathematical elegance, extensibility, causality in the theoretical transcription of data.

The connection between primary data (for which I shall henceforth use the symbol *P*) and the field of constructs (symbolized by *C*) is made in a variety of ways. In some instances one passes by long inurement automatically from a complex of data to a corresponding construct. This occurs when we 'reify' a group of immediate sensations into an external object. For the purposes of this paper the process of reification will be ignored or, more specifically, will be included within *P*-experience. Most important in science is a connection established by instrumental operations, operations which are often claimed to be the definitions of a construct in terms of data. I shall not discuss the propriety of the claim made by operationalists that such passages are definitions, nor the accompanying suggestion that all good scientific definitions must be of this type. My point is that they form a useful link between *P* and *C*. They form, as shown elsewhere,[1] one type of a very general epistemological relation called rules of correspondence, epistemic correspondences. To cite examples, the measurement of temperature by means of thermometers, force by the extension of springs, psychological responses by j.n.d. (just noticeable differences) reactions, public sentiment by statistical opinion polls are instrumental operations leading from vague, subjective $P$-experiences to rational constructs. The latter can then be joined and modified further by purely theoretical procedures which may take the scientist's concern far into the abstract domains of the $C$-field.

This situation is represented pictorially in Figure 1. Circles are drawn for 'individual' constructs even though it is clear to the logician that a concept is not often a discrete entity but, like any idea, is capable of 'decomposition' in many ways. The envelope around several connected concepts represents a theory $T$. Double lines are epistemic correspondences, whereas single lines are logico-mathematical relations, postulated for, or implied by, the concepts. For example, the constructs force, mass, and acceleration would appear as circles on the diagram; their operational definitions as double lines linking them with $P$, and Newton's law, $F = ma$, would form single-line connections between $F, m$, and $a$.

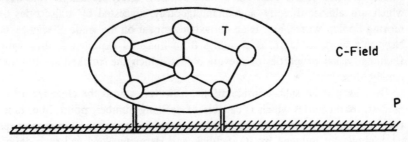

Fig. 1

## 2. EPISTEMIC CORRESPONDENCES

Instrumental correspondences render many constructs (usually called quantities or attributes in contradistinction to systems or substances) measurable. This permits the assignment of numbers to some of the lower-order $C$'s, namely, all those connected with $P$ by double lines, and mathematical treatment becomes possible. Numbers lead to functions, functions to operators, and so on, so that ultimately, as we proceed up into the $C$-field, almost all the devices of mathematics are carried into play. The question we wish to answer is whether their play is unique or can be expected to be unique at some future time.

To secure the answer we must note first a crucial fact about the epistemic correspondences: they are *not* implied or rigidly entailed by *P*- and *C*- experiences. They often *seem* to be, and we cite a few examples in which this appearance is strong.

The concept of number is very directly suggested by the operation of counting concrete and distinct objects, not clouds or ideas, to be sure, but fingers, toes, and pebbles. The story that some primitives take off their shoes to do arithmetic beyond the number 10 lends evidence to the contention that numbers are practically and perhaps uniquely determined by *P*-experience. Furthermore, historically, psychologically, ontogenetically, the process of counting is very probably prior to the number concept. None of these circumstances, however, sheds light on the logical or epistemological relation.

The mathematical ideas—point, line, square, triangle—were doubtless all gleaned from immediate observation of natural objects which appear to hold, as it were, the epistemic correspondence between triangular aspect and the construct triangle within their essence. There could be no doubt about the association of what looks like a line and the mathematical construct line; the correspondence seems implied. Similar remarks apply to simple functions which are almost directly and unambiguously suggested by trajectories of moving bodies, waves, etc. Even derivatives appear on the scene of science, in Newton's work at least, as gleanings from imageful experience; they were fluxions, visual properties of moving object, which the 'pricked symbol' ($\dot{x}$) merely 'describes'.

Over against all such considerations, however, stands the clear record of modern mathematics which is capable of defining number, point, line, function, and derivative in a fashion independent of *P*-experience (though perhaps psychologically induced by it). Schlick and Hermann Weyl call such definitions implicit; a most impressive one is that of area, which I shall set down here as an illustration.

Let a quantity $Q$ be defined by the following specifications:

1. $Q$ is a positive number.
2. If a 'piece' is dissected into two parts by line segments in its interior, then $Q$ for the whole 'piece' is equal to the sum of the $Q$'s for the parts.
3. Congruent 'pieces' have the same $Q$.

(For convenience, $Q$ will be thought of as associated with a piece of surface, to which the definitions above refer as the 'piece'!)

Remarkably, these simple propositions, devoid of reference to anything existing, give rise to a $Q$ which has all the analytic properties of measurable areas. They form a constitutive definition of the construct *area*. Experience shows that by establishing an epistemic correspondence, which associates $Q$ with the number of cm$^2$ contained in actual surfaces, one obtains a valid scientific theory. The constitutive definition alone gives no assurance that it will find application in the world. Logically, its success is a miracle.

Or take the implicit definition of $\pi$ as the ratio of circumference to diameter of an ideal circle. It certainly does not imply the correctness of the following observations.

If two numbers are written down at random, the probability that they will be prime to each other, i.e., will have no common factor, can be shown to be $6/\pi^2$. In a trial when fifty students wrote down five pairs each, it was found that 154 pairs were prime to each other: so from this trial we should get $6/\pi^2 = 154/250$, which gives $\pi = 3.12$. A trial on a much larger scale would doubtless give a much better approximation to the $\pi$ of Archimedes[2].

To establish the connection between the implicit definition of $\pi$ and its appearance as the square root of 6 divided by the probability that any two numbers will be prime to each other, one needs certain postulates of probability theory, plus the epistemic correspondence expressed in the frequency theory of probability. None of these are obvious, implied by constructs or enforced by data.

There are many instances in which formal or implicit treatments of sets of constructs *preceded* their application to observational science. Notable are Hilbert's elegant theory of function space, group theory, Boolean and matrix algebra. These have found profitable use in modern physics. Arising in science as formal constructs unconnected with the $P$-domain, they acquired epistemic connections as time went on. Yet no one would say that there is anything intuitively obvious or necessary about the association of vectors in Hilbert space with a set of points on a photographic plate which mark the incidence of an atomic particle. The lesson we draw is simple: epistemic correspondences are not entailed by mathematics or uniquely specified by observations. The picture drawn in Figure 1 can therefore lose and regenerate the double legs on which the constructs stand. It is precisely because the connections between $P$ and $C$ are not entailed by the theoretical components of the epistemological system that the success of mathematics in explaining physical facts so often seems like magic[3]. The phrase, God was a mathematician, partially expresses our present contention.

## 3. THE LIMITATION OF MATHEMATICAL THEORIES

Mathematical theories are limited in application through their rules of correspondence. Either this limitation is recognized from the beginning (a) or it develops with progressive refinement of observations (b).

(a) Instances in which the limitation was known—and the practical use of constructs outside the limitation was deliberately practiced in spite of it—are easy to find throughout formalized science. Boyle's law was always regarded as an idealization, corrections for the behavior of real gases, designed to extend its range of validity, were made at an early date. This means that the theory, like $T$ in Figure 1, had double lines extending only into a certain limited part of the $P$-domain, beyond which double lines rose to some other theory, although again, of course, the division of the domains is never sharp.

The concept of number and the theory of arithmetic were never invented to accommodate the flux of continuous phenomena; hence even this very fundamental and seemingly universal idea encounters limitations in its use.

Special relativity is based upon the postulate of the invariance of the speed of light in inertial systems, sometimes expressed by the equation $x^2 + y^2 + z^2 - c^2 t^2 = 0$, where the symbols are conventional. Now many students and some writers have registered great surprise upon learning that $c$ is the speed of light, and that the equation is not true when it is taken to be the speed of sound, for example. The point is simply that *this* aspect of the theory cannot be applied to the propagation of sound; even relativity has certain fundamental limitations. In another sense its very restriction to the velocity of light makes it applicable to the propagation of all signals, but in another form.

The recognition of limits for the valid use of mathematical ideas is not only of historical interest; lack of it can plague contemporary theory as well. I apologize for reiterating here a point already made at greater length in previous publications, maintaining however that it is relevant and still noteworthy. In his famous treatise[4] von Neumann introduced a mathematical operator, sometimes called a statistical matrix, sometimes a projection operator, $\rho$. This operator can be shown to have only two eigenvalues, 1 and 0, and it seemed admirably suited to represent the physical measuring process which interrogates nature through instruments and receives the answers yes (1) or no (0). This epistemic correspondence was most suggestive, and it has been accepted by many authors. Unfortunately, however, the definition of $\rho$ can be shown to entail another property besides the convenient one of possessing eigenvalues 1 and 0. When applied to a Hilbert state, it projects that state in a direction corresponding to the measured observable. If we hold to the rule of correspondence between $\rho$ and measurement, this latter fact can only be interpreted by saying that a measurement, when performed on the state of a physical system, leaves that system in an eigenstate, specified by the measured observable. This is clearly not generally true, for one can easily name many good measurements which do not produce that result. It is also possible to show that the infamous reduction of the wave packet, so widely misunderstood in quantum mechanics, arises from the retention of this rule of correspondence. To rectify the situation it is necessary to remember that an interesting and simple operator like $\rho$ may nevertheless have limited application to $P$-experience.

(b) We now cite instances where restrictions, unknown when constructs were introduced; developed in response to finer and richer observations. The first example is hackneyed and will be briefly recorded. Newton's inverse-square law of gravitation was thought to be universally valid; it was indeed called the law of universal gravitation. In our century, the deviationist motion of the planet Mercury, the precession of its perihelion, was discovered. Newton's law became inapplicable to Mercury; its limitations were recognized. At the same time a new construct, the metric of non-Euclidean space, was introduced. The correct features of Newton's theory became incorporated in it and a higher theory arose with wider rules of correspondence.

Until about 30 years ago chemical valence was explained in terms of electron bonds, a concept thought to be of universal scope. But certain ambiguities evolved, best remembered in connection with the benzene ring. Here, the simple concept of electron bond, single or double, could not be applied, and its range had to be restricted. But again a higher construct, the idea of quantum resonance, appeared, and its success, methodologically speaking, was twofold. It modified the simple bond theory slightly and it joined the

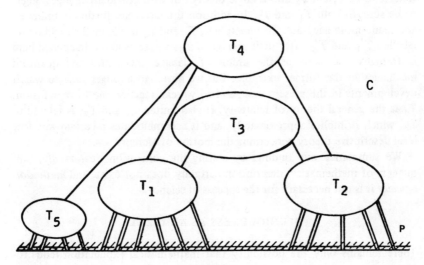

Fig. 2

conforming and the discrepant observations into consequences of a single theory with new rules of correspondence.

A third example, frequently discussed by philosophers of science, and only mentioned here, is provided by the transformation of Newtonian mechanics into special relativity. Limited applicability of the former was recognized in rapidly moving particles. A 'higher' theory arose which modified and unified the ingredients of classical mechanics.

Perhaps the most impressive evolution of this sort took place in physics during the last few decades, when quantum mechanics gained general acceptance. Classical mechanics, like the theory $T_1$ in Figure 2, was related satisfactorily to a certain part of $P$, and Bohr's theory of atoms $(T_2)$ to another. Quantum mechanics, symbolically $T_3$, joined $T_1$ and $T_2$, *modifying each in some respects*. It did this without changing the double lines to an appreciable extent, that is to say, it did *not* require new operational definitions of the low-order constructs which both the Bohr theory and classical mechanics involved. This is not always true; relativity theory, mentioned in the previous example, required a redefinition of mass and force.

A simpler, last example, may epitomize and extend the situation represented in Figure 2. Let us think of $T_1$ as Galileo's theory of falling bodies, clearly limited to observations near the surface of the earth. $T_2$ depicts Kepler's laws of planetary motion, designed to explain entirely different phenomena. Hence at this stage $T_1$ and $T_2$ are unconnected. $T_3$ is Newton's law of gravitation; it fuses $T_1$ and $T_2$ into a single theory but with attendant changes, slight to be sure, in both $T_1$ and $T_2$. In addition the latter are further illuminated and made more accurate. To be explicit, $T_1$ and $T_2$ in Figure 2 should be relabelled $T'_1$ and $T'_2$ after their merger in $T_3$. So far what has happened here is formally the same as the union of classical mechanics and quantum mechanics in the former example. But we have had a longer time to watch developments in the present case, and so we can register one further fusion. $T_4$ is the general theory of relativity. It also united $T_1$ and $T_2$, is related to $T_3$, which it implies approximately, and it accommodates a certain very low level descriptive theory concerning the motion of Mercury.

We summarize our findings by stating the conclusion: *historically,* uniqueness of mathematical description certainly does not exist, and *methodologically* it is not necessary for the success of science.

### 4. UNIQUENESS AS AN IDEAL

There remains only the possibility that mathematical explanation tends toward uniqueness as science advances, that diagrams such as Figure 2 will in time show concrescence below and convergence near the top. Very serious arguments can be advanced against this expectation. One of them is logical and is based on a peculiar interplay between the truth of mathematical theories and the span of the correlations with which they are equipped. More simply, a theory cannot be judged correct before it has a universal range of application; indeed it can be said to be erroneous if it is incomplete.

Consider Figure 3. If $T_1$ and $T_2$ are incompatible though not contradictory, i.e., have separate ranges of application, there will always be a region of $P$ labeled $X$, which remains unexplained. At the present time relativity and quantum mechanics are two such theories; $X$ contains observations on reactions between particles moving with speeds near that of light. I know of no case where $T_1$ and $T_2$ are separate and yet where $X$ is empty, that is to say, where theories manage neatly to coexist without incompatible claims on data. I believe one can prove that such situations cannot occur; that in the realm of science, at least, coexistence under incompatible premises is imposible.

Suppose now that $T_1$ and $T_2$ are joined into a supertheory $T_3$, as is the universal tendency we have discussed in section 3b. Now $T_3$ can cause $X$ to

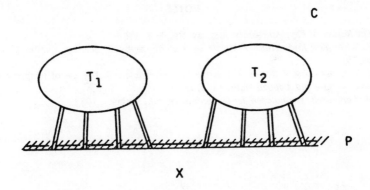

Fig. 3

vanish, but as we have seen, only at some sacrifice of truth in both $T_1$ and $T_2$. Each of them is modified. *Concrescence therefore destroys validity.* So long as we have separate, incompatible theories we can be fairly sure that they are wrong.

Because of this interplay between completeness and validity, the growth of science is not a simple process of continual accommodation; it is like the growth of a pile of large rocks when new ones are added at the top. Its structure suffers discontinuous readjustments of the entire pile at certain times, and it is difficult to assess its growth in terms of convergence to any rigorous sort of limit.

But these difficulties are aggravated by more general philosophical considerations. The continual enrichment of the $P$-domain by new observations cannot be described in clear and finite fashion. Wholly unexpected experiences may arise from the fertile ground of being. At the same time the $C$-field grows without limit, for there is no end to mathematical imagination. The creative drive of science will here primarily assert itself.

The dynamics of the scientific enterprise are therefore improperly described as a matching of a completable class of $P$-experiences against a finite class of theoretical constructs; it is not even like the solving of a giant picture puzzle with ever increasing boundaries. For the pieces we have placed do not stay put; moreover, they disappear and multiply, and the places themselves have the fluidity of ideas rather than the rigidity of geometric patterns. Science is an *open* field, and I see no way of understanding its convergence if it did occur, nor do I expect it. Indeed its threat would fill me with dismay.

## NOTES

[1] *The Nature of Physical Reality*, McGraw-Hill, N.Y. 1950.
[2] E. Whittaker, *Eddington's Principle in the Philosophy of Science*, Cambridge Univ. Press, 1951.
[3] E. Wigner speaks of this feature as the "unreasonable effectiveness of mathematics." *Comm. on Pure and Applied Math.* **13**, 1, 1960.
[4] *Mathematische Grundlagen der Quantenmechanik*. Springer, 1932.

# PART II

# FUNDAMENTAL PROBLEMS OF
# 20TH CENTURY PHYSICS

# PART II

A comment which concerns the entire book, but especially Parts II and III, seems called for here. As originally presented, the various chapters were independent essays and lectures, each attempting to convey its message fully. Some were addressed to general audiences, some to physicists, some to philosophers. This accounts for a feature which, though a matter of some regret to the author, seems almost inevitable: Since every chapter is a unit by itself, repetition of fundamental matters has occasionally been permitted. It is my hope that the reader will condone some redundancy of emphasis, indeed some repetition, on what is called 'decay of materialism,' a subject treated in a number of places.

# CHAPTER 7

# PROBABILITY, MANY-VALUED LOGICS, AND PHYSICS

## 1. THE RÔLE OF THEORY IN SCIENTIFIC PREDICTION

The present paper is concerned chiefly with the problem of scientific prediction. It aims at a factual analysis of the processes leading to prediction, and ventures an appraisal, in the light of this analysis, of some modern and unconventional theories of probability and truth. But although prediction is here chosen as the central issue of discussion, I do not wish to imply that, in its usual sense, it is the only or even the dominant issue of scientific research.

The reason for this cautious, though perhaps unpopular, attitude is not far to seek. In the first place an empirical survey of the various fields of human endeavor shows that the name science is not reserved for those speculative disciplines which do enable their followers to predict. There are sciences which do not make it their business to predict. Moreover, predictive ability does not of necessity constitute a body of knowledge or a set of rules as a science. If the procedures of astrology were some day found to be of "scientific validity," i.e. were shown by their nexus with established facts to be in some sense understandable, astrology would from that day on be classed as a science; but I believe that present astrological activities, while leading (presumably) to valid predictions, would even then be referred to as pre-scientific. The conclusion seems inescapable that science cannot be defined exhaustively as an endeavor to provide facility for prediction.

Logicians, however, may well observe that science is a denotative term given immediately in discourse, and that *every* definition of denotative terms engenders empirical exceptions. Thus, if a dog were defined as a mammal with four legs and many other characteristics, it would follow that a three-legged member of that species is not a dog. The preceding remarks are therefore reduced to the simple assertion: The term science does not coincide empirically with the criticised description of it in terms of predictive facilities. But the point to be noted is that it differs widely from it. The pragmatist is then still free to assert in a normative sense that we ought to call sciences those fields of activity which allow prediction though this is in many cases contradictory to convention.—To this we answer that the norm is likely to be rejected by most people actively engaged in making scientific predictions, for their concern is far more with the features of induction and analysis which precede prediction than with prediction itself. The discovery in any field of a new "principle" is heralded as a success far greater than the extent of the future it may possibly illuminate. The discovery may well illuminate the past and be no less important.

When we speak here of prediction we do not mean the word in its temporal sense. Every complete scientific procedure involves two sets of facts, fact which are ordinarily reduced to, or phrased as, observations. Between the two sets it formulates a sort of linkage, sometimes directly and very often indirectly, in such a way that when one set is known the other may be inferred. This inference is not always reciprocal not transitive.[1] It may involve a theory replete with symbolic constructs, or a theory of a more intuitive and simple kind, or none at all. An example will make the situation clear.

Suppose a physicist makes measurements of pressure ($P$) and volume ($V$) on a given concrete body of gas. He finds an approximate relation between different observations on $P$ and different observations on $V$, a relation well known as Boyle's law: $PV = k_1$. Trusting now to the principle of consistency or uniformity of nature, upon which are based certain rules of induction, he will soon be convinced of the validity of the relation and apply it to make inferences from a given or known $P$ to an unknown $V$. By scientific prediction we mean this *inference* from $P$ to $V$. It is important to note that no *theory*, as the word is understood in physics, is necessary to perform this passage. Required are only the operational rules leading to the numerical determination of $P$ and $V$ on the one hand, and a bit of arithmetic on the other.

But the physicist's interest transcends the behavior of the particular body of gas with which he has been experimenting. He will try another and arrive at the approximate relation $PV = k_2$. The new problem which now arises is to correlate $k_1$ and $k_2$. At this point the rôle of theory begins. An association is made between the operational $P$ and a certain *construct*, the thermodynamical pressure which is designated by the same symbol $P$. The fact that the same letter is universally used for the two entities has blinded many scientists to the recognition that they are *logically distinct*. An easy way to see this difference is to watch what is being done with the symbol $P$ in theoretical thermodynamics and then attempt to identify $P$ with observational manometer readings. If this fails to convince, we may point to the fact that the thermodynamical construct $P$ is altogether different from the construct $P$ employed in statistical mechanics. Gibbs' formalism in particular, in which $P$ is related to abstract properties of an abstract ensemble, provides a very forceful reminder of the logical distinction here drawn and of its inevitability. It is thus clear in the first place that the concept of pressure used in thermodynamics is different from that used in statistical mechanics. But if this is admitted it follows formally that at least one of them must differ in some way from the operational pressure, and a little further reflection will show that both, thermodynamical pressure and statistical pressure are concepts not to be identified with the operational $P$. One striking difference lies in the fact that, as a variable, the latter is capable of taking on discrete values only, whereas the former two vary continuously. In the present paper we shall follow the terminology previously proposed and refer to the former two concepts as (symbolic) constructs and to the latter as a "perceptible fact" or, more crudely, as "part of nature."[2]

Between perceptible facts and corresponding constructs there exist rules of correspondence of a very subtle kind, so subtle indeed that they are frequently taken for identities. For the physicist these rules amount for the most part to habits of thought; he knows instinctively, after years of training, how to correlate the result of his measurements on $P$ and $V$ with the thermodynamic or statistical notions $P$ and $V$ which function in his equations. The important property of these rules from the present point of view is that they are *unique*, provided the set of constructs to be used is specified. Thus from $P$ (observational) we may pass either to $P$ (thermodynamic) or to $P$ (statistical) or, if there were other theories, to other constructs. The return trip is without alternatives: it must always land at $P$ (observational).

An "explanation" of the constants $k_1$ and $k_2$ of the example under consideration involves a complete circuit of scientific inquiry. Starting with the observational relation $PV = k_1$ which is valid for a given experimental object, the physicist makes the transition to the field of constructs, using (almost unwittingly) the above mentioned rules of correspondence. Thus he associates with his object a great number of abstract properties which he is capable of ascribing also to a specifiable class of other bodies, properties which are not necessarily themselves observable but among which there are some which allow passage, by unique rules of correspondence, to observable facts. Among these properties are the constructs $P$ and $V$. He now finds himself in the field of symbolic construction in which his activities are not limited by experimental errors, least counts, practical feasibility, etc., but in which he is free to operate according to deductive procedures (embodied in his hypotheses) which he himself invents. The only requirement with respect to these procedures is that they be invariable and free from internal contradiction in the sense of the postulates of logic. These procedures plus the constructs they contain are either physical theories or idle speculations according to whether they lead, at *every* step, to situations amongst the constructs which allow transitions, by the rules of correspondence, to observable physical facts, or not. In the present instance the constructs introduced include the temperature $T$ and, if we refer to the statistical theory, the number of molecules, their momenta, etc. By well known reasoning in which the constructs are treated like continuous numerical variables (this is one of the components of the deductive procedure peculiar to the theory under discussion; it is not common to all theories; vide quantum mechanics where symbols are treated as operators!) we arrive at the relation $PV = RT$. From this the transition is easily made to the operational counterparts of the constructs appearing in the equation, and the landing is safely accomplished. Whether the theory is right or wrong depends of course on whether, by starting with $PV = k_1$, we are led correctly to the constant $k_2$ for some other body.

In the case of a single body we were able to infer $V$ from $P$, obviously without intervention of theory. But theory was employed in order to permit inference of $V$ from $P$ for an arbitrary experimental object. The sketch of the function of theory in this instance was in close alignment with the actual procedure of working physicists; it was not intended to assert that only theory could achieve the inference for an arbitrary body. To this question we return in a moment. We have established one fundamental thesis: the physicist and, presumably, every scientist, infers one set of observable facts from another set of observable facts. The inference may be accomplished with or without the use of theory. If without, the inquiry stays in every phase of the inference within what has been called nature; if with the aid of theory, the inference leads from nature into a field of symbolic construction and back to nature. This inference, regardless of the two alternate modes in which it may be conducted, will be called scientific prediction. I believe that most scientists who employ this word adopt tacitly the same definition.

In the example chosen above, could the use of theory be avoided? In principle this is possible indeed. For there is nothing to prevent the physicist from setting up a great number of relations between observed $P$'s and $V$'s for different gases under different conditions, and the experimentation could be carried on without regard for hypotheses until an approximate empirical equation between all the constants $k_1$ is found. In the course of this process $T$ would be introduced as an observed number, and the ultimate result, approached as a limit but never reached, would be a relation of the form $f(P, V, T) = 0$, where $P, V, T$ stand for measured numbers regarding which no assumption of continuity need be made. In one sense, the procedure differs only by the inclusion of a third variable from that leading to Boyle's law for a given body at a constant temperature: $f(P, V) = 0$, but in another sense it differs very widely: for the experimental problem of finding a relation among 3 variables is very much more complex and more difficult than among 2.

It is quite possible that theories first arose because of the peculiar and astounding enlargement of difficulties as one proceeds in the sorting of observations from two principles of division to three or more. If nature presented us with nothing but diadic relations the necessity for theories might never have been felt. It is useless, however, to decry the value of theories in science for that reason, for it simply happens that nature is not constituted in that way but in a manner which does impart relevance to theories, at least in a considerable number of aspects. It is idle to say that prediction of facts is the only important goal of science and that theories are of ancillary service in achieving this end, because the striking and a priori unpredictable usefulness of theory in ordering and assorting observations is to the open-minded just as significant a feature of reality as the ordered facts. If we couple these points with the undisputable circumstance that the class of scientists displays an intense interest in successful theory, all grounds for the pragmatic assertion of the all-importance of observation in the fields of endeavor called sciences are removed.—Later, in studying the problem of empirical correlation, we shall discover further justification for the claim that theory, i.e. symbolic construction, be an integral part of scientific analysis. At present we shift our attention to the two endpoints of the circuit of prediction: the two sets of observation which it connects, losing sight for the moment of whether a theory is operative in the process or not.

## 2. CORRELATION VERSUS SYMBOLIC EXPLANATION

The set of observations from which prediction starts, being the result of measurements, is a group of definite numbers. The same is true of the set by which the prediction is verified or refuted. Thus arises the question: If a prediction is to be valid, must it agree with the measured set absolutely, in the manner of point coincidence? All practical experience demands the answer no. We arrive at the problem of how to soften the rigidity of prediction without destroying its essence. One thing is certain: whatever procedure we adopt to diffuse one set of observations must, by symmetry, be applied to the other as well, for the process of prediction may in special cases be reciprocal. In science this diffusion is accomplished by two well known methods: (1) the method of statistical correlation; (2) the method of introducing errors of observation. As the following analysis will show, the second method may be regarded as a limiting case of the first.

Prediction is more than the simple act of inferring one set of data from another set. Since it makes use of one of the two methods mentioned it takes on the complications entailed by them. In order to apply either method it is necessary to generate a *series* of sets of data at both ends of the predictive chain. Thus, to revert to our example, prediction passes from a single $P$ to a single $V$ (the set here reduces to a single number; we might also have linked $P$, $V$ at one temperature with $P\,V$ at another for the same gas, in which case the set would have consisted of two numbers), whereas correlation is meaningless unless we assume a series of $P$'s as well as a series of $V$'s. The generation of sequences of this sort and the assumptions to be made about them form an important aspect of the problem of scientific induction, which, however, will not be dealt with *in extenso* here.

The methods of correlation and of errors are relevant to the main issue of this paper and will therefore be subjected to closer scrutiny. We start with the former. In *Philosophy of Science,* **5**, 237 (1938), C. H. Prescott gave an interesting survey of numerous features of the problem of correlation, and we desire to refer the reader to his treatment for some parts of the subject which we can not treat adequately in this report. We state, however, that we are not at all in accord with his thesis that the ultimate domination of nature will be achieved by establishing extensive correlations, as may be seen from the remarks in the preceding section. Rather, we view the establishment of correlations as a procedure incidental to the success of science in many fields, but subordinate to the explanatory domination of nature in terms of the symbolic constructs introduced earlier. We also point out that the rules mentioned by Prescott for determining degrees of correlation, while ideally sound, are not those commonly employed by working scientists.

Let us discuss the matter in its simplest form, restricting our attention to cases involving two variables, such as $P$ and $V$ measured for a given body at a constant temperature. Let us call these variables $x$ and $y$. To obtain the necessary series, numerous simultaneous observations, $x_i$ and $y_i$ are made. Statisticians express the degree of correlation between the observations on $x$ and those on $y$ by means of certain numbers called correlation coefficients. Numerous recipes for obtaining such numbers are on the market, all based on sound mathematical reasoning, but measuring correlation on different scales. Clearly, the suitability of the scale depends on the problem to be treated, which accounts for the fact that actuaries may use different coefficients for their correlations than biologists do. Only one of the various possibilities will here be chosen for illustration. It has, to be sure, several disadvantages, in particular that of yielding a value equal to unity for the correlation coefficient only if the law combining the variables is a linear one. This, however, will be no source of difficulty for the present purposes.

The rule for finding the number in question, $r$, is this.[3] List all observations, $x_i$ and $y_i$, in pairs, and let $\bar{x}$ and $\bar{y}$ be their mean values. If $i$ runs from 1 to $n$, we define

(1)
$$r = \frac{\sum_{i=1}^{n} (x_i - \bar{x})(y_i - \bar{y})}{\left[ \sum_{i=1}^{n} (x_i - \bar{x})^2 \cdot \sum_{n=1}^{n} (y_i - \bar{y})^2 \right]^{\frac{1}{2}}}.$$

One can see immediately what the limiting values of $r$ will be. For let us suppose, in the first place, that the various $x_i$ and $y_i$ are entirely "independent." This means that the sum $\Sigma_{i=1}^{n} x_i y_i$ approaches to value $n\bar{x}\bar{y}$. In that case the numerator of (1) is zero while the denominator (barring the case of zero dispersion which is uninteresting) remains finite. Hence, for statistical ("stochastic") independence, $r = 0$. Next, let us take the case in which there exists a linear relationship between $x$ and $y$, for instance $y_i = ax_i + b$. Then $y_i - \bar{y} = a(x_i - \bar{x})$, so that numerator and denominator of (1) become equal. Therefore, if $x_i$ and $y_i$ are linearly related, $r = 1$. (if $a$ were negative, $r$ would be $-1$.) The absolute value of the correlation coefficient, $r$, is confined between 0 and 1, 0 indicating *no* correlation (stochastic independence), 1 indicating exact functional (in our case linear) relationship.

But why was an expression of the type (1) seized upon to represent $r$? The only way to validate this or similar choices is to go back to certain axioms of probability theory. They may be summarized by saying that the series of measurements must be assumed to be a probability aggregate (Von Mises' "Kollective", or something equivalent to it in other theories), and this involves a drastic element of metaphysics. To avoid it seems to me impossible; for even if we were willing to regard (1) merely as a conventional formula which happens to have proved helpful in practice, and by which we agree to judge the relatedness of our observations, the use which is made of $r$ in scientific work, and the conclusions drawn from it, would be completely unjustified. If the fundamental axioms in question are not retained, the value of $r$ has no bearing even upon the problem of prediction which has initially led us to investigate this matter of correlations. These axioms are definitely *metaphysical* in the sense that observation does not dictate their adoption. It is a rather amusing commentary on the efforts of those scientists and philosophers who wish to banish symbolic constructs from science, and who therefore talk of nothing but correlated facts for fear of permitting entrance to metaphysics, that one look behind their screen of correlations shows metaphysics galore.

The axioms referred to may be stated in several different ways. The clearest is probably the formulation of von Mises in terms of two component postulates: Existence of the limit of relative frequencies, and "Regellosigkeit." It is true that there are difficulties connected with this formulation, especially the contradictory character of the limit postulate if the term limit is used in its mathematical sense, and the non-constructibility of the series in view of its randomness. Whether these may be resolved or not is a question affecting the validity of this specific scheme but not the fact that it necessarily transcends observations. An interesting alternative formulation which has not yet enjoyed wide-spread study among philosophers and physicists is that of Kolmogeroff[4] who treats the subject of probability and correlations as a branch of the theory of set functions, thus introducing the axioms of completely additive sets. We point out that correlations between *observations*, to which this treatment and, as we hope to show, all scientific uses of probability notions, are restricted, can be founded entirely upon a *frequency* interpretation of probabilities and involves no concept so problematic as the probability of a single event. For a review of the various probability theories a paper by Bures[5] may be consulted.

It will be instructive to apply formula (1) to a few simple instances. As our first illustration we consider the correlation between the second digits ($x_i$) of the numbers and the second digits $y_i$ of their squares, the numbers running from 11 to 30. The reader can easily make a table from which the results here given may be verified. In the first column of this table are written the numbers from 11 to 30; in the second the $x_i$ which run from 1 to 9, then from 0 to 9, the final entry being a 0, obviously. The third column contains the $y_i$: 2,4,6,9,2 etc. From these one finds at once: $\bar{x}$ = 4.5, $\bar{y}$ = 4.4. In the next two columns are written $x - 4.5$ and $y - 4.4$. This shows in detail the customary procedure. One thus obtains for the numerator of (1) the number 57.6, for the denominator $\sqrt{570} \times 540 = 555$. Hence $r = 0.104$. This result would be regarded as indicating a very low correlation, almost an absence of relatedness between the $x_i$ and the $y_i$.

Next, we consider the correlation between the volume of water carried in the rivers of a certain district in Sweden per year, ($x_i$), and the yearly amount of precipitation ($y_i$) in that area.[6] The procedure outlined leads in this case to the value $r = 0.705$, indicating quite definitely the presence of some "connection."

Finally, we will take a typical series of fairly accurate measurements on Boyle's law; the series is listed below for the reader's inspection (in arbitrary units).

| $P_i$: | 5.69 | 2.94 | 8.64 | 12.83 | 4.80 | 6.01 | 9.10 | 7.42 | 10.29 | 3.88 | 5.30 |
|---|---|---|---|---|---|---|---|---|---|---|---|
| $1/V_i$: | 10.82 | 5.70 | 16.40 | 24.39 | 9.08 | 11.44 | 17.28 | 14.12 | 19.60 | 7.37 | 10.11 |

If the value of *r* is computed for these data, it turns out to be 0.9999918.

Now there is absolutely no a priori cause for suspecting that, because in the last example chosen the correlation coefficient is so much nearer unity than in the others, the relationship between the data expressed is of a differend *kind*. So far as the observations and the method of procedure are concerned no difference at all is present. But this circumstance must not predispose the philosopher of science against the clear realization that, in encountering cases of observations like the one in the last example, the human mind has been struck by the closeness of the correlation, by the frequency of occurrence of these cases, and that it has chosen in regard to them a new type of analysis, namely that in terms of what is called a theory. Our concern is not with the question whether theories are unavoidable—we have already emphasized that nature can in principle be completely explored by phenomenologic correlations—but with their methodological analysis when they are present. The attitude which views them as unimportant by-products of an essentially factual description of the universe is nothing but an affected naivety, a grotesque *nihil mirari* which as it were shrugs its shoulders at the sight of new and astonishing types of growth, casually suggesting that science is not concerned with them, but only with the fertility of the soil that produced them.

The transition from correlational to symbolic or theoretic procedure is made, as was indicated before, by definitely leaving the field of nature and later returning to it. This is not an arbitrary undertaking, permissible whenever the investigator pleases to indulge in it. The criterion for its permissibility involves the notion of experimental *error*. Regarding this, it is only necessary to say at present that there are well known and perfectly standardized methods for determining the experimental error of a series of observations. When this has benn ascertained, the following criterion has meaning: If the defect of correlation $(1 - r)$ is smaller than a certain measure[7] of the experimental error, the scientist takes or may take the liberty of abandoning the correlational procedure and to invoke the method of symbolic explanation in terms of theory. Although I know of no place where this rule is stated, I am quite certain that almost all scientists adhere to it.

Both correlation and symbolic explanation have been alive throughout the history of human thought. Ancient association between the ascent of stars and the seasons was a primitive form of correlation, mythology an early attempt at symbolic explanation in a crude way. Aristotle's endeavors were chiefly correlational, while the atomists were clearly proponents of symbolic explanation. This example is particularly interesting because Aristotle was *aware* of the fundamental distinction between the two modes of analysis, for he saw the incompatibility which existed between his philosophy and that of the atomists. At least in this respect was his thought clearer, I believe, than that of the modern pragmatist or positivist who accuses him of having retarded the growth of science. At the present time, sciences may be divided into two classes. In one, correlation of data predominates, in the other the method of symbolic explanation. Physics and chemistry belong to the latter class, descriptive biology and statistical economic science to the former. Most success has been achieved in the fields depending on symbolic explanation, and there seems to be a general feeling, expressed forcefully by the groping efforts seeking to introduce physical methods of analysis into the realm of human behavior, that symbolic explanation is a superior tool of research. The assertion here implied of a methodological similarity between mythology and modern physical analysis may not be palatable to physicists. Yet it must be evident to the unbiased thinker that the electron and the ancient gods have very much in common as regards lack of direct observability, their function as foci from which flow numerous observable phenomena, and as ultimate termini of inquiry. The reason, however, why mythology is not a science is found in a rather simple but significant detail: The deductive use of the constructs is neither invariable nor free from internal contradiction. (Cf. section 1) Unbridled practice of symbolic explanation (speculation) does not make a field a science.

After these lengthy excursions, let us return to the problem of prediction, distinguishing now case a) (pure phenomenologic correlation) from case b) (symbolic explanation). When a correlation coefficient $r$ is known, a set of observations to which the coefficient relates may be predicted with a certain probability $p$, $p$ being a function of $r$. Thus in case a) prediction passes directly from one set of observations to another, the relation between the two sets involving probabilities. In case b) a theory is interposed. This theory links a sharp set of observations with another sharp set of observations; the theory itself has no diffusing effect; its elements are as clear and sharp as *logical* constructs, and its usefulness depends on this clarity. But diffusion is introduced in case b) by admitting experimental errors in both sets of observations, so that the net effect of prediction in case b) is the same as in case a): one set of data is linked by probabilities with another set.

### 3. THE "PROBABILITY OF THEORIES"

The stage has now been set for a discussion of one of the most interesting problems of the philosophy of science, the problem of scientific induction.

This, however, would lead us more deeply into epistemology than I feel competent to go. The remainder of the paper is negative in its constructive scope, inasmuch as it attempts to analyze in what respects the premises of certain proposed solutions of the problem of induction fail to correspond to the foregoing conditions. One of the most noteworthy recent proposals of the sort in question is embodied in Reichenbach's well known book Wahrscheinlichkeitslehre[8] and in other interesting writings by the same author. The book represents a farreaching attempt of clarifying and solving age-old problems by rearing them on the basis of a new probability interpretation, and by endowing the probability concept with a universality of application which, in our opinion, it does not possess. The merit of the work in question as a speculative approach to the problems of philosophy is beyond doubt, and unaffected indeed by the present criticism; but the claim of its author[9] that his system is part and parcel of present investigations in the strict sciences, particularly in physics, must meet with emphatic rejection. Its success can not be regarded as assured in view of current scientific procedures; it can only be granted if it modifies and improves them in the future.

Reichenbach's assertion that the probability implication applies generally to the relation between theory and facts is incompatible with physical methodology, and it is at the same time the thesis which lands him ultimately in the midst of many-valued logics. If, therefore, it is possible to refute it, most of the strangeness of his system—and perhaps its uniqueness—disappears. Let us, therefore, concentrate on the probability of theories. It is true, of course, that physicists occasionally, though rarely, speak of the probability of a theory. But this is hardly sufficient cause for widening the meaning of the term in an elaborate way, no more so certainly than there is occasion for extending physical theories because philosophers occasionally indulge in a loose use of technical scientific terms. Yet, as was already pointed out by Nagel,[10] they never attach significance to such statements. Let us rather see what physicists actually do when they apply their theories. From Reichenbach's point of view hypotheses have probabilities even before they are linked with facts. Since his definition of probabilities is of the frequency type, the hypothesis in question must form a member of a certain class or aggregate of hypotheses in which successful ones must be distinguishable from failures. This leads, in Reichenbach's own terminology, to a probability of the "second form." The details of the manner in which he proposes to count off relative frequencies among comparison theories and of their selection, are not of great interest in this connection; astounding perhaps is only the fact that such procedure is actually deemed descriptive of what physicists are doing. But aside from our inability of finding a single instance of it in physics, we do not see how it could possibly work if it were tried, and this for the following simple reason.

Reichenbach's own endeavor is to produce probabilities which, in the face of a given body of evidence, are unique and measurable. To achieve this result a frequency theory is needed. Unfortunately, however, even a frequency theory will fail to give unique probabilities if the elements to which it is applied are not clearly defined as single entities. A theory cannot be defined quantitatively as an entity. As an example, take the quantum theory. From one point of view it is certainly a single theory, but every physicist knows that Schrödinger's approach is in every respect different from that of Heisenberg, although the two may now be shown to be isomorphic. Is the quantum theory one entity, or does it amount to two? Moreover, it is perfectly easy to break up each of the components into a number of other theories again. This property of resolvability is peculiar to all physical theories inasmuch as every theory permits an infinite number of singular statements. The trouble here is not that the use of the term theory *has* not been standardized, but that it *cannot* be standardized. In view of this the counting of theories is meaningless; one count is as good as any other, and probabilities may be made—within due limits—anything the counter pleases.

Next we pass to Reichenbach's "probability of the first form" for theories.[11] This concerns the truth frequency of theories and is to be determined in the following way. One lists all the testable propositions to which a theory gives rise, and then divides the number of those which are experimentally verified by the total number. Now it is evident that this prescription is entirely without value when the number of testable propositions is infinite. This, as we have already seen, is always the case. Since the number of tested and hence that of verified propositions is finite the "probability of first form" is zero for every theory. This criticism is so simple and obvious that we feel the prescription, given by Reichenbach in this form, must have been intended differently. Indeed the example chosen by him for illustration suggests that the relative frequency is to be taken not with respect to all *testable*, but to all *tested* propositions. This example concerns the probability of the quantum theory and is very illuminating. (Cf. last reference, where the symbols we shall now use are defined.) $\psi$ is the quantum theory, $\varphi_i$ is a class of tested propositions having a truth-frequency $W(\varphi_i)$. Then, according to Reichenbach,

$$W(\psi) = \prod_{i=1}^{n} W(\varphi_i).$$

Now it was already pointed out that no theoretical proposition (prediction) can ever agree exactly with experiment; hence $W(\varphi_i) < 1$. The number of tested classes of propositions is n. We are thus immediately led to the result

$$\lim_{n\to\infty} W(\psi) = 0$$

In words: the probability of any thoery may be made smaller and smaller by testing it long enough.

But this is not the only difficulty. It is in fact easy to show that, according to Reichenbach's definition, the probability of the quantum theory is zero at present. Quantum theory requires a certain intensity distribution in the continuous spectrum beyond the limit of, say, the Lyman series in hydrogen. This intensity distribution has been photographed. In this instance, then, theory produces an infinite number of propositions of the type $\varphi_{1\lambda}$ where $\lambda$ is a continuous variable, any value of which may be tested. $\varphi_{1\lambda}$ means: At the wave length $\lambda$ there is to be found an intensity of such and such a value. Now the reduced photometer trace agrees with predictions only at a few points, disagrees everywhere else. This makes $W(\varphi_1) = 0$ and therefore $W(\psi) = 0$.—It is idle to object that we must not require exact agreement because the theory predicts only with a certain probability; this would mean putting the cart before the horse, since the present procedure is supposed to lead to a determination of that probability. The physicist knows, of course, where the trouble lies. His observations themselves are subject to error in an *empirically determinable* way. This leaves his theories sharp and connections can readily be established.

The present criticism to Reichenbach's prescription for finding the probability of a theory (of the first form) is not *in general* valid if, in carrying out his procedure, we operate with experimental errors. When the "softening" of data by means of experimental uncertainties is permitted, however, it is extremely difficult to see what methodological advantage could be gained by softening up theories as well. True, there is no rule against it, but Occam's razor has thus far prevented physicists from practicing such unnecessary diffusion.

There is a rather deep seated difference between Reichenbach's conception of a physical thoery and that which we are advocating as representing more truly the physicist's view. According to the former a theory is the logical product of proposition sequences, e.g. in the case of the quantum theory $\psi = \varphi_1, \varphi_2, \varphi_3 \ldots \varphi_r$. This definition is untenable. In the first place, if the product contains a finite number of factors referring to tested sequences, the definition is certainly wrong, for every theory implies much that is not tested. In the second place, if the factors stand for all possible sequences, the logical product is an infinite one and therefore very troublesome if not meaningless. The main inadequacy of Reichenbach's conception—and this criticism may be made of the entire Vienna circle as well—is that if fails to recognize the inherent power of every theory of *generating* testable propositions ad libitum. To this most significant circumstance justice can only be done by formulating the meaning of theories in terms of symbolic construction in the way we have attempted.

### 4. WHEN IS A THEORY TRUE?

In his reply to Nagel's competent critical review[12] of his Wahrscheinlichkeitslehre, Reichenbach further elaborates his views on the rôle of physical theories in a manner most vulnerable to criticism by physicists. We shall again use his own example to clarify the fundamental difference between his point of view and that which is in our opinion almost universally adopted by working scientists. The characteristic feature of Reichenbach's thesis is that theories make predictions *only* with a certain tolerance of error. There appears to be no close relation between his "probability of a theory" discussed in the last section, and this tolerance, as indeed one frequently misses the exact link connecting Reichenbach's various pronouncements with regard to the function of theories. But they have in common the implication of a certain internal diffuseness which ultimately leads to many-valued logic. The example to be considered is the theory of the constancy of the speed of light in free space in inertial systems. This theory, Reichenbach asserts, in essence is not a unique affair, but resolvable into an infinite number of statements of the following sort: c is constant within an accuracy $\pm \triangle$ with a probability greater than $p$. $p$ varies presumably from 0 to 1, and $p$ is a function of $\triangle$. Thus if $\triangle = .0015\%$, $p = 2/3$; if $\triangle = .0052\%$, $p = .9999$. If one looks into the computation of these figures one finds that the partial probabilities $p$ of the constancy theory are obtained by the rules for treating the divergence of observations in a given set experiments, such as those of Michelson and Morley, and that they have little if anything to do with the probabilities of theories of either the first or the second form as previously defined. However we are here concerned with more than the consistency of Reichenbach's own procedures; we wish to exhibit their contradiction with well established physical methods.

When the physicist speaks of the constancy of the velocity of light he means this in a far more absolute sense. His theory is either true or false; a single experimental contradiction (and this has often been pointed out before) invalidates it completely. But by contradiction is not meant a divergence from prediction, however slight. The degree of permitted divergence has nothing at all to do with the theory and is by no means arbitrary; it is derivable by an analysis of observations. Suppose that there exist a number of "good" measurements of $c$ in different inertial systems (the qualifying adjective is clear to physicists; it bars measurements containing discernible systematic errors), and let $c_{1i}$ be a sequence of measured values in one system, $c_{2i}$ in another. Finally, denote by $\bar{c}_1$ and $\bar{c}_2$ their arithmetic mean values and by $\triangle c_1$ and $\triangle c_2$ their standard deviations. The theory is then said to be true if the ranges $\bar{c}_1 \pm a \triangle c_1$ and $\bar{c}_2 \pm a \triangle c_2$ overlap, false otherwise. If *any* new series of measurements produces a range $\bar{c}_n \pm a \triangle c_n$ which does not overlap all others, the theory is declared false and abandoned. The constant $a$ is of order of magnitude unity and is inserted as a concession to a convention which has not completely crystallized. It is furthermore supposed that every set of measurements is sufficiently large to be susceptible to statistical treatment.

The "justification" of this criterion may be questioned from several philosophic points of view. If, for instance, a theory were regarded as part of nature's permanent and transcendental constitution one might well ask whether a set of observations, judged sufficiently large to be a fair sample, may not be a statistical freak and thus accidentally fail to satisfy the criterion just formulated. The physicist's answer is simply that he is not concerned with any theory independent of his observations and that in case of such failure the unfortunate theory falls by the board, accidentally or not.

The important point is that lack of agreement with facts never reduces the probability of a theory, but causes the scientist to abandon it. Striking indeed is the readiness with which physical theories are abandoned or modified. It may even be said that the clean truth-false alternative peculiar to the validity of theories is the main spring of physical progress. For if a contradiction merely lessened their probability the incentive to try modifications would not be great, and it is precisely the constant temptation to adjust theories to facts which is responsible for the rapid development of modern physical research. *Theories* are subject to change, and not their probabilities.

A case in point is the recent discovery of the heavy electron. Had Street, Stevenson, and Anderson applied consistently the method of Reichenbach, their findings would merely have reduced the probability that the electron's rest mass is $9.115 \times 10^{-28}$ gm. Instead, they concluded essentially a failure of the theory which attributes equal masses to all electrons, and suggested a new construct, the heavy electron.

At first sight, the argument maintaining the fixity of a theory's propositions might appear indefensible in view of such specimens as the statistical theory of gases or quantum mechanics, which do operate with probabilities. Under certain conditions (non-commutability of observables) quantum mechanics allows only inferences of the type invisaged by Reichenbach in connection with the constancy of $c$, although with one important difference. Quantum mechanics predicts all sorts of curious probability distributions for the observed data, not only Gaussian ones, and in that respect goes far beyond Reichenbach's interpretation. If this situation were to be described in his manner, it would lead at once to probabilities of higher order; i.e. the theory would have to be construed as follows: Theory says that the probability of finding an electron at a given place is $p_1$, and the probability that this statement be true is $p_2$. This, again, is not in conformity with the physicist's interpretation, which is much the same as in the former case and permits only the values 0 and 1 for $p_2$. It seems that the order of Reichenbach's probability is always higher by one than the physicist's.—One cannot say that quantum mechanics proves Reichenbach's point, because this theory definitely permits conditions of perfectly sharp inference, in fact it does so in most of the interesting experimental cases (sharp spectral lines, etc.). In earlier papers[13] I have treated these matters at greater length. For the present it suffices to state that even theories, such as quantum mechanics, whose sharp inferences ($\psi$ functions) are related to probability distributions on the side of nature in a somewhat complicated way, are always found either true or false by the criterion discussed.

## 5. MANY-VALUED LOGICS

After the critical remarks on the preceding section we hasten to record again that they were not intended to damage the analytic structure of Reichenbach's probability theory, which, I feel, would be improved if the notion of probability of theories were dropped. Very interesting and fruitful is his replacement of v. Mises' condition of randomness by normalness, which renders probability aggregates constructible. His approach from the side of symbolic logic is illuminating and the formalism suggestive, although in places perhaps unnecessarily complex. Most note-worthy is Reichenbach's attempt of getting around the limit difficulty by appealing to probabilities of higher order. On this matter the last word has of course not yet been spoken, and one hopes its author himself will return to it with further clarifying thoughts. But all these matters stand even if those features which contradict scientific practice are dropped. Why, then, were they inserted?

We venture to suggest a two-fold answer. The first adverts to an attitude closely related to the philosophy of the Vienna circle. In terms of the developments of this paper, it is the failure to distinguish between the practice of correlations and that of symbolic explanation. Positivism is prone to look down upon the latter as meaningless and because of this never gets to the heart of any exact science. But Reichenbach obviously does not wish to go to the logical limit of denying the importance of theories. On the other hand, not venturing to concede to them the methodological distinction they deserve, he has to treat them by the method of correlation as though they were themselves observations, or a part of what we have called nature. Correlation, as we have seen, imparts diffuseness to the correlates, and hence there emerges the diffuseness, or probability, of theories.

The other aspect of our answer regards the scope of Reichenbach's system. It is clearly larger than that of all other probability theories, and it is made so by the universality implanted in the probability concept. This implantation, it is here maintained, is not in keeping with actual methods of physical procedure. But this universality lands its proponent finally in many-valued logic. The fact should be of interest to the philosopher that this destination is reached because of the retention in the system of elements, such as the probability of theories, which have no scientific status, and that their removal would not lead to many-valued logics.

Lest the reader misconstrue our attitude as fanatical enmity against many-valued logics, we make these concluding remarks. New systems of logic are highly interesting and useful in their own right; their establishment need not be justified by the allegation that they are implied by scientific procedures at present. Their status is much the same as the work of Riemann, Bolyai and others on non-Euclidean geometry before concrete use was made of it in the theory of general relativity. Many-valued logics may someday revolutionize science. But their potential value would be impaired if confusion as to their bearing on contemporary methods were allowed to creep into their very making. For that reason the physicst's "optimum wager," to use Reichenbach's term, is to wait until the structure of many-valued logics is more fully developed, until their relation to two-valued logic is no longer in dispute, and until rules are given for applying that calculus to observable facts.

I record here my sincere appreciation of the benefits derived from discussions with Professor F. S. C. Northrop.

## NOTES

[1] The reader is asked to pardon the vagueness of the terms. The matter is discussed more adequately and fully in my papers on Methodology, *Phil. of Science*, 2, 49; 2, 164, 1935.

[2] I am fully aware that the distinction just made meets with widespread disapproval among logical positivists and physicists. To clarify my position I may perhaps be permitted to say that I admire the successes achieved by the syntactic analysis of the Vienna circle in clarifying the calculus of propositions. But I am forced to regard their fundamental proposals as an interesting "Philosophie des Als-Ob" which proceeds on the tentative rule that we ignore all "pseudo problems." This very statement implies that there must be a deeper sense in which the problems, arbitrarily ignored, still persist. Positivistic ignoration is a useful artifice indeed for it promises to solve the remaining problems with greater ease. However, methodology of science, unlike some parts of science itself, automatically breaks across positivistic landmarks, as the present distinction shows. In deference to the merits of positivistic procedures I should like to call propositions whose meaning is not clear in the formal idiom (Cf. Carnap) *forbidden* propositions, distinguishing them from "meaningless" ones which may be defined in other ways.

[3] Cf. for instance, R. v. Mises, *Wahrscheinlichkeitsrechnung*.

[4] A. Kolmogeroff, *Grundbegriffe der Wahrscheinlichkeitsrechnung*. H. Cramér, *Random Variables and Probability Distributions*.

[5] C. E. Bures, *Phil. of Science*, 5, 1, 1938.

[6] This and many similar examples may be found in C.V.L. Charlier, *Vorlesungen über die Grundzuge der mathematischen Statistik*.

[7] This phrase is being used to avoid fixing the exact meaning of experimental error, which may be taken to be the mean deviation from the mean, the standard deviation for measurements under "identical" conditions, or any other index used by scientists for such purposes. The following relation is illuminating in this connection: If we define

$$y_i = a(x_i + \triangle_i); S(x) = \Sigma_i(x_i - \bar{x})^2; M = \Sigma_i(x_i - \bar{x})(\triangle_i - \bar{\triangle});$$

then formula (1) may be seen to be identical with

$$r = 1 + \left[\frac{S(x)S(\triangle) - M^2}{(S(x) + M)^2}\right]^{-\frac{1}{2}}$$

If $M \ll S(x)$ and $S(\triangle) \ll S(x)$, $1 - r = \frac{1}{2}\frac{S(\triangle)}{S(x)}$

[8] H. Reichenbach, Leiden, 1935.

[9] H. Reichenbach, *Phil. of Science*, 5, 21, 1938.

[10] E. Nagel, *Mind*, 45, 501, 1936.

[11] H. Reichenbach, *Erkenntnis*, 1935, p. 267.

[12] E. Nagel, *Mind*, 45, 501, 1936.
See also his *The Structure of Science*, Harcourt Brace and World, New York, 1961.

[13] H. Margenau, *Phil. of Science*, 1, 133, (1934); 4, 337, (1937).

CHAPTER 8

# ON THE FREQUENCY THEORY OF PROBABILITY

In his impressive critique of frequency theories,[1] Professor Williams accomplishes one-half of a worthwhile task: he discredits the claim that frequency is the sole meaning of probability. This is indeed a service to humanity, for it is true that frequentists have been strutting about proclaiming their gospel dogmatically, and with an unwholesome indifference to other points of view. For this reason, if for no other, Williams' attack is likely to afford the objective reader a decent measure of retaliative satisfaction; but it is to be questioned whether it satisfies completely his intellectual appetites. Having deflated the frequentist, the author might have gone on to dispel the ghosts from other haunts, to survey and examine the field of his study with the equanimity and detachement which normally settle upon the critic when his destructive deed is done. Unfortunately, however, this second synthetic part of his task remains unfinished. We are offered, as a substitute for the frequency theory, the hoary formula of Laplace. To use a simple metaphor, a modern edifice has been torn down because of a few cracks in its facade, and it has been replaced by a cave dwelling.

Having indicated my main position as succinctly (and perhaps as offensively) as seemed feasible I shall endeavor in the following discussion to present some of the evidence on which it was reached. Unable to differ with the larger sweep of the arguments in the paper I may first leave its criticism intact and develop the alternatives untouched in Williams' treatment, reserving for later consideration those minor aspects in which that analysis seems to be defective. The point to be made is that there are many divergent definitions of probability besides the one characteristic of the frequency view, definitions which it is neither accurate nor profitable to subsume under the Laplacean doctrine. With all these the frequency interpretation enters into a unique relation, a relation which confers validity upon the specific theories wherein the divergent definitions are embedded. The frequency postulate, while powerless alone, is a necessary complement designed to impart substance to the other views.

Before embarking on this principal theme it seems desirable to raise an issue which, though recognized in the paper, has not perhaps been given the emphasis it deserves. It has to do with the *practical* shortcomings of Laplace's formula. No final decision will here be based on pragmatic considerations, which may be felt to be out of place in any logical analysis; but the fact is that the preference of people who use probabilities is at least psychologically rooted in factual matters such as the following.

A dicer is instinctively ready to accept Laplace's dictum in placing his bet on the appearance of a five. Perhaps he would prefer to know more about the die than that it has six faces, but the simplicity of the formula appeals to him. If, before the game, he had a choice of either counting the number of faces or of inspecting the results of a long series throws, there is no doubt about what he would do. But suppose he is prevented from ascertaining relative frequencies. If the game goes on and the player is reasonably satisfied with its progress he will harbor no grudge and speculate no further. If not, he will demand to know more about the die. If he is a confirmed Laplacean he will try to modify his computation of the number of favorable cases, perhaps by determining the center of gravity of his die or by examining its magnetic behavior. In that case, Mr. Laplace will let him down rudely, for *he specifies no method whereby his rule can be adapted to special physical circumstances*. The frequentist, on the other hand, is not embarrassed, for his recipe is designed to cope with situations of this sort.

In practice, therefore, the player must at this junction either turn frequentist or abstract logician, who says: my original estimate of the probability is still correct, only it does not apply to this particular die. To aver that it did apply, but becomes modified in view of further evidence inherent in subsequent throws of the die, is to make a noise in hopes of confusing the critic, or, in more charitable interpretation, to shift the meaning of probability to a new sense that is of no interest to the player.[2] If, then, Laplace's rule must be maintained, the loaded die is not an object to which probability considerations can be applied. This rule drastically limits the range of application of the probability calculus and thereby reveals the distinguishing character of any *specific scientific theory*, which in fact it is.

Let us discount the objection that the frequentist also employs Laplace's rule in arriving at his measure of probability. Of course in a sense he does; but to admit it would wipe out the distinction between the views which are being contrasted; it would run counter to the current usage of terms and would be about as significant as the assertion that Democritus is responsible for modern physics because he introduced the atom. The frequency theory here under discussion is a very specific device, developed through the collaborative efforts of philosophers and mathematicians and certainly distinct from Laplace's early views.

Sharing to a large extent the repugnance which Williams holds for purely pragmatic considerations, I shall attach to the preceding deliberations no more weight than is commensurate with their bare factualness: They indicate that there are situations in which Laplace's rule cannot be used, or in which its use becomes equivocal. Moreover, it should be added that in science these situations are very numerous, a fact to which anybody engaged in actuarial, statistical, or biometrical research will testify. The frequentist somehow always manages to float when the Laplacean calls for help.—After these preliminary remarks, we proceed more directly to the heart of the problem.

The roots of the conflict between the two views under discussion go deeply into scientific methodology; indeed the conflict is nothing but a minor phase of the battle between the empiricist and the rationalist, fought in a terrain that does not lend itself to direct and open warfare because its topography is not completely known. By this we mean that probability still contains a few mysteries of a technical, mathematical sort. With the reader's indulgence, we shall therefore transfer the conflict to another, more familiar arena, and watch it there as mere bystanders.

A rationalist (R) and an empiricist (E) wonder about the volume of an object that looks very much like a sphere. R handles it, admires its smoothness, guesses at its radius and says: "the volume of this object is $4\pi r^3/3$," which he quickly computes to be, let us say, 5000 cc. Thereupon E, with a superior smile, takes the object, immerses it in a tank of water and measures the overflow. Soon be announces: "You are quite wrong, my friend; the volume of this object is 3459 cc." This puts R clearly on the defensive.

But not for long. "Well," he says, "I had no idea you were such a stickler for accuracy." Producing a caliper, he proceeds to meausre the diameter of the object. While doing so, he protests: "Now don't take this to be a concession to your operational philosophy; I never said I could find the volume of this thing in *a priori* fashion. I merely maintain that I need not fuss with anything, like your loathesome water bath, that will measure not volume but the water displaced by the object." Then, with just pride in his accomplishment, he announces: "the volume is 3596 cc."

"Oh, no," says E, whose quick eye has already apprized him of a defect in the object's supposedly spherical form. "This thing is not a sphere, it is slightly ellipsoidal." R, after suppressing an irksome feeling, almost voiced, that the blasted thing really had no business departing so far from ideality, says: "Oh, I see," uses his caliper again and declares "You are right. I measure two different diameters in two different directions." Then, producing pencil and paper, he calculates $V = 4\pi r_1{}^2 r_2/3 = 3475$ cc.

Here the real battle starts. E accuses R of ignorance as to the actual shape of the ball, pointing out that it is improper to use the formula for the volume of an ellipsoid. But then R begins using his heaviest ammunition; he challenges his opponent to do his messy water experiment over again and see if he gets the same value for the volume as before. E admits that he does not. And here R declares with finality: "If you get a different value every time you perform your experiment, pray tell me which of them *is* the *true* value of the volume of this object." Now E is driven into a corner; he admits meekly that he uses a rather arbitrary procedure, namely, that of taking the arithmetical mean of his results, to fix the true value. It is at this juncture that R exploits his triumph by making the devastating remark: "And you cannot even be sure that, for an infinitely long run of experiments, this arithmetical mean converges to a limit." E says "No" and grows pensive.

After some cogitation, E performs a series of measurements, each time immersing the object in the water tank. He takes the mean and finds it to be 3465.4 cc. All his individual measurements are included between the limits 3454 and 3479 cc, and he declares, in true scientific spirit, "the volume of this object is 3465.4 ± 14 cc." The plus-minus ambiguity puzzles the rationalist who insists that, surely, the volume of an object is unique.

Now E launches his offensive. "You declare that your value was unique," says he. But how did you arrive at it? Did you not start with measured values of $r_1$ and $r_2$? If you were to measure these quantities over again, you, too, would obtain different results and arrive at different values for your computed volume. From the point of view of uniqueness your scheme has no advantage over mine."

Incidentally, it is *not* recorded that R replied: "My friend, operations never determine anything. When I first computed $V$ = 5000 cc, that *was* the volume of the ball. Of course, this volume *changed* on further evidence."

Being unable to arrive at a satisfactory *measure* of the object's volume, the philosophers turned to the question of the *meaning* of the concept, volume. E defined it in terms of his immersion experiment. To R it was something to be found in Euclid. E made the point that speculative definitions à la Euclid are often dangerous and useless (here he invoked relativity and non-Euclidean geometry), but R was able to show that many applicable formulas may be derived from his definition (so why was it useless?). "Yes," said E, "but you never know when these formulas are applicable. You stumbled into that trap when you computed the volume of the ball." "And you," said R, "have no right to apply the term 'volume' to anything but the result of this specific experiment. What is usually called the volume of the earth is something utterly different from what you term volume; for the earth cannot be immersed in your water bath." And so on.

A similar dialogue could have been conducted about every physical quantity that is measurable in a scientific sense.

More significantly, what has here been said about volume refers, with minor modifications, to the issues surrounding the concept of probability. The reader need not be told that R was the Laplacean and E the frequentist. He may, however, have wondered whether it is true that the Laplacean probability is not unique. For according to widespread belief, both the number of favorable cases and the total number of "equipossible cases" are definite integers, hence their ratio is a unique rational fraction. It is commonly admitted that the matter of weights is troublesome, introducing uncertainties which can only be eliminated by the principle of ignorance. The point here to be made is that even the number of alternatives, favorable and total, is not uniquely specifiable in any concrete instance. A penny need not fall heads or tails, it can stand on edge. A die has more than six faces, as microscopic examination will reveal. This lack of precision is just as significant, and just as detractive to perfection, as the failure of a frequency series to possess a limit (of relative frequencies in the mathematical sense).

Having, thus far, thrown up paradoxes at random with an obvious bias in favor of the frequency thesis—and be it said that this partiality was displayed mainly to counteract Williams' aversion—it behooves us to create order out of the general chaos.

Since the core of the problem is how to define probability, we should first specify what should be demanded of a good definition. Many philosophers, and most logicians, require that the definition be explanatory of the widest usage of the term. To demand adequacy of the definition in this inclusive sense would, in the case of probability, involve an investigation of what the word *generally* means in our particular language. Then, if consent of many investigators can be secured, that meaning, or collection of meanings, is standardized. The results of this procedure favor the Laplacean interpretation. They justify the conclusion that probability is a "rational expectation," or a "degree of belief based on evidence," that a single occurrence has a probability; in fact they identify probability with any kind of likelihood. The type of probability that is *measurable, verifiable* and highly quantitative may be included as a special case. From this point of view, our problem has been solved admirably by the Oxford dictionary. The solution, however, does great violence to science and is, we feel, a trifle archaic for the purposes of philosophy. This situation is usually recognized in other examples: in analyzing the meaning of force the philosopher of science is not greatly concerned with the police force.

An alternative procedure leading to the stabilization of meanings involves an investigation of that special brand of learning in which the term in question has attained its greatest refinement. In the case of probability, whose history has long surpassed its early concern with gambling, *the field thus specified is science*. It is in frank espousal of this second alternative that the following remarks are offered, and the reader who feels that science has no prereogative in this connection will derive no illumination from them. To him, Williams' challenge must remain an irrefutable indictment of the messiness of language.

But even when the peculiar competence of science in this special realm is granted there lurks a danger of miscarriage for our venture in the well-known fact that scientists are frequently careless in their language. To see how scientists use the term may lead to all the infelicities which were to be avoided by an appeal to science. It is necessary to see what they *do* with it, to study their method rather than their language. We feel that, when this is done, there results a view which is clear of the pitfalls noted and described in the paper under discussion.

Science deals with certain regular aspects of experience, not with what the philosopher likes to call being. To be sure, the scientist ascribes existence to external things; in fact many do so in the uncritical attitude of primitive realism. But the mood is changing as indeed it must in view of recent discoveries in the field of atomic physics: scientific existence is coming to mean nothing more than a set of invariant features in our way of describing experience. Experience itself is fluid and defies anyone who wishes to confine it in the straight jacket of invariance, but the elements which we invent and place in unique correspondence with aspects of experience must and do satisfy methodological principles which regularize their use. These constructed elements, such as volume, mass, valence, price—and, for that matter, the Laplacean "probability" for the appearance of a five uppermost when a die is thrown—are sharp and clear in contrast with sensory experience. How, then, are they related to experience?

In answering this question reference will first be made to the *exact* sciences like applied mathematics, astronomy, physics, and chemistry. They have evolved a methodological scheme, rarely exhibited in science books, whereby a passage can be made from perception to theoretical description via certain variform *rules of correspondence*. Thus the physicist knows, often instinctively, what in a given set of observations corresponds to volume, for example, or to mass. Knowing the mass, he can *reason* about the phenomenon with the aid of theories in which the idea of mass is embedded. The validity of the whole procedure obviously rests on two conditions: (*a*) that he be able to "extract" the construct mass from his observations; (*b*) that he be able to manipulate it in his theories. The question we posed at the end of the last paragraph is simply: How does he achieve condition (*a*)? With it there is now associated the other: how does he achieve (*b*)?

The answer to both questions is: by the use of proper *definitions*. To wit, the rules of correspondence alluded to in question (*a*) are tantamount to, or found within, *operational* definitions, and the logical fertility of constructs implied by (*b*) is brought about by utilization of *constitutive* definitions—to use a terminology I have previously proposed. In the example of mass, the operational definition has reference to balances; it prescribes a standard process or perhaps many standard processes, i.e. empirical tests through which mass can be quantitatively determined. On the other hand, the physical entity can be defined constitutively by saying: it is the symbol $m$ which appears in Newton's second law of motion. *When both definitions are at hand* science can set its intricate machinery in motion and perform its amazing feats of predicting the mass of the moon, or the motion of Jupiter, or the radiation emitted by atomic electrons.

Now it is a fundamental principle of exact science that every quantity, to be scientifically adequate or valid (notice that we do not say logically meaningful), must be capable of both constitutive and operational definition. To be sure, this is not the only requirement for validity but certainly the *conditio sine qua non*. Implied here are both the claim of the rationalist and that of the positivist, but not in mutual exclusion. Operationalism turns out to be one facet of the product which the scientist offers; but the intriguing luster it has recently attained by virtue of much philosophic polishing should not misguide the spectator to a belief in its sole supremacy.

Nor should one overlook the fact that every physical quantity possesses in general many operational and many constitutive definitions. The number of the former is primarily a matter of convenience, convention, and manipulative skill; the number of constitutive definitions is equal to that of the theories in which the quantity plays a major rôle. Moreover, a verified theory often generates a new operational definition, and situations may arise in which it is difficult to state whether a given definition is of one type or the other. But such a possibility, while baffling to the logician whose interest is confined to static fact in contrast with the elusive, dynamic forms of modern theory, merely confirms the fruitfulness of the interplay between the two fundamental forms of scientific definition.

We have now come face to face with the crucial issue: Is probability a physical quantity? Professor Williams appears to say it is not. I should prefer to answer in the affirmative; the reasons for this position have been stated. If probability *is* a physical quantity it must partake of the duality of definitions: its operational definition is in terms of relative frequencies, the constitutive aspects are largely, though not exclusively, presented by Laplace's rule. It is through the frequency interpretation that we apply the concept to experience; the other formulations permit us to reason about it, to transfer probabilities from one aggregate of observations to another via theory.

The failure of frequencies to be unique, or to tend uniformly toward a limit, is no longer disastrous when it is realized that all operational procedures yield fluctuating results without destroying science. Measurements never fix the value of any physical quantity; probability would be an exception to a general rule, and hence a monstrosity, if its operational definition permitted ultimate precision. The irreducible haziness with which every quantity is surrounded on the empirical side would indeed be embarrassing to science if it were a purely inductive affair, never passing beyond the flux of sensory experience. This, however, is not the case: since physical quantities have also theoretical aspects which are not beclouded by uncertainties, stabilization of meanings can and does occur.

Descending from this plane of generality to particulars, consider again the example of mass. Measurements of the mass of a given object are spread about a certain mean; we grant that if such a set of operations is used to define mass, uncertainties adhere to the concept. Using now the constitutive definition of the term by the aid of which we are able to place it in relation to other mechanical quantities, i.e., the symbol $m$ in the laws of motion, we *identify* the mean value with the symbol $m$ and proceed to solve the equation. Although the mean is sometimes called the *true* value of the mass, this terminology has no metaphysical importance and had better be disregarded. The solution of the equation yields values of other quantities, such as position and velocity of the mass at some other time, and these values are *sharp*, for theories themselves never impart diffuseness to predictions. The physicist then examines, this time using the operational definitions of position and velocity, the solutions thus obtained. Clearly, he will *not* find them *sharply* verified, but encounter again a certain spread among his observations. If the range of spread is "compatible"[3] with the spread of the original mass determination and includes the predicted value, the theory is said to be correct in this instance.

Similarly with respect to probabilities. Using the frequency definition, the probability distribution for drawing the several colors of marbles in a bag may be ascertained. From this, though it be slightly uncertain knowledge, other quantities can be predicted, as, for example, the probability of drawing one white and one red marble in succession from the same bag. This is possible only on the basis of some constitutive definition of probability, usually of Laplace's form. Such a definition then forms part of a *theory*, and the theory is violated or confirmed by empirical reference to the frequency with which the successive drawings do occur. The situation with regard to spread is the same as above.

Science is not always engaged in verification of theory; it may use theory to avoid observation. If, for example, I possess certain empirical knowledge (operationally obtained) about a material object, let us say its acceleration and the force applied to it, I may appeal at once to a constitutive definition of mass and obtain its numerical value. Here again, an exact parallel is found in the field of probability where knowledge of the physical properties of the marbles in a bag permits one to compute, by Laplace's rule, the probability of drawing a red specimen. In both instances the calculation may be wrong. Laplace's rule here might correspond to Newton's laws of motion; both have well-known limitations. The physicist freely admits that there are cases where Newton's laws do not apply. Should not the probability philosopher be equally open-minded about Laplace's rule?

Thus far our concern has been solely with the exact or theoretical sciences. A brief glance at biology, psychology, economics, and sociology will further confirm the thesis here advanced, namely, that the normal duality of definition with respect to all measurable scientific quantities is reflected in probability theory in the complementarity, often construed as contrast, between so-called *a priori* and *frequency* interpretations. For in the last-named sciences, which are devoid of quantitative theory and hence barren in constitutive definitions, the investigator is forced to use frequencies almost exclusively. His procedures remain on the level of correlations; unaided by Laplace's rule or an equivalent, he must measure relative frequencies in various aggregates of phenomena and content himself with the ascertainment of correlation coefficients. Only when a *theory* is given do the words "favorable event" and "equipossible event" have any meaning. The use of the two types of probability in the various sciences is thus in complete accord with their respective functions, as they were here analyzed.

We have assigned to the frequency definition and to Laplace's rule what we believe to be their proper places in the scheme of scientific things. Let us now extend our view and observe whether these two definitions are the only ones in their respective categories. In general, a quantity capable of one operational and one constitutive definition is of limited interest. Probability is no sterile concept.

In the first place, there hide behind the frequency proposition a number of specific operations which further scrutiny reveals as differing essentially from one another. A few examples will make this clear. The relative frequency $n_i/n$ or its limit, may refer to a coexisting aggregate of observations like a sample of the ages of a group of living people, or the behavior of a group of molecules. But it may also concern a temporal sequence of ordered events, such as the outcomes of successive throws of a die, or the behavior of one molecule on numerous different occasions. Clearly the operational rules involved in these two interpretations are highly disparate, and the frequency theory alone offers no justification whatever for identifying the constructs to which they lead. Scientists are keenly aware of this situation; their endeavors to provide such justification on the constitutive side have produced some of the most abstract formalisms of modern mathematics (e.g., the ergodic hypothesis). Furthermore, there is in use an operational device which can hardly be classified as a frequency measure: in actuarial work probabilities are often determined by noting what assessment of odds makes or loses money.

On the other hand, one easily perceives that Laplace's rule characterizes but a very small portion of the vast range in which probability is constitutively significant. It merely represents that theory which is most successful in connection with games of chance. Any aggregate which presents a continuum of properties defies its direct application. Statistical mechanics, the field in which the ideas of probability have reached their highest quantitative perfection, is dominated by at least three major theories (those of Boltzmann, Gibbs, and Darwin-Fowler) performing duties similar to Laplace's rule. In quantum theory, probabilities are computed by solving differential equations. A study of all these constitutive apparatus is quite indispensable to the student who wishes to arrive at a full and objective appraisal of the meaning of probability. But the present occasion forbids their consideration, and even the mention of a host of others operative in different fields.

To summarize: what appears to Professor Williams as a conflict between two rival formulations of the probability idea, one of which he wishes to abandon, or in some cases to see reduced to the other, turns out on further scrutiny as the highly fruitful interaction between two forms of scientific definitions. What he deplores as a logical contradiction emerges as the dynamic element energizing all of the theoretical science.

This more organic view, however, demands a sacrifice if it is to be attained. Just as the follower of Newton, who revolutionized the meaning of force, had to abandon Kepler's cherished notion of the *vis a tergo*, so the modern probability philosopher must decide to forego the use of the term "probability" in reference to intrinsically untestable situations like single events. Decisions of this sort were never made on the basis of logic. Logically, Newton's force is not superior to Aristotle's, nor does it have a firmer basis in any metaphysical essence of things. Dialecticians could argue endless debates on this issue and have done it; the crucial point is that Newton's deductive theory, which science has pronounced correct by its own specific methods, entails the new definition of force. The *vis a tergo* of the seventeenth century has become the "probability of a single event" of today.

Of course men are still interested in rationality of expectations. They would like to know the chance that, in a single throw of a die, they will cast a six. They ponder with heartbreaking fear over the chance that *their son* will die in battle. There is a large domain of experience not amenable, or at least not yet amenable, to scientific treatment and this domain will continue to be of concern to man. But so in Newton's day: the planet still moved as if propelled by Kepler's *vis a tergo;* people no longer called the agency a force, they described it by the vaguer term "inertia."

At one time[4] I was hopeful that a similar device might work in the present situation. To provide a distinction in language corresponding to the conceptual difference between a probability that is testable and one that is not, it was suggested that the word "probability" be used for the former, "likelihood" for the latter. Present usage, being indifferent to the distinction here intended, lends it no support and hence, I suppose, the proposal must remain futile. A case could be made for it on the basis of etymology. The word "probable" did, at one time, connote provable, while "likely" has always had the flavor of subjective "seeming." The English language has an advantage over others in possessing *two* words to convey this distinction, and it would be a pity to keep them blunted by promiscuous use.[5]

Like a single event, a scientific theory or hypothesis does not possess a testable probability but only a likelihood (if anything). The question as to the probability of a theory is a curious one because of the very lop-sided interest it enjoys. Probability theorists grapple with it most valiantly; they seem to feel their reputation at stake unless they are able, in final desperation, to wring some meaning out of the probability of an hypothesis. The exact scientist, on the other hand, is alsmost wholly unconcerned. To him, a theory or an hypothesis is never *true* in the sense of being ultimate. It is valid if no present or past experience contradicts the deductions which flow from it. If these deductions have not been tested, the scientist does not go around computing probabilities but bends his efforts to design crucial tests. A theory which by its nature does not expose itself to test is not scientific. Finally, if a theory is contravened by a *single* experience, it becomes wrong or invalid and is rejected; to say that its probability had decreased would be distinctly amateurish.

To be sure, a theory may be in error and nevertheless be useful as an approximation. This situation is in fact a very common one; note, however, that the degree of approximation is never expressed in the form of a truth probability but in more suitable mathematical ways. All this should be clear to the observer who realizes that science, instead of seeking to establish some static truth, is in fact a dynamic concern furnishing reliable knowledge. If it functions in a way displeasing to the probability logician he may condemn science; but he must not glory in his presumed feat of having made his probability theory more scientific by being able to define the probability of a scientific hypothesis.

To say that scientists profess no interest in an idea does not, of course, make the idea worthless. We asserted that the likelihood of a theory is not testable, and the burden of the proof is still upon us. To be testable, a probability must permit the assignment of a fixed numerical measure. In connection with a theory, the probability therefore relates in some way to the fractional number of valid predictions. The point is that this number is not unique. The difficulty lies not so much in the fact that the number of valid and the total number of possible propositions are both infinite, this trouble being encountered and resolved in other instances. It resides in the circumstance that each concrete proposition implies an infinite number of others, a circumstance which makes it impossible to say what propositions are atomic and hence of equal weight. In mathematical terms, the probability of a theory cannot be normalized. To impart meaning to it the logician must first invent an adequate theory of the continuum of logical propositions, to rank with the much simpler theory of the arithmetical continuum—and this is a frightening challenge.

Conversely, specific formulations of the probability of theories may be shown to lead to absurdities when fully analyzed.[6] But space here forbids this undertaking. Thus, to be in keeping with our former suggestion, we should speak of the likelihood of theories before they are tested, expressing thereby their degrees of subjective appeal. These degrees may be expressed on some numberless scale, similar perhaps to a scale of esthetic values which, I suppose, are not amenable to numerical fixation.

There is, however, one sense in which theories may be said to have testable probabilities, namely, that which assigns the value 1 to a correct, 0 to a wrong, theory. But this interpretation is trivial, rarely adopted, and may therefore be dismissed without further comment.

In insisting that a ban be placed on intrinsically untestable probabilities, the meaning of the adverb must be clear. Legitimate uses of probabilities which cannot be measured *directly* are numerous everywhere. The physicist freely speaks of the probability of light emission from a single atom, using a construct which, in that particular setting, he could not validate by frequency determinations. Nevertheless this probability is not *intrinsically* untestable because it leads to other propositions which can be vertified. Probabilities of single events and of theories do not possess this measure of fruitfulness.

Having now outlined my general position, I shall turn to a few specific points in Professor Williams' paper. What has here been called frequency theory without further qualification should be regarded as identical with the "collective" theory in Williams' terminology. Concerning it he states, correctly, "that in describing probability as something which can never be encountered in experience, it makes a probability something which can never be inferred from experience, and something *by* which nothing else can be inferred from experience." Be it noted, however, that science is not in any such predicament because it uses the operational collective definition in conjunction with those here called constitutive. The quoted statement, if it were taken as a criticism of the use of the collective idea, would apply with equal stringency to every physical quantity defined operationally. It would be most damaging if such a definition functioned in isolation, which it never does.

One might even go farther than this. It is my own belief that an exact science never *infers* from experience and that the whole logical problem of inference is of minor importance in this connection. Theories, though frequently suggested by experience through psychological associations that are difficult to specify, are fertile in a *deductive* sense. They are essentially postulates whose *consequences* are verified. Thus it is quite true, as has often been stated, that a theory can never be completely confirmed; but to regard this as an indictment of scientific method is to mistake the deductive rôle of theory for an inductive one. Nagel[7] points out the further fact that a probability statement, in its frequency sense, can never be refuted by finite experience. But this too is a property it shares with the predictions of every other scientific theory if precise logical refutation is meant. These observations are not disastrous to the frequency formulation of probability any more than to theoretical science.

Professor Williams adverts to the difficulty inherent in an operational test of the randomness of a probability aggregate. This is a technical point recognized by almost all collectivists. Directions for avoiding this difficulty[8] have already been given, albeit they may not appeal to all mathematicians. My own opinion is that this problem will no doubt be solved, and that it is far less serious than the puzzles posed by the principle of indifference which arise when the Laplacean theory is detached from the frequency formulation.

In footnote 53 and the paragraph to which it refers, Professor Williams admits the possibility that one probability for a given aggregate, obtained with the use of incomplete evidence, may be superseded by another, based on more relevant evidence. This, he feels, "is not without difficulties." From the point of view here presented all difficulties vanish; the two ways of computing the probability correspond to the use of two different theories, the first being an approximation to the second. To make this clear, let us consider an analogous situation in science. The chemist wishes to determine the pressure of a gas, the pressure being defined operationally by reference to gauges. Instead of measuring it he uses a theory to compute it. If he knows the volume and the temperature of the gas, and nothing more, he will employ the perfect gas law in his calculation. The resulting pressure will be wrong, for the theory is known to be inaccurate. But under the circumstances it may pass as a satisfactory approximation. If, however, the chemist happens to know certain constants pertaining to the gas, he will use Van der Waal's equation and obtain a more accurate value.

To describe this state of affairs, he will not say that the *pressure* has changed as a result of new evidence, though he will use this phrase with respect to his *knowledge* of the pressure. Nor would he attach any meaning to the statement, found in the paragraph just cited in reference to probabilities, that "the former (pressure) was an equally genuine (pressure) while it lasted."

We merely observe in instances of this sort how the use of different constitutive theories may lead to different numerical values for a specified probability: which theory is more adequate can be decided only by reference to frequencies.

Professor Williams' criticism of the use of the qualifier *a priori* as applied preferentially to Laplace's theory is much to the point. The word is one which has taken on a very specific meaning in epistemology, and this should be respected by scientists.

Reluctantly I forego a detailed reply to the important and timely query of the author's last paragraph, hoping that the preceding developments provide some sort of answer. An adequate world view, it seems, is not a conglomerate of empirical knowledge; it is a postulational thesis whose demonstrative consequences are confirmed in experience.

## NOTES

[1] *Philosophy and Phenomenological Research*, V, 449–484.
[2] Later in this article we shall examine with greater care whether and how probability can change with evidence and thereby provide further substantiation of this indictment.
[3] Here the logician will balk and demand to know the meaning of compatible. My answer is not likely to please him: Compatibility is a matter of convention, a tacit understanding among scientists. More evasive answers, plausibility arguments can be presented,

but this bare fact remains. Science is, after all, a man-made discipline which derives its distinction, however great, from acceptance of a specific methodology. This methodology includes logically vague but practically decisive maxims concerning compatibility of measurements (cf. any treatise on the theory of errors).

[4] Cf. *American Journal of Physics* Vol. 10, (1942), p. 232.

[5] It is quite possible that the naïve literal sense of the German *wahrscheinlich*, which has no reference at all to the methodology of science, has prepossessed investigators and obscured many issues.

[6] See, for example, H. Margenau, *Philosophy of Science*, Vol. 6, (1939), p. 65.

[7] E. Nagel, *International Encyclopedia of Unified Science*; Vol. I, No. 6.

[8] See, for example, an early and interesting one by A. H. Copeland, *American Journal of Mathematics*, Vol. 50, (1928), p. 535.

CHAPTER 9

# CAN TIME FLOW BACKWARDS?

The nature of time is one of the crucial problems in the philosophy of science, and it cannot be solved by an appraisal of past formulations of the time concept, nor by introspective examination of our awareness of time. Among the philosopher's tasks is the seemingly thankless one of scrutinizing the advance of modern science for significant facts and ideas, and to integrate these into the larger notions he has formed of time. Recent physics bears suggestions of peculiar interest in this regard; chief among them is the theory of quantum electrodynamies developed by Feynman which involves reversals in the course of time and thereby cherishes, in the minds of many, an age-old phantasy of more than scientific appeal. To appraise that theory in philosophic terms is the purpose of this note.

In part 1 I attempt to give, not a popular exposition of Feynman's theory, but an account of it suitable for the philosopher of science who, while conversant with the elements of mathematical analysis, is unfamiliar with the techniques and presuppositions of the original publications.[1] It goes without saying that a knowledge of the scientific basis is indispensable for correct philosophic understanding. Thus, although I regret the inclusion in this paper of so much material that is second-hand, it seems to be a better course than its omission.

Part 2 contains the philosophic analysis and shows that Feynman's innovations are entirely compatible with accepted rules of scientific methodology.

### 1. THE PHYSICAL SITUATION.

The physical theory under review results from the union of two propositions, 1a and 1b. The first of these provides a novel, or at any rate an unusual method for solving quantum mechanical problems; the other is a known theorem of relativity physics.

*1a. Quantum mechanics describes motion as a series of scattering processes.* Let us recall one of the principal lessons which physicists have learned from the study of atomic behavior. Put simply, it is this. Some questions that arise quite naturally in our experience with bodies of ordinary size, questions which our facile imagination insists must have an answer, should not be asked concerning the entities of the physical microcosm. *Where an electron is* at a given instant, for example, is not in general knowable either by observation or theoretical analysis. Such is the fundamental meaning of the well known uncertainty principle in one of its aspects. A more modest question about the same state of affairs, however, *is* of significance. We may properly ask, and expect an answer to, the question: *What is the probability* that, at a given time $t$, and electron will be found at a specified point $x$? This probability is both measurable (through numerous observations) and predictable by theory. That theory will now be sketched.

Let $w(x, t)$ be the probability in question. If its value is 1/3, for example, then the odds are 2 to 1 against my finding it when I arrange to look for it with appropriate experimental means at the point $x$ and at the moment $t$. To be explicit, the point of space in question should be designated by all its coordinates, $xyz$; we shall for brevity include these in the single symbol $x$. The positive quantity $w$ can be expressed as the square (accurately, the square of the absolute value) of another quantity $\psi(x,t)$, so that $w = |\psi|^2$. Knowledge of $\psi$ entails knowledge of $w$, even though the reverse is not true; $\psi$ is called "probability amplitude" or *state function*.

Schrödinger discovered that $\psi$ satisfied the equation

(1) $$i\frac{\partial \psi}{\partial t} = H\psi$$

which governs the causal development of the state function. The italicised proposition at the beginning of this section adverts to an interesting way of solving this basic equation. But before we proceed to discuss it, let us see what that equation means.

That left side represents the local rate of change of the state function, multiplied by $\sqrt{-1}$. On the right stands a new function, resulting from $\psi$ by an application of the *operator*, $H$, which is called the Hamiltonian. This operator consists of two parts,

(2) $$H = T + V$$

which symbolize, respectively, kinetic and potential energy. $T$ and $V$ have different forms for different types of particle and for different versions of the quantum theory; if as in this article the particle under study is an electron, these forms are known from Dirac's theory.

Eq. 1 has no determinate solution which can be written down without further specification. But if the function $\psi$ is known for all space points at a given time $t_1$, the equation allows it to be calculated at any other time $t_2$. In other words, it tells how a given state unfolds or develops in time.

There are many ways of describing this process mathematically. Feynman's method is to represent the later state as an integral over the former with the use of what is called "Green's function":

(3) $\quad \psi(x_2 t_2) = \int K(x_2 t_2, x_1 t_1) \psi(x_1 t_1) \, dx_1$

Here $\psi(x_1 t_1)$ is the known initial state function ($t_1$ being the initial time); $\psi(x_2 t_2)$ is the state at the later time $t_2$. Both are functions of the coordinates $x$, and it may seem perplexing that these coordinates are labeled by a subscript 1 on the right and by 2 on the left. This is necessary only for the sake of book-keeping; the integral appearing on the right extends over all coordinates, and to keep the variables of integration separate from those which appear in the resultant function on the left, they must be distinguished by suitable subscripts. $K(x_2 t_2, x_1 t_1)$ is Green's function; it depends on both sets of coordinates and on both times.

Eq. 3 is a rather formal way of stating the solution of Eq. 1; it is useful only when $K$ is known. That function is determined by the natures of $T$ and $V$ in Eq. 2. Now $T$ is the same for all electrons, but the potential energy $V$ has different forms for different force fields in which the electron moves. For a *free* electron, $V$ is zero everywhere, and in this special case Green's function has a well-known form which will be called $K^\circ(x_2 t_2, x_1 t_1)$. Our present purposes do not require that we write it down explicitly.

Suppose that the electron is *nearly* free, that it moves in a field of force which is either of limited spatial extent, or of short duration, or is weak everywhere during its existence. One may then expect $K^\circ$ to be a first approximation to the exact $K$, and the correct Green function to be a series of the type

(4) $\quad K(2, 1) = K^{(0)}(2, 1) + K^{(1)}(2, 1) + K^{(2)}(2, 1) + \cdots$

where the arguments $(x_2 t_2, x_1 t_1)$ are abbreviated to $(2, 1)$. The different members of the $K$-series can in fact be calculated without difficulty; they turn out to be

$$K^{(1)}(2, 1) = -i \int K^{(0)}(2, 3) V(3) K^{(0)}(3, 1) \, dx_3 \, dt_3 \quad \text{(a)}$$

(5) $\quad K^{(2)}(2, 1) = (-i)^2 \int K^{(0)}(2, 4) V(4) K^{(0)}(4, 3) V(3) K^{(0)}(3, 1)$

$$dx_3 \, dt_3 \, dx_4 \, dt_4 \text{ etc.} \quad \text{(b)}$$

These formidable expressions are in truth quite simple and have a most interesting interpretation. Before discussing it, let us note carefully their mathematical structure.

In formula (5a) for $K^{(1)}$ there occurs one set of "dummy" variables, namely $x_3$ and $t_3$, and these are abbreviated to 3. They appear in both $K^{(0)}$-functions and also in $V$. When the integration is performed, only the variables

$x_2 t_2$ and $x_1 t_1$ are left. If $V$ vanishes, all $K^{(i)}$ except $K^{(0)}$ are zero. If $V$ has a small effect on the problem, $K^{(2)}$ will in general be of lesser importance than $K^{(1)}$ because it contains the product of two $V$-functions while $K^{(1)}$ has only one; for a similar reason the terms in the series (4) decrease in magnitude as the series progresses.

The pleasing and simple visual interpretation of the quantities $K$ will be explained after a brief excursion into more familiar territory. In a certain country there are observation stations where ornithologists watch the flight of birds. They take note of (a) the number of birds of a given species present in the neighborhood of each station; (b) the probability (or relative frequency) of flights from the neighborhood of one station to that of any other. If the stations are numbered $1, 2, 3 \cdots i \cdots$, the data of type (a) can be represented as a sequence of numbers $n_1, n_2 \ldots n_i \ldots, n_i$ being the number of birds normally present at station $i$. The probability of flights from station $i$ to station $j$ may be denoted by $k_{ji}$.

These quantities are not entirely arbitrary, but satisfy a sort of continuity equation which allows the observer at station $i$ to predict $n_i$ if he knows all other data. For the birds present in the vicinity must come from somewhere, and if the observer multiplies the number of birds normally present at station 1 by the probability that a bird will fly from 1 to $i$, adds to this product the number at station 2 times the probability that a bird flies from 2 to $i$ and so forth, he clearly obtains $n_i$. In symbols,

$$n_i = \sum_j k_{ij} n_j$$

Assume now that the observations are not made at specific stations but at all points in the area. The number $n$ then becomes a function of position, $n(x)$, and $k$ a function of two coordinates, $x_1$ and $x_2$; the summation takes the form of an integral and we have

$$n(x_2) = \int k(x_2 x_1) n(x_1) \, dx_1,$$

an equation of exactly the same from as (3). The number of birds corresponds to the state function, and $k(x_2, x_1)$ to Green's function.

Equation (3) refers to electrons and not to birds; nonetheless our analogy suggests the following picture for $K(x_2 t_2, x_1 t_1)$: it represents the probability that an electron makes a passage from (the "world point") $x_1 t_1$ to $x_2 t_2$; or, more strictly, the probability that an electron, observed at $x_1$ at time $t_1$, will be found at the point $x_2$ at the time $t_2$. Unfortunately, however, there is something wrong with this interpretation.

Probabilities cannot be negative numbers; the $\psi$'s and $K$'s of Eq. 3 may take on negative values when a solution of Schrödinger's equation is sought.

In this respect, then, our analogy is imperfect. We can remedy the situation by altering our language; retaining the indicated picturesque suggestion we shall cahnge the word probability to "tendency" and speak of $K(2, 1)$ as the *tendency of passage* from $x_1$ to $x_2$.[2]

If the significance of our analogy stopped at this point one might not be greatly impressed by it. The remarkable fact is that it goes right on and permits a simple interpretation not only of Eq. 3 but also of Eq.'s 4 and 5. What is meant by a passage from one point $x_1$ to another, $x_2$? Is it a direct passage along a straight line? Or does the particle go first to intermediate points and later to $x_2$? Clearly, there are many possibilities.

(0) It can pass directly from 1 to 2. This "tendency" is given by $K^{(0)}$ which, as we have seen, describes the behavior of *free* particles whose character it is to move along straight lines.

(1) It can go from 1 to some intermediate point, let us say $x_3$ or, briefly, 3, not necessarily on the line joining 1 and 2; from there it moves to 2. This change of direction and/or speed can only take place if forces are present, which is to say, in a region having potential energy $V$. The "tendency of passage" from 1 to 2 with a single scattering event on the way (this event taking place at 3) is proportional to three things: the "tendency of passage" from 1 to 3; the "power" of the field to produce scattering at the point 3 which may be expected to be $V(3)$; and the "tendency of passage" from 3 to 2. Since the scattering can take place at any point within the field, the coordinate $x_3$ is a variable, and an integration over it must be performed.— All this is summed up by equation 5a for $K^{(1)}$, which speaks as follows: $K^{(1)}$ (2, 1) is proportional (never mind the $-i$) to the "tendency of passage" from 1 to 2 with a single scattering process taking place somewhere on the way. A pictorial illustration of $K^{(1)}$ is figure 1.

As an afterthought we add a last refinement. The integration in Eq. 5a is over all times as well as all points of space. The sentence above describing $K^{(1)}$ should therefore end "somewhere, sometime on the way." This will henceforth be understood but not explicitly said.

(2) The electron can go from 1 to 2 and suffer two intermediate deflections. An extension of the preceding consideration, which the reader can easily perform, will show that the "tendency of passage" under these conditions is exactly proportional to $K^{(2)}$ (2, 1) as defined in 5b. An illustration of the meaning of $K^{(2)}$ is found in figure 2.

The terms of the sequence (4) are thus identified. $K^{(i)}$ (2, 1) is the "tendency of passage" of an electron from 1 to 2 with $i$ acts of scattering on the way. After all, it is reasonable to classify the paths of birds in accordance with the number of times they have changed the direction of their flight.

Feynman suggests that this also be done for electrons and he shows what practical advantages accrue from such a procedure.

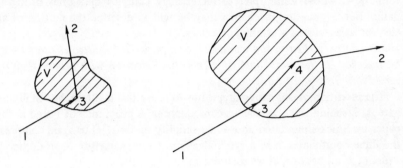

Fig. 1. Pictorial illustration of $K^{(1)}$. A particle going from point 1 to point 2 is scattered once on the way by a field of force of potential energy $V$ extending over the shaded region.

Fig. 2. Pictorial illustration of $K^{(2)}$. A particle going from point 1 to point 2 is scattered twice on the way by a field of force of potential energy $V$ extending over the shaded region.

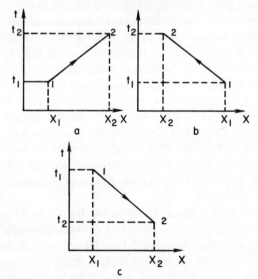

Fig. 3a. World line of an electron moving from left to right.
3b. World line of an electron moving from right to left.
3c. World line of an electron doing the impossible, moving from future into past.

The specific sense of our original assertion will now be clear. One of the principal tasks of quantum mechanics is to solve Eq. 1. This is done through formula 4, whose terms refer to increasingly complex categories of zigzag paths. Hence quantum mechanics may be said to describe the motion of an electron as a series of scattering processes.

1b. *In relativity theory, the world line of a positron (positive electron) may be regarded as the world line of a (negative) electron with proper time reversed.*

The intention of our exposition is not to prove this theorem, but to illuminate its meaning and consider its consequences. A proof may be found in the paper by Stückelberg cited above. To simplify matters, let us again compress the three coordinates $x$, $y$, $z$ symbolically into one, namely $x$. A world line is a plot of $x$ against $t$ or, as we shall draw it, of $t$ against $x$.

If an electron moves uniformly from $x_1$ to $x_2$ in the time interval from $t_1$ to $t_2$ its world line is that depicted in Figure 3a. Notice that the arrow on the $t$-axis points upward, into the future, so that $t_2$ is later than $t_1$. Similarly, $x_2 > x_1$, which means the electron moves to the right.

If we leave the time coordinates $t_1$ and $t_2$ unchanged but reverse $x_1$ and $x_2$, as in figure 3b, the resulting world line represents a perfectly reasonable motion: it goes from right to left, in the interval from $t_1$ to $t_2$.

The converse procedure reveals the strange asymmetry of time. If we leave $x_1$ and $x_2$ unchanged but reverse $t_1$ and $t_2$ as in figure 3c, the diagram has no correlate in ordinary experience, for the electron would have to move from the future into the past in order to make it meaningful.[3]

This type of situation, wherein old rules of correspondence suddenly fail when concepts are innocently extended, occurs quite often in science. It happened, for instance when the square-root sign was applied to negative numbers. The result appeared at first to be nonsense, but on further study new correlates were found, and while complex numbers are no longer related to immediate experience by the simple correspondence rule called counting, their meaning with respect to many classes of physical phenomena is nonetheless quite clear.

And so it is with time reversals. It was found that the world line of Figure 3c, meaningless for electrons, corresponds to a perfectly reasonable motion of a positron, namely motion from point 2 to point 1. The world line of a positron going from $x_1$ to $x_2$ in the interval from $t_1$ to $t_2$ ($t_2 > t_1$) is identical with that of an electron going from $x_2$ to $x_1$ in the inverse interval, from $t_2$ to $t_1$. The same theory describes positrons and electrons, provided *both time and space coordinates are reversed* when it is applied to positrons. But to reverse both sets of coordinates is simply to reverse the arrow on a world line, is to trace it backwards.

One may properly ask why this peculiar identification of reversed-electron-world lines with the motion of positrons, even if it is formally valid, should be interesting. Would it not be preferable to discard such anomalous world lines and draw normal ones, extending from past to future, for positrons as one does for all other particles? The answer is a clear and certain no. The relativity theory of electrons naturally leads to, and involves, world lines with proper times reversed. These appear in it automatically, and their artificial proscription would mar, indeed destroy the theory. When the identification in question is made the theory not only remains satisfying and complete: it also tells the empirical truth about positrons. Without this recognition modern physics would be in the condition of Greek mathematics after it had discovered the irrationals and before it understood their meaning.

Let us restate the practical content of theorem 1b: whenever you see an electron's world line extending *downward* (from future to past), reverse its arrow and assign it to a positron.

This rule allows a simple pictorial representation of many atomic processes. Fig. 4a shows an electron first going to the right and then scattered to the left at time $t_2$; in 4b the electron goes first to the left and is scattered to the right. Figures 4c and 4d represent these same processes for a positron.

In fig. 5a a novel effect occurs. The left leg of the diagram clearly depicts an electron proceeding from $t_1$ to $t_2$, the right a positron coming from another point

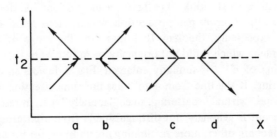

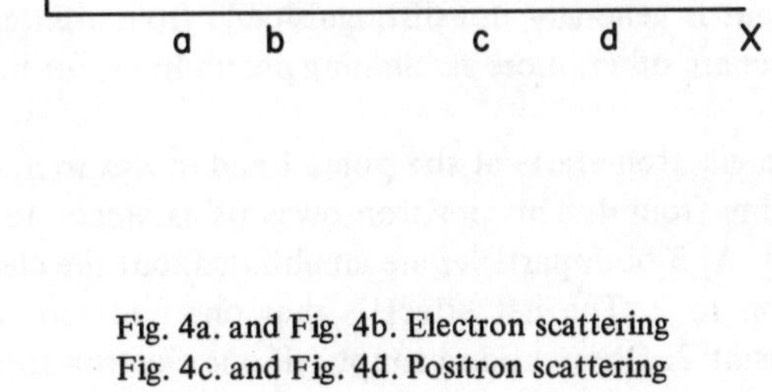

Fig. 4a. and Fig. 4b. Electron scattering
Fig. 4c. and Fig. 4d. Positron scattering

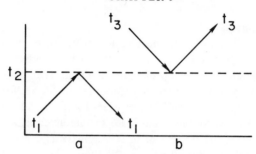

Fig. 5a. Pair annihilation; 5b, Pair production

at $t_1$ and moving to the left. The two particles meet at $t_2$, and the absence of lines beyond $t_2$ indicates that both have disappeared. The event is called "annihilation of a pair." Fig. 5b shows pair production. An electron created at $t_2$ moves to the right, a positron to the left. Prior to $t_2$ no particles existed.

The versatility of the new interpretation and of such diagrams is evident. The individual figures 4a to 5b may be regarded as single world lines—the corners can be rounded. Shape and orientation of that one line give rise to a rich variety of seemingly unrelated physical effects. In the light of these considerations the integrals $K^{(1)}$, $K^{(2)}$ etc. of section 1a, to which we now return, take on a larger meaning than was implicit in their definition.

Fig. 1 shows only one possible version of the quantity $K^{(1)}$. The mathematics of relativistic quantum mechanics, unless artificially restrained, generates others such as fig. 6, for the mathematics is indifferent to the direction of world lines. Indeed $K^{(1)}$ denotes "tendency of passage" for the class of all world lines with a single kink. But fig. 6 means pair annihilation under the influence of $V$; fig. 7 means pair production, which is also described by $K^{(1)}$. Somewhat unexpectedly, theorem 1b brings into the range of theorem 1a several phenomena which it did not originally seem to govern.

The meaning of $K^{(2)}$ is similarly enlarged. Fig. 2 depicts an event called double scattering. If the line from 4 to 2 has the same direction as that from 1 to 3 it denotes "virtual" scattering, since "actually", i.e., in terms of observations, this event is generally not distinguishable from uninterrupted propagation. But there are other, more fascinating possibilities, for instance the one drawn in fig. 8.

Here a single electron starts at the point 1 and moves to 3, where it meets a positron coming from 4. This positron owes its existence to an act of pair production at 4. At 3 both particles are annihilated, but the electron originating at 4 goes on to 2. The net effect is that one electron starts out at 1, another emerges at 2. Observation cannot tell one electron from another and will only reveal a change in the motion of one electron. Hence the occurrence of pair production remains hidden, and the process is called virtual pair production.

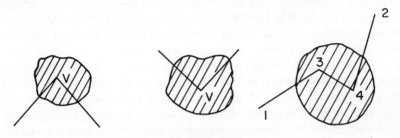

Fig. 6. Pair annihilation in a field
Fig. 7. Pair production in a field
Fig. 8. *Virtual pair production.* An electron moves from 1 to 3. Meanwhile a positron, created together with another electron at 4, collides with the first electron at 3 and both are annihilated. The other electron moves on to 2.

Let us note here in passing that the multiplicity of particles is not always reflected in a multiplicity of world lines; in the present example a single line—albeit with changes in direction—describes the fate of *three* physical entities.

To summarize: Quantum mechanics can be said to portray the motion of electrons as a series of broken passages, zigzag transits from one point to another with calculable probabilities assigned to each possible leg of a journey. But unless the doctrine of time reversals is adopted many legs remain devoid of meaning and encumber an otherwise beautiful theory by their unwanted presence. Allowing time reversals, and interpreting them as indicated, makes the theory singularly significant, powerful and true.

## 2. THE PHILOSOPHY INVOLVED

We face the philosophic question whether, and to what extent, the new quantum electrodynamics entertain departures from accepted methodological practice, in particular the question chosen as the title of this paper. The analysis of scientific method and the terminology I have offered in *The Nature of Physical Reality* provide suitable means for arriving at answers and will be employed. Here then, are the points of importance for the present inquiry in brief reiteration.

Physical theory is not obliged to confine itself to the use of directly observable entities. It may introduce speculative constructs freely, provided it states the connections which these constructs have with observations. These connections, though they must ultimately lead to matters of immediate perception, may be lengthy chains involving many steps of reasoning, imagination and experiment. And the constructs of a theory are valid, become verifacts, when scientific scrutiny moving along these specified connections affirms the designated observations or perceptions. The verifacts, or approved constructs, then become part of physical reality. In the book cited the data against which the scientist ultimately checks his theories, that final boundary of his cognitive experience, were said to form the $P$- (for perception) plane; it constitutes his last instance of appeal and contains his measurements, observations and whatever else he deems empirically incontrovertible. The chains connecting data with constructs were called rules of correspondence, and the constructs together with their logical and mathematical relations compose an hypothesis before, a theory after verification. To repeat: observations, $P$-plane facts have to be taken as they are; the constructs which correspond to them are at our rational disposal.

The words we use in science often have double meanings between which the labels subjective and objective are meant to discriminate. This distinction is, however, unfortunate and ineffectual, for the difference between these meanings resides not in their degree of objectivity but in their referral, once to $P$-plane facts and once to constructs. Thus the word *time*, in its perceptory sense, means temporal awareness, time immediately experienced; in the other sense it means the time of physical theory, stabilized by axioms of structure and by a special choice of clocks. To say that the two are the same will be recognized as nonsense; the two are connected by rules of correspondence.

Quantum electrodynamies does not ask us to experience time backwards, nor even does it suggest that we change our habit of reading clocks—a procedure that leads to the simplest form of "constructed" time (which is really a mixture of perceptory and constructional elements). In fact, the new theory leaves the $P$-plane entirely unchanged with respect to temporal experience. (If it attempted to do otherwise it would have to appeal to us in the manner of Rudolf Steiner's "Wie erlange ich Erkenntnisse höherer Welten," or of the advocates of a perceptible fourth dimension, an effort which it wisely restrains.) Instead, it takes a construct, time of flight of an electron, which already occurs as such in previous theories and which is linked by clear rules of correspondence with empirical observations, modifies it suitably and sets up a new rule of correspondence for the modified construct. To wit: Let $\triangle t_o$ be the time referring to an electron; this is an already established construct. The modification amounts to affixing a minus sign, and the new rule of correspondence says, briefly, let $-\triangle t_o$ be related to the motion of a positron in the same manner as $+\triangle t_o$ is related to the motion of an electron.

One can state the situation in a more formal way. $\Delta t$ is a time interval. The subscript $e$ refers it to an electron, $p$ to a positron. The superscript $P$ indicates that it is to be taken in its perceptory or immediate sense, $C$ identifies it as a construct. The symbol $\sim$ means "corresponds to." Then, if the old theory said

(6a) $\quad \Delta t_e^C \sim \Delta t_e^P$

the new says

(6b) $\quad -\Delta t_e^C \sim \Delta t_p^P$

All this is perfectly in line with theoretical practices of long standing, and the uneasy feeling which befalls us when we contemplate the backward flow of time associates itself with almost all redefinitions of familiar terms in science. The naive mind must have balked at the introduction of negative numbers because new rules of correspondence were required to make them meaningful, and a similar psychological predisposition makes some men hesitate now in interpreting negative entropy as information. If, as was argued, constructs need not be accessible to *direct* observational experience, the doctrine under review does not depart from standard scientific methodology when it operates with "proper times" reversed. Indeed it may be added that the employment of time reversals is no more strange than the customary and universal use of $\psi$-functions in quantum mechanies, which cannot be directly inspected either.

The direction of time cannot be fixed by the elementary processes of physics, as is well known. Only irreversible processes—among which the philosopher may count the flow of consciousness if he desires—can put an arrow on time. It is therefore unwarranted to take Feynman's theory as asserting anything at all about the "true," experienced direction of time. It speaks of constructs, and these are made to conform to $P$-plane time, which is regular and in accord with the direction of increasing entropy, by the rules 6a and 6b. I emphasize this in order to prepare for the solution of a paradox which is sure to occur to some imaginative reader.

There is nothing mathematically absurd about a *closed* world line, like that in fig. 10; in fact, this is included in $K^{(3)}$. It implies that a pair is born at 1, that electron and positron are scattered at 2 and 4, and that the same pair is annihilated at 3. But the diagram suggests a periodic process, an indefinite repetition of birth and annihilation. The preceding analysis takes the mystery —and perhaps the charm—out of this perplexing situation. For since the direction of time is established uniquely by a reference to phenomena other than those depicted in fig. 10, and since that figure must be translated into obser-

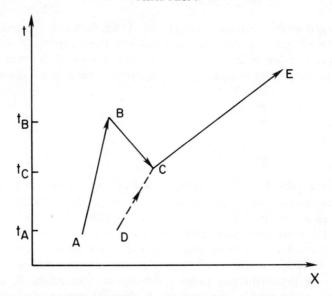

Fig. 9. The events represented by this world line, when regarded from the point of view of classical mechanics, do not form a causal sequence.

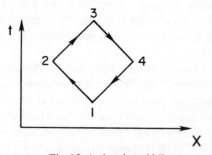

Fig. 10. A closed world line

vational experience by rules 6, the process cannot be recurrent since there is nothing to take the *observer* back from $t_3$ to $t_1$. Fig. 10, therefore, represents a single process of pair creation and death, non-periodic and devoid of all the mystic suggestions one may at first see in it.

Another philosophic problem is implied by Feynman's ingenious manipulation of word lines, a deeper problem not at once in evidence. It arises in reflection upon the curious causal aspects of diagrams like fig. 8. To fix attention, I review briefly what may be meant by causality.[4]

The word is applicable to a description of phenomena which, having recognized clearly definable and reproducible objects or systems (particles, fields, machines, perhaps organisms, groups of people, etc.), assigns to them *states* capable of temporal variation in such a way that a state at time $t$ allows calculation or prediction, via laws of nature, of a state at a later time $t_2$.—I hasten to exemplify this heavy definition.

A satisfactory physical system is a particle, e.g., an electron or a positron. Its state might be the position $x$ and the velocity $v$ it has at time $t$. Then, as is well known from ordinary mechanics, Newton's law makes possible the calculation of $x$ and $v$ at time $t_2$ when $x$ and $v$ at any other time $t_1$ are known. Causality holds because this is true; it holds, to state the case more sharply, because the fate of the particle is wholly determined by its complete state at the time $t_1$, that is, by $x_1$ and $v_1$.

Let us look again at fig. 8 or, better, at an elaboration of it shown as fig. 9. The state of the electron at the point $A$, its position and velocity, is completely specified by the $x$-component and the slope of its world line at $A$. The fact that at a time $t_c$, which is later than $t_A$, a pair is produced with a positron moving to annihilate our electron at $B$ is not implied by the world line at $A$, nor could it have been predicted from any property possessed by the electron at $A$. The birth of a pair at $t_c$ and the motion of our electron at $t_A$ are unrelated events. Hence the diagram does not describe the deterministic behavior of a single electron starting out at $A$. Causality in this sense, which is suggested by ordinary (or classical) mechanics, thus clearly fails.

One might try to restore it by enlarging the physical system, saying that fig. 9 describes the motion not of one particle but of three. If the state of the electron at $A$ *and* that of the pair in terms of the two lines starting at $C$ were known, the rest of the diagram could be filled in. This is true. But causality asks that we specify the state of a system at one time, while the procedure in question refers to two instants, $t_A$ and $t_c$. As it stands, the diagram is still not causal; to make it so we must cut off its lower portion along a horizontal line through $C$—a part which our theory is not willing to surrender.

Finally, it might be supposed that the diagram is incomplete in failing to state the cause from which the pair results. Perhaps it should include another feature, the presence of a photon at $D$ moving toward $C$. For it may have been this photon that produced the pair at $C$. Yes, it may have been; but nothing in the theory and no possible observation can tell us, at time $t_A$, that the photon actually will produce a pair at $C$. The photon may create a pair anywhere along its path or none at all. The device of introducing a photon satisfies our desire to know the complete state at time $t_A$, but this state is not connected with a determinate future state through Newton's law or, indeed, through any law we know of. Hence, the principle of causality still fails.

The language we have spoken is the familiar language of classical mechanics, the states we used are Newtonian states, corresponding to single world lines. Three things in Feynman's theory should have led us to suspect the applicability of such states: 1. That theory does not operate with single world lines but with many. The quantities $K^{(i)}$ are indiscriminate in their reference to specific diagrams like 9 for they involve an integration over all of them. They do not say which of them actually occurs. 2. Quantum electrodynamics assigns *probabilities* (we spoke of tendencies) to the different world lines. 3. It never uses Newton's laws but adopts Schrödinger's eq. 1 as its basic formula.

These facts remind us of a need for revision of the causal scheme, a revision demanded not only by the present theory but also by the simpler forms of quantum mechanics.[4] If the function which appears in eq. 1, and whose square is a probability, is taken to be the complete specification of an electron's or a positron's state, Schrödinger's equation becomes the law of nature which links states determinately in time, and theorem 1a shows a way to make this causal linkage explicit. The sacrifice we have to bring in order to restore causality, if it be a sacrifice, is a renunciation of the desire to see motions definitely mapped in space and time, to think of an electron as a billiard ball. Instead, it becomes necessary to acknowledge as ultimate protocols of —ultimate so long as the present theory is held to be true—the shadowy *probabilities* that an electron exist in one or the other of Newton's states. I do not think this reinstates the allegory of Plato's cave with its suggestion of realities beyond the shadows; the evidence is very strong that the shadows comprise reality.

Does this view, then, deprive the world lines of their meaning? Are the diagrams with all their imaginative connotations false? Far from it! To speak significantly of the probability of a particle's position one must have a clear conception of the points of space; to understand the $K$'s in Feynman's theory one must visualize all world lines as possible paths for a particle's motion. The diagrams, correctly interpreted, do not tell what actually happens but are samples of what *might* happen; they are visual structures, indefinite by themselves yet serving as instances to which probabilities can be assigned.

When taken in the light of the preceding considerations, quantum electrodynamics does not cause time to flow backwards, nor does it violate the principle of causality.

## NOTES

[1] The important papers are: R. P. Feynman, *Phys. Rev.* **76**, 749, 1949; Stückelberg, *Helv. Phys. Acta* **15**, 23, 1942. A preliminary attack upon these problems which forms the background for the present theory is J. Wheeler and R. P. Feynman, *Rev. Mod. Phys.* **17**, 157, 1945.

[2] Physicists, being conservative in their habits of speech, sometimes refer to $K$ as the "probability amplitude" for motion. It is my desire here to relieve the innocent word *amplitude* from one of the many duties it is forced to perform in modern physics. Tenddencies can be negative, indeed complex.

[3] This paper is not concerned with the older aspects of time reversibility, which are hardly problematic any longer and are well understood. Nevertheless, a brief summary of the "classical" theory of time reversibility may be helpful here.

Distance and time, although they appear conjoined in the theory of relativity (as long as it neglects phenomena involving entropy), do not have identical empirical attributes. Distance can be traversed in either direction simultaneously by two bodies, or at different times by one body. This property I have called "two-wayness" of distance in ref. 4; it might be symbolized by

$x$

Time, on the other hand, can flow only one way, and this way is called "into the future". A formal demonstration of this asymmetry of space and time was given in ref. 4, p. 160. Hence the "one-wayness" of time. But while time, in our experience, flows in only one direction, *the basic laws of mechanics* (in so far as they do not contain odd derivatives or odd functions of $t$) *are invariant with respect to time reversal*. If $t' = -t$ is substituted in them, they remain unchanged. This means that to every solution for which $t$ goes from 0 to $r$, there is another for which $t$ goes from 0 to $-r$. If a direction of $t$ is not previously fixed, the only ordering relation which has meaning is the conjunction of times with distances, i.e.

$$(x_1, 0), (x_2, t_2), (x_3, t_3) \cdots (x_n, r)$$

for the first solution,

$$(x_1, 0), (x_2, -t_2), (x_3, -t_3), \ldots (x_n, -r)$$

for the second solution. If we read the second series backwards, as we may because the direction of $t$ is not fixed, we have a motion different from the first but equally possible. Both can be observed, yet never simultaneously. Hence we can chosse the direction of time *as we please so far as the laws of mechanics are concerned*. The mechanistic character of time might be termed, perhaps awkwardly, "either-wayness" and symbolized by

$t$

Only the irreversible laws of nature allow the elimination of one of these arrows, for they represent tell-tale changes superposed on the indifferent mechanical laws. They force us to symbolize time in the one-way manner

$\longrightarrow$
$t$

Our intuitive experience seems to convey this final conclusion directly, probably because the "flow of consciousness" is intimately connected with irreversible organic phenomena.

[4] See in this connection an interesting critical paper by M. Bunge, *Methodos*, 295, 1955.

CHAPTER 10

# CAUSALITY IN QUANTUM ELECTRODYNAMICS

Quantum mechanics, even in its early and simple phases, has often been regarded as a non-causal discipline. The argument supporting this view cites the uncertainty principle as prohibiting the ascertainment of complete knowledge concerning physical states upon which causal prediction could be based. Recent developments in atomic physics have added new and puzzling features to the problem of causality insofar as they operate, not only with intrinsically unmeasurable states, but also with time reversals which have been interpreted to mean that the effect can be prior to the cause. Feynman's theory of quantum electrodynamics is particularly rich in unorthodox suggestions which tantalise philosophers. The purpose of the present paper is to exhibit them, appraise their methodological function and see in what manner they violate the rules of causal description. This purpose, it seems, is best achieved by a sequential discussion of three questions: What does causality mean in physics? What is the new method of quantum electrodynamics? Is this new method compatible with the causal doctrine in some satisfactory form?

## 1. THE MEANING OF CAUSALITY

When modern science speaks of causal connexions, it has reference to one or the other of two quite different relations between events or observations. The first is illustrated by such a chain of events as this: appearance of a cloud in the sky—darkening of the sun—lowering of the temperature on the earth—people putting on coats, etc.; or another, perhaps more scientific: emission of light from a star—propagation of an electromagnetic wave through space—absorption of light by a metal—ejection of an electron, etc. The events composing these chains form a sequence of *continuous action*; we know precisely, in terms of visible or postulated agencies, how the appearance of a cloud leads to the obscuring of the sun, how this in turn brings about a lower temperature, how this makes people uncomfortable and induces them to put on their coats. The other sequence can be traced in a similar manner; the events it connects, while occurring in widely different places and in totally different objects, are linked by some continuous action, some pervasive influence the details of which are understood. If there were gaps in this understanding, missing links in the chain of continuous action, the term causal would not be applied to it.

I shall speak of that meaning of causality which these examples illustrate as continuous action in time and space; it adverts to little more than relatedness by scientific agents and therefore makes causality tautologically equivalent to scientific understanding. It is a variant of the Humean doctrine of invariable sequence, refined by the inclusion of connective agencies between the members of that sequence. This interpretation of causality is large and generous, enjoys favour chiefly in the non-physical sciences and, of course, in everyday language; it is the stock-in-trade of lawyers and biologists. However, it is difficult to formulate with precision, and the difficulty resides in the circumstance that the view at issue places no restrictions upon the location and the kinds of events which are connected into a single chain. The emission of light can be on Sirius, the absorption can take place in some photocell on earth, and so forth. The only supposition is that the effect is later than the cause.

Physics, while at times espousing the continuous action view (often without being aware of the difference which I am exposing), is partial to another meaning of causality, a meaning first clearly formulated by Kant and Laplace. To wit: A physical system is described in terms of *states* which change in time. For example, the state of a body undergoing thermal changes may involve its temperature, its volume and perhaps its phase, and these variables are said to be variables of state, or variables defining a state. Another physical system, called an elastic body undergoing deformations, has states which are defined through stresses and strains; an electro-magnetic field (which is also a physical system in our sense of the word, for physical systems need not be material!) assumes states specifiable by an electric and a magnetic vector. Common to all these instances is the supposition that the variables of state, however defined, change in time in a manner conforming to certain equations which are ordinarily called laws of nature. Future states are therefore predictable if a complete present state is known. A prior state of a given system is called the cause of a later state, the later state the effect of a former.

The principle of causality, in this sense, asserts the existence of a determinate temporal continuum of evolving states, all referable to the same physical system. Between a given cause and a given effect there is an infinity of other causes of the same effect, though only one cause at one time. The advantage of this view, which I shall designate by the label of *unfolding states*, resides in its greater logical precision and in the uniqueness it confers upon the causal relation. When a cloud appears in the sky, that cause (in the former sense) has *many* effects at a given later time, one of which is a lowering of the temperature; when an elastic body has a given distribution of stresses and strains, a *single* definite distribution at a specified time is its effect. The continuous action view permits many causal chains, the unfolding state view only a single train of evolution.[1]

The simplest physical system, indeed the one for which a causal theory of the latter type was first developed, is the moving mass point. Its states are pairs of variables, positions and momenta, and the law of nature governing their evolution is Newton's second law. The latter is a differential *equation of the second order* requiring two constants of integration in its complete solution. Position and momentum of the particle at a fixed time can serve as constants of integration and therefore determine the solution of all times. States and laws of a causal theory must always have this internal affinity. The states must be so chosen that they provide the information demanded by the initial conditions that make the solution of the law complete. It follows from this circumstance that the definition of states in a causal theory cannot be arbitrarily altered without corresponding changes in the law of nature, and a change in the law will generally necessitate a redefinition of the state of a system.

Newton's theory of the motion of particles is the prototype of all causal description, and the laws and states it demands, rather than its formalism, have come to be regarded as essential elements of causality. This misunderstanding, or, at any rate, this inflexible identification of states, has led to the belief that quantum mechanics is no longer a causal discipline. Let us recall the important details: Newton's law was found to fail for atomic particles, and Schrödinger succeeded in replacing it by a new equation. But that equation did not have solutions specifiable by the old positions and momenta. Heisenberg discovered through his uncertainty principle that these variables had furthermore lost their soundness as universally observable quantities, for they cannot both be known with precision. This seemed to spell the doom of causality because the states it involved are neither theoretically significant nor observationally available.

It is not idle, however, to ask the question whether Schrödinger's law selects, or is compatible with, states in terms of other variables, and whether these variables permit a causal description in the second, more formal sense of our principle. That is, in fact, the case; only it is the misfortune of these variables, or rather of the states which they define, to be somewhat strange and elusive when judged from Newton's familiar standpoint. They turn out to be probabilities.

We shall need a little of the mathematical context of the quantum theory in the next section and, therefore, do well to be explicit at once. Schrödinger's equation is

(1) $$i\frac{\delta \psi}{\delta t} = H\psi$$

$H$, the Hamiltonian operator,[2] contains co-ordinates and the time, and is in general of the form $T + V$, symbolising kinetic plus potential energy. The arguments of the function $\psi$ are therefore likewise the space coordinates and the time. If the functional form of $\psi$ were known at any time $t_1$, it could be computed from equation (1) for all other times. In other words. If $\psi$ can be regarded as a state of the particle, quantum mechanical description is causal.

The identification of $\psi$ with observable matters was made by Born and Jordan: $|\psi(x,y,z,t)|^2$ is the probability that, when the particle is looked for in suitable ways, it will be found at $x,y,z,t$. If probabilities are not decent physical variables—and many physicists do regard them as loathsome—then this interpretation of states and its causality must be rejected. Further probing into why probabilities are objectionable reveals that they cannot be determined by a single measurement but require many observations. The position of a particle, or its momentum, can each be measured in a single act. But if the condition of the particle is unspecifiable by statements saying where it is and how fast it is going; if it is found sometimes here and sometimes there, then an aggregate of measurements must be performed, and the interesting information is the relative frequency, i.e., the probability, with which it is found here or there. It is difficult to see why an observable should lack the fitness to serve as a variable of state if its determination requires more than one measurement. These measurements can be performed, it should be added, in a way that will yield the probability at a definite time, so that the reference to an instant, which is crucial to the causal sequence, is not lost.

Subject to the acceptance of probabilities as physically meaningful states, quantum mechanics remains a causal theory. Henceforth we shall take this stand and proceed to show that the latest advances in quantum electrodynamics leave this status unchanged.

## 2. SUMMARY OF SOME RECENT INNOVATIONS

I shall attempt to sketch here the theory proposed by Feynman, partly because of its successes and partly because of its richness in stimuli for philosophic reflection. Among the many approximation methods for solving equation (1) Feynman selects one which represents $\psi$ as an integral over the initial state with the use of a Green function or kernel, $K$:

(2) $$\psi(\lambda_2 t_2) = \int K(\lambda_2 t_2, \lambda_1 t_1)\, \psi(\lambda_1 t_1)\, d\lambda_1$$

For simplicity, we have abbreviated the co-ordinates $x,y,z$ into the single symbol, $\lambda$. the kernel $K$ can be found if the form of $H=T+V$ is known.

For a free electron, $V=O$, and $K$ therefore has a definite representation which we call $K^{(0)}$. As we shall make no explicit use of it we need not write it down.

If $V$ is small but finite (it is understood to be a function of $x$ and $t$) $K^{(0)}$ is its dominant part, and it is possible to write a series

(3) $\qquad K = K^{(0)} + K^{(1)} + K^{(2)} + \cdots$

in which successive members decrease in magnitude because they involve $V$ in increasingly higher powers. Each constituent of the kernel is, of course, a function of two sets of co-ordinates, $K^{(i)} = K^{(i)}(\lambda_2 t_2, \lambda_1 t_1)$ and we shall now abbreviate it by writing $K^{(i)}(2, 1)$. Feynman shows that the terms in series (3) can be computed as follows:

(3a) $\qquad K^{(1)}(2,1) = -i \int K^{(0)}(2,3) V(3) K^{(0)}(3,1) \, d\lambda_3 dt_3$

$\qquad K^{(2)}(2,1) = (-i)^2 \iint K^{(0)}(2,4) V(4) K^{(0)}(4,3) V(3) K^{(0)}(3,1)$

(3b) $\qquad dt_3 dx_4 dt_4$ etc.

These formulas have an interesting and suggestive interpretation.

If $\psi(\lambda, t)$ were the probability that an electron be situated at the world point $x$, $t$, and $K(\lambda_2 t_2, \lambda_1 t_1)$ the probability that an electron makes a passage from $x$, $t$, to $x_2 t_2$, equation (2) would be the relation connecting these quantities, as a little reflection will show. To be sure, $\psi$ is not a probability but a 'probability amplitude' (since $|\psi|^2$ plays the role of a probability) and the same must be said about $K$. To carry through the interpretation, however, we will ignore this distinction.

We have recognised $K(2, 1)$ as the probability of passage of an electron from point 1 to point 2. Such a passage need not be direct. Indeed it is reasonable to classify all passages according to the number of times the electron changes its direction of motion. Thus we define $L^{(i)}(2, 1)$ to be a path leading from 1 to 2 and having $i$ corners. $L^{(0)}(2, 1)$ is the characteristic path of a free particle, i.e., of an electron in the absence of a potential energy $V$. A possible $L^{(1)}(2, 1)$ is drawn in fig. 1; the corner at the point (3) lies within the space-time region in which $V$ is finite, since otherwise a deflection could not occur. What is the probability of $L^{(1)}(2, 1)$? It is proportional to three quantities: (i) the probability of passage from 1 to 3; (ii) the probability of a deflection at 3, and this might be supposed to be proportional to $V(3)$; (iii) the probability of passage from 3 to 2. When these three quantities are multiplied together and integrated over all intermediate points 3, the result is the

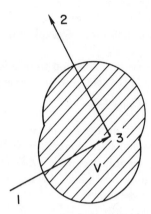

Fig. 1. Pictorial illustration of $K^{(1)}$. An electron going from point 1 to point 2 is scattered once on the way by the potential $V$ extending over the shaded region.

probability that the electron will go from 1 to 2 via *any* path $L^{(1)}$ (2, 1). But the indicated operations are exactly those defining $K^{(1)}$ (2, 1), equation 3a (except for the factor $-i$, which need not concern us here). We conclude, therefore, that $K^{(1)}$ represnts the probability of any passage in which the electron suffered *one* deflection.

An extension of this reasoning serves to show that $K^{(2)}$ (2, 1), as defined in 3b and as illustrated in fig. 2, represents the probability of *any* passage from 1 to 2 in which the electron suffered two deflections.

Equation 3a makes no specifications as to where the points 1, 2, and 3 shall be; it asks us to compute values of $K^{(1)}$ (2, 1) for all possible points and includes paths via all intermediate points 3. One *mathematically* possible path is depicted in fig. 3, another in fig. 4. But what is the physical meaning of such diagrams?

In fig. 3 an electron starts out at 1 and goes to 3, moving to the right. At 3 it continues to the right but *goes backward in time*.

In fig. 4 it starts going backward in time and at the point 3 it takes on a reasonable behaviour. A path on which time is reversed, i.e., a leg of a diagram that is directed downward, seems to be obvious nonsense.

At this juncture, however, the theory of relativity has something important to say: *The world line of an electron moving backwards in time represents a positron moving forward*. Thus, if we reverse the arrow on the nonsense leg off fig. 3 and direct it from 2 to 3, that leg represents a positron

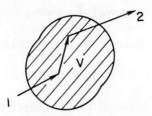

Fig. 2. Pictorial illustration of $K^{(2)}$. An electron going from point 1 to point 2 is scattered twice on the way by the potential $V$ extending over the shaded region.

moving from 2 to 3. The whole diagram, then, depicts an electron and a positron converging toward 3 where, since there are no lines at times later than $t_3$, they cease to exist. The diagram corresponds to the annihilation of a pair. The reader will have no difficulty in seeing that fig. 4 represents pair production.

The contents of this section, when briefly summarised, might be put as follows. Quantum electrodynamics portrays the motion of electrons as a series of broken passages, zigzag transits from one point to another, with calculable probabilities assigned to each possible leg of a journey. It interprets

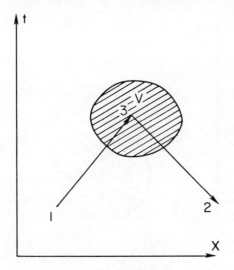

Fig. 3. Pair annihilation.

182                                  CHAPTER 10

world lines with times reversed as belonging to positrons. Unless this interpretation is made, many legs whose consideration is required by the mathematical formalism remain devoid of meaning and encumber an otherwise satisfactory theory by their unwanted presence. When time reversals are allowed, the theory becomes correct and powerful in its predictions.

### 3. THE CAUSAL STATUS OF THE THEORY

Diagrams such as those in figures 1 to 4 present no problem to causal analysis. But there are others, encountered in the study of $L^{(2)}$ (2, 1), which are hopelessly acausal when viewed from the standpoint of classical, or Newtonian physics. Consider, for example, fig. 5. A single electron starts from point $A$ at time $t_A$ and moves to $B$, where it meets a positron and is annihilated. This positron was created, together with a second electron, at $C$; the time of its birth was later than $t_A$. The second electron moves on toward $D$. It is interesting to note that the single world line $ABCD$ (the fact that it has sharp corners is not significant, for they can be rounded without detriment to our interpretation) represents the fate of three different particles. Between $t_A$ and $t_C$ there exists but one electron, between $t_C$ and $t_B$ there are three particles; after $t_B$ again only one. Observation may not disclose all these events. It will in general tell that an electron starts at $A$ and emerges at $B$, for it cannot distinguish one electron from another. The occurrence of pair

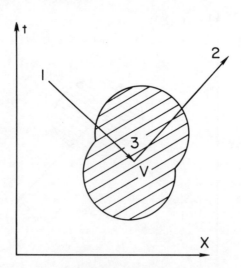

Fig. 4. Pair production.

production and annihilation remains hidden, and the whole process is called 'virtual pair production'.

The Newtonian state of the electron at point $A$, its position and momentum, is completely specified by the component and the slope of its world line at $A$. But the fact that at a time $t_C$, later than $t_A$, a pair was produced with a positron moving to annihilate our electron at B, is not implied by the state at $A$, nor could it have been predicted from any property possessed by the electron at $A$. The birth of a pair at $t_C$ and the motion of an electron at $t_A$ are unrelated events. Hence the diagram does not describe the deterministic behaviour of a single electron starting out at $A$. Causality in the Newtonian sense thus clearly fails.

At this point some of the comments in the preceding article become relevant. If one tries to restore causality by saying that Figure 5 implies the existence of three particles instead of one, there might seem to be a mechanistic solution. Knowledge of the state of electron $A$ *and* of the pair at $C$ would allow the remainder of the diagram to be filled in. The principle of causality, however, requires that we specify the state of the system at *one* time only, while the suggested procedure refers to *two* instants, $t_A$ and $t_C$. Therefore the diagram is still not causal. For we may not cut off the diagram along a horizontal line through $C$.

We also pointed out that the diagram is incomplete in failing to state the cause from which the pair results. Perhaps it should include another feature, the presence of a photon at $E$ moving toward $C$. For it may have been this photon that produced the pair at $C$. While this is a possible conjecture, neither theory nor observation can tell us, at time $t_A$, that the photon actually will produce a pair at $C$. The photon may indeed create a pair anywhere along its path or none at all. It is clear that the device of introducing a photon into the diagram, while satisfying our desire for a more complete description of the state at time $t_A$, does not yield a state in Newton's sense which is connected with a determinate future state through Newton's law or, indeed, through any law we know of. Hence the principle of causality still fails.

No such disaster occurs if the probabilistic interpretation of states is adopted. In that case a single world line says nothing about a positive, actual occurrence; it merely presents a sample of what might occur, a hypothetical instance to which probabilities can be attached. World lines do not lose their meaning, any more than points of space lose their meaning in the more orthodox form of quantum theory. For while this latter theory denies that under certain conditions an electron can be said to be at $(x,y,z)$, it nevertheless needs that point as a peg for its probabilities. In the same way, the

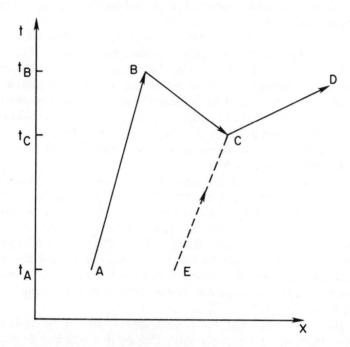

Fig. 5. The events represented by this world line, when regarded from the point of view of classical mechanics, do not form a causal sequence.

Feynman theory denies that a given diagram is a positive portrayal of reality, but it needs that world line as a carrier for its probabilities. Nothing more is implied in the use it makes of the $K$ integrals; it integrates over all $K$s to get a state in accordance with equation (2); it regards them as possible samples of what might occur without committing itself in Newtonian fashion as to actual paths. Equation 1 remains its basic law.

That equation induced a new definition of physical states, as we have seen. Quantum electrodynamics leaves this definition unchanged, and those who adopt it must regard it as a causal discipline.

## NOTES

[1] Our survey of the meanings of causality is not quite complete. One deficiency lies in its failure to analyse further the laws which connect the states. They must in some sense be invariable, or time-free. This point has been discussed in my book, *The Nature of Physical Reality* where further reflections concerning the suitableness of probabilities to function as state variables will be found.

Also omitted has been a version of causality which, though extremely limited, has found its way into the technical literature under the label 'causality conditions' (see, e.g., Van Kampen, *Phys. Rev.*, 89, 1072, 1953). It is nothing more than the requirement of relativity limiting the speed of a wave packet to the speed of light and says, in effect, that a cause at one point at time $t$ cannot produce an effect at another point, a distance $r$ from the first, at a time earlier than $t + r/c$.

In extensive treatments of the causality problem see Ernst Cassirer, *Determinism and Indeterminism in Modern Physics*, Yale Univ. Press 1956. Mario Bunge, *Causality*, Harvard University Press, Cambridge, 1959. Max Born, *Natural Philosophy of Cause and Chance* Dover, N.Y. 1964.

[2] Here and everywhere else in this paper, energies are understood to be frequencies, i.e., every energy is divided by Planck's constant.

CHAPTER 11

# RELATIVITY: AN EPISTEMOLOGICAL APPRAISAL*

Physicists today readily accept the principle of relativity and would probably feel uncomfortable with a theory of nature which did not possess the extension, the range of applicability, and the sweep of thought which a relativistic theory provides. It is furthermore a great satisfaction to the physicist to know that he can rely on the heuristic power of an idea which is so thoroughly agreeable with respect to symmetry and simplicity. Yet it is difficult to find an account of the principle of relativity which is clear and exact while at the same time providing a just characterization of its wider meaning. One may easily state an uncertain generalization having some resemblance to the intention expressed in Einstein's Theory of Relativity; or alternatively, a narrow and overly technical definition which in effect coincides with Einstein's explicit directives in their final form. But neither of these provides the perspective that makes Einstein's early efforts intelligible, or guarantees that the principle will serve as a guide to further fruitful thought. More serious is the fact that scientific effort is wasted and understanding is lost if we cannot clearly state the principles which are involved in the theory, and, furthermore, distinguish them from the particularities of an equation or a fact. The present paper is an attempt to do this, an attempt preceding a more extensive effort to axiomatize relativity.

### 1. PRELIMINARY ORIENTATION.

Attention must first be drawn to the domain of theoretical constructs, apart from their empirical validity, for only here is it possible to focus on the full meaning of the relativity principle. We shall see that its spirit is to provide a structural restriction on the kind and scope of physical theories to be entertained. When a formally complete theoretical structure, such as the logico-mathematical content (C-field) of Einstein's special relativity theory, has been proposed, and the rules of correspondence[1] have been specified, it is possible to speak of the empirical adequacy of the theory. But the character of physical theory is such that empirical evidence alone does not dictate a unique theoretical structure. In particular, we will show that neither the principle of relativity nor the Einsteinian axiom of the constancy of the velocity of light is subject to experimental proof in the sense that we cannot imagine a theoretical alternative satisfying the experimental requirements. For this reason, the meaning to be attributed to them is primarily methodological. Their interest lies in their heuristic desirability, their simplicity, and their suggestiveness of further theoretical research.

It must also be recognized that the principle of relativity, as it is understood in the scientific tradition, conveys multiple meanings and connotations which cannot all be displayed as part of a single proposition. For this reason, it is helpful to distinguish two aspects of the principle which are logically distinct. They are (1) *the older meaning* of the idea of relativity, and (2) a more recent meaning which has gained prominence in the theories of Einstem, and which we will call the *modern form* of the principle of relativity.

1. Historically, relativity is associated with problems arising out of the need to provide a reference for particle motion. The question which philosophers have asked is: Should quantities like particle position, velocity and acceleration be referred to an absolute, primitively given spatial frame of reference; or should the kinematical notions which fix the state of the particle have meaning relative only to other particles? Those philosophers who believed that the relata of kinematics must be particles are said to have supported the principle of relativity. The essays of Leibnitz, Mach, the young Einstein, and others who generally endorsed the idea of relativity make it clear that this is their meaning. If we may put the matter differently, the historically older form of the principle of relativity challenges the frame of reference as a suitable foundation for the elaboration of scientific theory, and attempts to subordinate it to material particles for definition and meaning. The logical structure of a theory of nature which succeeds in this attempt would, accordingly, employ the particle as a primitive idea or at least as a concept prior to metrical space.[1]

The old principle of relativity therefore revolves about the question of priority which may be assigned to scientific constructs in the logic of physical theory, and focuses this question on the primitive or non-primitive status of particles and frames of references. We have noted that Einstein spoke in defense of the idea that material particles provide the basis of kinematical meanings and hence define the subordinate notion frame of reference. But in his theoretical research, Einstein actually proceeded to solve another kind of problem. Newton's physics not only supposes a prior meaning for frame of reference, but specifically requires an inertial frame of reference relative to which its propositions are valid. Newton chose the inertial frame from among the class of all frames of reference to perform a special task in his mechanics. On the other hand, Einstein labored in his general theory to eliminate the privileged role of this restricted sub-class of frames of reference[3] from the foundations of physical theory. His success has been startling and of immense interest to philosophy of science, but the Einstein theory does not settle the question of the logical priority of the wider class of coordinates (reference frames) which that theory calls for. This issue remains, although its disposal does not radically effect scientific interest in the methods developed by Einstein, or significantly change the new meaning which has become associated with the principle of relativity.

2. The modern form of the principle is not concerned with the status (primitive or defined) of the concepts it treats, but rather with the extensibility of the axioms over the range of individuals included in the axiomatic structure. It requires the elimination of special or preferred individuals. "Individual" here refers to membership in any given or generated collection of constructs of a specified kind. Clearly, we advert here primarily to individuals which are particular frames of reference contained in the class of all frames of reference; but other examples are contingently given events, physical objects, etc.

The remainder of this paper is concerned primarily with the modern form of the principle of relativity. We will therefore largely disregard these elements of the older question which tend to distract attention from a full formulation of the modern principle, and from an examination of its methodological foundations.

## 2. A FORMULATION OF THE MODERN PRINCIPLE OF RELATIVITY.

A physical theory includes a formal mathematical structure, plus rules of correspondence which identify selected parts of the structure with immediate experience. The structure is composed of constructs, or formally related scientific concepts which have logico-mathematical meaning. The collection of these constructs, comprising (together with the relations between them) the formal aspects of physical theory, are represented by the matrix of circles in fig. 1, and are called a C-field. Each circle is a construct, or formal scientific entity, and its relation to the other constructs is represented by connecting lines. Rules of correspondence are represented in that figure by double lines joining the more immediate constructs with a straight vertical line, thought of as a projection of the domain of empirical or datal experiences called the P-plane.

As we proceed away from the P-plane into the C-field, we encounter constructs of greater and greater generality and abstraction, and it becomes increasingly difficult to maintain rules of correspondence joining the P-plane

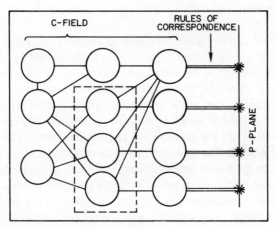

Fig. 1

directly with these constructs. As our progress continues away from that which is empirically identifiable, we find that we must rely increasingly on the formal properties of the C-field alone. It is in this domain of physical theory that the modern principle of relativity begins to perform its important functions.

The modern principle of relativity is a requirement of form and symmetry, binding on a group of constructs in the C-field. Referring again to fig. 1, we may say that certain constructs such as those enclosed within the dotted lines are most centrally affected by the requirement. However, the principle is not so restricted in its meaning as to affect sharply only a narrow group of constructs without any bearing on the others. The principle has extensions which reach to the left into the more abstract regions of the C-field, and extensions to the right which touch on more immediate constructs. For this reason, a satisfactory exposition of the principle will progress in several stages ranging from its broadest features to its particular implications close to the P-plane.

We will present the principle in three stages. The first (A) illuminates the sense in which the principle of relativity draws on symmetry requirements of a most general nature. The second stage (B) features the Einstein Principle of relativity, which is the central requirement as it is currently maintained. And the third stage (C) will develop a narrower, somewhat more technical statement of the principle, leading to the completely specific and unambiguous formulation of Einstein's special theory. Before proceeding with our enumeration we recall that the individuals of an axiomatic structure are said to be members of a given or generated collection of constructs of a specified kind.

## CHAPTER 11

### A. General Premise

No proper sub-set of individuals belonging to a constructed set of individuals of a certain kind will be treated primitively or axiomatically in preference to the remaining individuals of the same kind.

This premise, resembling in some ways the pronouncements of Leibnitz, requres a great deal of care in its application. The hazard of misuse seems to be minimized, however, if we allow ourselves to be guided by an application already before us in the Einstein theory, and by other applications found in the history of physics. The premise then leads us to an attempt to cast the laws of Nature in their most universal form, and to avoid fundamental dependence on a special individual construct such as a preferred frame of reference. The premise serves equally well to disqualify the Sun, or the Body Alpha as regal objects of Nature. The difficulty with the requirement as it stands is the ambiguity in the notion "same kind of individuals". For this reason, the premise does not settle very much by itself but must be used in conjunction with other expectations we may have as to the place of a particular construct in scientific thinking. For instance, if we believe that an inertial frame and, say, a rotating frame of reference should be viewed as the same kind of thing, then the sense of Einstein's Principle of Relativity flows with ease from the general premise[4]. In examining that Principle, we must remember that although all frames of reference are given an equal status, we are not told anything about that status.

### B. Einstein's Principle of Relativity

The laws of nature (axioms of the structure) are of the same fundamental form relative to all frames of reference.

There are two important aspects of this proposition which require some elaboration. One is the meaning of transformations between coordinate frames, and another is the notion of invariance under transormation.

A frame of refernce is best understood to be a numbering or ordering of physical events. It may be viewed as a language in terms of which propositions relating those events are stated. To effect a mathematical transformation from one frame of reference to another is therefore the same as switching from one language to another. The dictionary which provides the rules for translating is the relevant transformation equation. We normally define or generate the set of frames of reference which are to be admitted into the formalism by specifying the character of its transformation equations. Therefore, when we speak of 'all frames of reference' in the above sense, we mean those frames of reference which are connected in the broadest transformation group permitted by our theory[5]. The transormation equations therefore serve two purposes. They define the set of coordinates systems and provide the means whereby one may translate from one coordinate language into another.

The notion of invariance of physical law under coordinate transformation is perhaps the most significant feature of the relativity principle of Einstein. Examples of invariance may be cited from fields other than physics, and may help to clarify its meaning. In music, for instance, we speak of the invariance of a composition with respect to scale transpositions. A tune which is played in a major scale will have the same musical form when transposed to another major scale. However, the tune will not be preserved if it is transposed from a major to a minor scale. The tune is invariant with respect to some, but not all transpositions. This selectivity is often a property of invariant things: they preserve their character for certain classes of transformations, but not for all transformations.

In geometry, we speak of the invariance of a circle under rotation about its origin. Suppose a person lived in a round house which had no windows or interior furnishings except a compass which permitted him to maintain a coordinate system with one coordinate always pointing North. If, during his sleep, some outside agent gave the house a 45° turn, then the occupant would not know of the change when using this coordinate system to examine his environment in the morning. The inability of the occupant to detect the rotation provides an example of geometric invariance applied to a circle. By contrast, if a square house having no windows or interior furnishings were subjected to a 45° rotation, then a person living in that house would know about the rotation by reference to a compass coordinate system; for the compass needle which first pointed toward a wall, would point toward a corner after the rotation. The geometry of a square house is therefore not invariant under a 45° rotation, although the square house is invariant under 90° rotation. These examples illustrate two points: a) Every claim of invariance must specify the "operations" (transpositions, rotations, or in general, transformations) that are permitted in order to be meaningful; b) The invariant entity (Tune, geometrical figure or, in general, mathematical relation) cannot be used to discriminate between the operations with respect to which it is invariant. This is the reason why the laws of Newtonian mechanics can not distinguish absolute uniform motion from rest. The lack of discrimination involved in the notion of invariance is part of the meaning of relativity.

There are a great many illustrations of invariance from geometry, projective geometry and topology, so that the versatility of the idea may readily be recognized. Einstein's principle of relativity is to be identified with the idea of invariance applied to physical law. This feature of relativity theory is guaranteed by the use of tensor equations in formulating special propositions. Physical law, stated in tensor terminology, remains invariant under coordinate transformation in the same way that the geometry of a circle remains invariant under rotation. For this reason, the principle has considerable appeal, as it brings symmetry and simplicity of high order to the basic postulates of physics.

The third stage of our definition will deal with the more explicit technical instructions which follow from the previous stages. At this point, we will confine ourselves to the Special Theory of Relativity. This theory does not conform to the Einstein principle in the generality it has under B because it is restricted to a limited set of coordinate system (i.e.: those connected by the Lorentz group), a well-known defect which Einstein's General Theory endeavors to remove. But if this limitation is overlooked, we will find the special theory adequate for displaying a most interesting and successful application of the principle of relativity.

## C. Principle of Special Relativity

The laws of Nature, including those of light propagation, must be expressed in a form which is invariant with respect to transformations between inertial systems.

However restricitive the above requirement may be, it does not tell us what law most appropriately meets the demands of experimental physics. We must, therefore, add a hypothesis as to the precise mathematical expression which is to remain form-invariant under transformation. In special relativity, the hypothesis which secured the desired results was an expression guaranteeing the constancy of the velocity of light in all inertial frames. (It is written below and labelled *Inv*.) This choice was suggested by the unexpected results of the Michelson Morley Experiment, and has since been found empirically satisfactory in a great variety of ways. But it is important to understand that the results of experiments of this kind, however exhaustive, cannot prove either a) the principle of relativity or b) the hypothesis of the constancy of the velocity of light. The sense of this statement will now be clarified.

a) It will be helpful to recall Lorentz's electromagnetic theory which takes as its primary postulate the physical contraction of bodies, and the retardation of clocks which move relative to a luminiferous ether. His program is experimentally and mathematically as complete as special relativity. It consists in choosing one privileged inertial frame (the one at rest with respect to the supposed ether) and formulating a theory relative to that frame alone. While Lorentz's theory contains some of the now familiar features of the Special Theory of Relativity (contraction and retardation), these features are regarded as causally related to the physical agent "ether" which defines the privileged frame of reference. That Lorentz's non-relativistic theory succeeded (from the point of view of experiment) is not surprising when we recall the meaning of a relativistic theory specified under B. If the laws of nature are of the same form with respect to all frames of reference, then they are of that form in any chosen frame of reference. The laws may therefore be initially stated with respect to the preferred frame, and known in other frames of reference by a suitable transformation (in this case, the Lorentz transformation). Clearly then it does not make sense to speak of settling the matter between Einstein and Lorentz by experiment. What is significant about Einstein's principle is that it *can* be maintained, not that it *must* be maintained. In rejecting it, Lorentz did not commit any technical error, but he did forego its conceptual and heuristic advantages.

b) For somewhat different reasons, the hypothesis of the constancy of the velocity of light is also removed from the possibility of direct measurement. It is of such fundamental character that it is impossible to imagine an experiment which avoids question-begging assumptions, and at the same time provides a proof of the full content of the hypothesis. In the Special Theory, light signals perform a metric-defining role by providing a criterion of simultaneity of spatially separated events[6]. To speak of proving experimentally that the velocity of light is rightly or wrongly conceived as a constant appropriate to this role is to presume that we are in possession of a more fundamental metric which may serve as a basis of the proof. If the Special Theory is correct, we have no claim to a more worthy kinematic starting point. A condition on measuring the velocity of light is therefore that the instruments of measure be calibrated with respect to the velocity of light. The circularity implicit in such an experiment therefore prevents us from calling the hypothesis of Einstein an inductive generalization of experimental facts.

How are we to suppose that a theory can have synthetic content if it begins with a formal requirement, and leads to a postulate which is critical to our definition of a space-time metric? Fortunately, although the complete and literal content of the theory is not subject to experimental determination, there are ways in which the theory can be exposed to empirical tests. This is done by first stating the rules of correspondence which identify selected components of the theoretically constructed universe with the empirical clocks and rods of the laboratory. Then by appropriately conceived experiment (i.e.: conceived with respect to those theoretical features of special relativity which distinguish it from Newton's physics, a theory using similar rules of

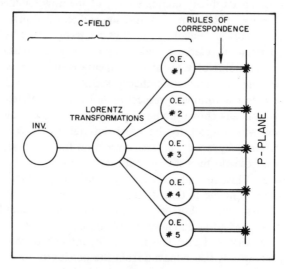

Fig. 2

correspondence), it is possible to disqualify one of the contending theoretical possibilities. An example of this kind of empirical filtering is the Michelson-Morley Experiment. Michelson by no means proved Einstein's hypothesis, but he did satisfactorily demonstrate the inadequacy of the theoretical thinking of his day.

### 3. EPISTEMOLOGICAL STRUCTURE OF SPECIAL RELATIVITY.

At this point it may be well to present a succinct, pictorial summary of the epistemological situation in the special theory of relativity. We shall use the terms and the schema of reference 1, particularly the distinction between the field of constructs (C-field), protocol experiences (P-plane, comprising immediate data, observations and anything taken to be epistemologically primary within the science in question—in this instance physics) and rules of correspondence (linkages between constructs and P-experiences).

Consider fig. 2. The proposition expressing invariance, often called law or principle of invariance, has the form:

(Inv.) $\quad ds^2 = dx^2 + dy^2 + dz^2 - c^2 dt^2$

and is of the nature of a postulate. It lies far to the left in the C-field and being a postulate, permits passages only to the right, none to the left. For to the left lies greater generality which, in the special theory, is not attainable. The general theory of relativity, if it could be successfully formulated, would present constructs to the left of (Inv.) in the C-field.

The postulate of invariance connects the constructs d$s$, d$x$ . . . $c$ and d$t$ in unique fashion. These constructs, which satisfy all metaphysical requirements (cf. ref. 1) and are contained in various circuits of empirical verification, are individually connected by formal relations with other constructs and each finally makes contact with the P-plane by some rule of correspondence. We have not drawn these individual connections between the constructs in (Inv.) and the P-plane, which are partly constitutive and partly epistemic, in fig. 2; for they are *not* unique and cannot establish the full meaning of the postulate.

To illustrate the lack of uniqueness one might attempt to define dx. The procedure is well known, and it exhibits the vagueness here claimed. The interval dx can have reference to solid rods (which do not exist), to the path of a light ray or to any of an infinite variety of signals. Or it might take its meaning from a previously established metric; indeed it could be defined in terms of dt. The postulate Inv. is empirically empty because of this bewildering variety of paths connecting its component constructs with immediate experience.

Its meaning is acquired in the following way. By virtue of its formal properties, the postulate permits the deduction of certain consequences called the Lorentz transformations (L.T.):

L.T.  $x' = (1 - \beta^2)^{-1/2}(x - vt)$, etc.

We shall not trouble here to explain the symbols, which have their customary meaning. While the Lorentz transformations appear to make reference to inertial frames, nobody is able to guarantee that such frames exist or that they are interesting. All the Lorentz transformations say is that, if Inv. is to be affirmed, *they* must be affirmed. To be sure, the deductive passage from Inv. to L.T. is not entirely unique; more general transformations are known which can also be "deduced", in the sense just stated, from Inv.[7] But these are nonlinear and are ruled out by various (metaphysical!) considerations, chiefly the requirements of simplicity and elegance. For practical purposes, then, the formal passage from Inv. to L.T. is the only one known to be acceptable.

But the Lorentz transformations are still empirically empty! They mediate between different inertial frames, which may or may not exist. And if the frames are called into being by the establishment of suitable rules of correspondence, there is still the question whether what the Lorentz transformations claim is verifiable under retention of any of the customary definitions of $x$, $v$ and $t$. Indeed, closer inspection shows that these transformations cannot be verified in their usual form at all, chiefly because they talk about coordinates and times, and not about intervals. They represent nothing more than a dictionary permitting translation of terms from one language (inertial frame) to another, and there is no assurance that both the languages between which the dictionary mediates are actually spoken in the world of our experience.

Those who try to demonstrate, or derive inductively, anything like the transformation equations from physical observations (and this group includes many teachers of physics), seem to us misguided and succeed only in creating confusion.

The way out of this state of flotation is via another stage of deduction. By assuming the "existence" of the reference systems to which L.T. refer (i.e. generating—or, really, accepting previously formulated-suitable rules of correspondence) one can *derive* from the Lorentz transformations 5 socalled observable effects; these are propositions between constructs lying rather close to the P-plane, constructs like length of a physical object and interval of time, for which reasonable rules of correspondence in the form of operational definitions are already at hand. True, these need to be refined by subtle considerations not called for in prerelativistic physics; we note for example the special care which must be used in the synchronization of clocks. But these changes from established practice are slight, natural and occasion neither surprise nor embarrassment.

The observable effects in question, labelled O.E. No. 1 to 5 in figure 2, are the contraction of moving objects, retardation of moving clocks, relativistic velocity addition, mass increase of moving material objects, and the mass-energy equivalence. It is at this stage of the epistemological scheme, several steps to the right of the invariance postulate as we have portrayed the process, that effective verification takes place. The "prediction" of all 5 effects has met with astounding success; and this success, though strictly confined to the region between the 5 observable effects and the P-plane where we have drawn double lines meeting the P-plane in stellar fashion, reflects the glory of empirical truth upon all the constructs and theories to the left of the observable effects. This truth is of course methodological, tentative and time bound, as is the truth of all relations between constructs. For no one can guarantee that the stars on P may not some day be met in even more glorious fashion by a theory having a different postulational start, perhaps beginning several stages further to the left and terminating successfully, not in the 5 stars here shown but in a dozen more which have not yet been spotted in the field of observation.

To complete the story, we shall now indicate briefly in what manner the 5 effect were verified, since this is not done systematically in any treatment of special relativity we know of. Physical confirmation is rarely direct. In the present instance, special experiments sometimes provide full evidence for one, sometimes hypothical evidence involving two effects. The isomorphic situation depicted in the figure is idealized beyond the historical course of the verifying process; the double lines can only be drawn in ex-postfacto manner after all the evidence has been surveyed.

Fairly direct is the test of the *lengthening of time intervals* in moving systems. During the last 20 years nature has supplied the physicist with particles carrying "clocks" and moving very fast. They are the mesons. The clocks mark off only two instants of time, namely their birth and their death; for these particles have a finite mean life period, and thus the particles themselves act as clocks. In one of the earliest experiments Rossi and Hall[8] found that the mean life of fast and slow $\mu$-mesons differed by a factor 3. This discrepancy disappeared when the retardation of the meson's clock was taken into consideration: with this correction all $\mu$-mesons have the same mean life span regardless of their speeds, and this must be regarded as strong confirmation of the retardation effect.[9]

Even more direct is the observation of the *relativistic mass increase*, first accomplished by Bucherer[10] and now a commonplace in every accelerator laboratory. It involves sending swift charged particles through a magnetic field and observing their deflections.

Confirmation of the *contraction of lengths* is most easily afforded by a study of the Doppler effect. The formula for this effect is well known and is confirmed in numerous observations. Its theoretical derivation, however, involves two assumptions: retardation of clocks and contraction of space intervals. The first of these, unless coupled with the second, leads to an empirically erroneous answer. Since the first assumption is independently confirmed, the Doppler effect serves to confirm the second.

In 1851 Fizeau measured the velocity of light through moving water[11]. His result did not conform to expectations based on Newtonian mechanics, i.e. on the ordinary vector-addition formulas for velocities. Hence, Fizeau concluded from his observation that the flowing water partially "entrains" the luminiferous ether in which light was supposed to be propogated. Today we know this conclusion to be erroneous, indeed the ether hypothesis of Fizeau's day has been abandoned; but Fizeau's observations are exactly those required by the relativistic *formula for velocity additions*.

The 5th effect is called the mass-energy equivalence, $E = mc^2$. It is corroborated in so many instances, from the early Compton effect to the release of energy in nuclear explosions, that further comment on it in the present context seems unnecessary.

It is through these 5 effects and their combination that the constructs to the left of the P-plane in fig. 2, and particularly the postulate Inv., attain their *empirical* validity. Their *formal* merit depends on the idea of relativity in the various forms here discussed.

## NOTES

\* Co-authored with Richard A. Mould.
[1] H. Margenau, *The Nature of Physical Reality*. Carnap, "The Methodological Character of Theoretical Concepts." *The Foundations of Science and the Concepts of Psychoanalysis*. Minnesota Press, 1956.
[2] The distinction between the metric of space and a reference frame was not generally recognized in the historical debate. This distinction follows from the fact that one may designate a great variety of metrical relations corresponding to the possible geometries, while retaining a single frame of reference (or coordinate system) for the purpose of locating and identifying points.
[3] For a concise statement of the change which took place in Einstein's view of coordinate systems, and how this change effected his principle of relativity, see p. 117 in *The Principle of Relativity* by Einstein and others. Dover Publications, Inc. Translated from Die Grundlage der allgemeinen Relativitätstheorie, Ann. der Phyk *49* (1916).
[4] Ibid.
[5] These include all Gaussian coordinates. See A. Einstein, *Relativity, the Special and General Theory*, Henry Holt and Co. New York.
[6] A. Grünbaum, *Amer. Jour. of Phys.* **23**, 7 450 (1955) p. 456.
[7] Lindsay and Margenau, *Foundations of Physics* John Wiley and Sons, p. 335.
[8] B. Rossi and D.H. Hall, *Phys. Rev.* **59**, 223 (1941)
[9] When a more careful analysis of this experiment is made the interpretation is seen to involve the relativistic mass change along with the time retardation. Strictly speaking, therefore, only the conjunction of the Rossi-Hall experiment with the Bucherer experiment demonstrates the retardation effect.
[10] A. H. Bucherer, *Phys. Zeitschr.* **9**, 755 (1908).
[11] H. Fizeau, *C.R.* **33**, 349 (1851); *Ann d. Phys und Chem. Erg* **3**, 457 (1853).

CHAPTER 12

# PHILOSOPHICAL PROBLEMS CONCERNING THE MEANING OF MEASUREMENT IN PHYSICS

1. The trouble with the idea of measurement is its seeming clarity, its obviousness, its implicit claim to finality in any inquisitory discourse. Its status in philosophy of science is taken to be utterly primitive; hence the difficulties it embodies, if any, tend to escape detection and scrutiny. Yet it cannot be primitive in the sense of being exempt from analysis; for if it were, every measurement would require to be simply accepted as a protocol of truth, and one should never ask which of two conflicting measurements is correct, or preferable. Such questions are continually being asked, and their propriety in science indicates that even measurement, with its implication of simplicity and adroitness, points beyond itself to other matters of importance on which it relies for validation.

Measurement stands, in fact, at the critical junction between theory and the kind of experience often called sensory, immediate or datal. The coverall term for this latter type of experience is, most unfortunately and misleadingly, "observation". This word is vague enough to hide a variety of problems; its penumbra of meaning overlaps that of measurement and the two are often confused, a circumstance which further aggravates the analysis here to be conducted. What should be clear upon very little critical inspection is the following: If observation denotes what is coercively given in sensation, that which forms the last instance of appeal in every scientific explanation or prediction, and if theory is the constructive rationale serving to understand and regularize observations, then measurement is the process that mediates between the two, the conversion of the immediate into constructs via number or, viewed the other way, the contact of reason with Nature.

Theories are welded in two places to the $P$-plane (if I may continue to use a term introduced previously to designate the "perceptory" or "protocol" phase of experience, i.e. the kind just called observation), and both unions are measurements. In the simplest instance certain quantities (e.g. position and velocity of a moving object) are measured; the results are then fed into a theory (e.g. Newtonian mechanics); here, through logical and mathematical transformations a new set of numbers arises (e.g. position and velocity, or some other variable relating to the object at some other time) and these are finally, again through measurement, confronted with $P$-facts. No scientific theory can have but a single contact with the $P$-plane—if it makes that claim it is called magic. To change the metaphor, measurement enables both embarkation and debarkation of a theoretical traveller at the shore of empirical fact. Ordinarily, these operations are without difficulty and without interest. But when the sea is rough they present problems and require special consideration. Fundamental reorganizations of theory, like seismic disturbances at the bottom of an ocean, produce troubled seas, and nowhere in modern physics has there occurred a greater revolution of thought than in quantum mechanics. Here the landing has been difficult, and the problem of measurement clamors for understanding and solution with particular urgency.

2. This paper attempts to prepare an understanding, but makes no claim of providing a complete solution of the problem. In this attempt, the first step must be to clear away the debris of older misconceptions. Most philosophers and many scientists regard measurement as a simple "look-and-see" procedure, requiring at the most a careful description of apparatus and the recording of a number. In doing so they ignore two things. First the relevance of the number obtained, its reference to something that is to be measured, its physical dimension. For the apparatus and the act alone do not tell us that the measured number represents a length or an energy or a frequency; this identification involves a use of certain rules of correspondence with preformed theoretical constructs which greatly complicates the meaning of measurement. In the second place, a single measured number is devoid of significance except as a tentative indication, acceptable only under the duress of conditions which forbid the repetition of a measurement. Generally, measurements must form an aggregate to be of importance in science.

Eddington's persuasive claim of the reducibility of all measurements to pointer readings on a scale is equally fallacious. It is contradicted by the obvious possibility, indeed the increasingly prevalent method, of counting events without use of pointer or scale; by the existence of yes-or-no measurements performed while watching a signal. Merely to see whether a spectral line occurs in a given region of a photographic plate may, in certain cases, constitute an important measurement. Clearly, one must beware of oversimplifying the meaning of that term.

3. Let us look briefly at the collective aspect of measurement. As was said, a measured number by itself signifies nothing that could safely be interpreted by means of rational constructs. If an aggregate is at hand, and only then, can the theoretical significance of the measurements be assessed. For in that case only do we have facilities for determining the error, or the measure of precision, of the results and can know what to do with them theoretically. But the discernment of errors raises further problems which need to be discussed.

An empirically "true" value of a measured quantity does not exist. What passes for truth among the results of measurement is maximum likehood, a concept that attains meaning if a sufficient statistical sample of differing measured values is available. When such a sample is obtained the physicist can plot a distribution curve which represents the quintessence of the intended measurement, since this curve reveals and determines the answer sought. It tells in the first place whether the set of values under inspection is trustworthy or whether it is to be rejected because of some manifest bias of the distribution. A simple though not always applicable test for acceptibility is to see if the distribution is Gaussian. But to justify this or any other test is to invoke some sort of uniformity of nature, to appeal to "randomness" of the observations (a term so far not susceptible of rigorous mathematical definition); in short, it introduces the entire group of annoyances known to philosophers as the problem of induction.

When the absence of bias has been established the search begins for that value which is to be regarded as the most acceptable result of the series of measurements. Ordinarily one chooses for this distinction the value at the top of the distribution curve, reasoning that if an infinite number of measurements were available that value would occur most often. But this, presumably most popular, value is not in reality among the measured set, and its selection is attended by some uncertainty. For it is possible to draw an infinite number of error curves to approximate the finite collection of measurements under treatment, each error curve having a slightly different maximum. The choice of this maximum again introduces a need for considerations transcending any simple meaning of measurement.

4. The difficulties thus raised culminate in two questions regarding the manner in which sequences of measured values approach the ideal of truth, in so far as that ideal is revealed through measurement. The first concerns internal, the second external convergence.

To explain the first, let me suppose that a measurement is repeated $N$ times with the same apparatus. The $N$ results enable the construction of an error curve which fits them best according to some mathematical criterion, and the curve has a maximum, $M_N$, as well as a certain width at half maximum, $W_N$, called the half width. As a matter of experience, $W_N$ remains approximately constant as $N$ increases, and may therefore be considered as a sort of instrumental uncertainty attached to the apparatus employed. The quantity $M_N$ will fluctuate as $N$ increases, and the question of internal convergence of the measurements asks whether a limit, $\lim_{N \to \infty} M_N$ exists. Experience answers this question affirmatively: we know of no instance where internal convergence fails. It is true that the meaning of the term "limit" must be changed from its strict classical understanding to the modern stochastic one in order to justify the foregoing statement; but this is a small price for a most satisfying nod of nature.

External convergence has to do with the behavior of $W_N$ when different measuring apparatus are employed in a sequence of sets of measurements. Is it possible, at least within reasonable limits, to choose different devices of increasing instrumental precision in such a way that $W$ becomes smaller and smaller, falling each time within the range of all preceding $W$'s? In other words, does $\lim_{S \to \infty} W^S$ approach 0? In writing this formula we have omitted the subscript $N$ because, as we have seen, $W$ does not depend on it; but we have added the superscript $S$ to designate the $S$th measuring apparatus employed, these different apparatus being arranged in the order of increasing instrumental precision. Thus, if we are to measure a length, $S = 1$ might designate a carpenter's rule, $S = 2$ a carefully calibrated yardstick, $S = 3$ a vernier caliper, $S = 4$ a travelling microscope, $S = 5$ an interferometer device, etc. To be sure, external convergence cannot be tested in as simple and exhaustive a way as internal convergence because apparatus are not infinitely available. The interesting fact, however, is that despite this difficulty we already are aware of an important failure of external convergence: $\lim_{S \to \infty} W^S$ does not approach zero when the measurements involve atomic systems. An accurate account of this failure is given in Heisenberg's uncertainty or indeterminacy principle of quantum mechanics, which we are thus led to consider.

5. According to the textbook version[1], two canonically conjugate quantities, like position and momentum of a particle, or the energy of a physical system and the time at which it possesses this energy, cannot be measured simultaneously with unlimited precision. The story is that a measurement of one "inevitably" disturbs the other, and the argument then becomes inductive, appealing to a profusion of experimental situations in which the physical effect of a position measurement is to *alter* the momentum of the particle. Just why a change in the magnitude of the momentum should preclude its simultaneous measurement is supposed to be obvious, or at any rate is deemed a question too silly for the physicist to answer. This stereotyped attitude with its logical myopia has not been dislodged by the clear evidence that one is often able to measure rapidly varying quantities with considerable success, nor by the patent possibility of making simultaneous measurements upon position and momentum of any particle, including an electron, with actually existing apparatus. Even the famous gamma-ray microscope, the pièce de résistance against every doubt afflicting the argument just offered, permits simultaneous measurements of both position and momentum, for there is

no reason whatever why I cannot bombard an electron at the same time with many short and long wave gamma rays and wait until I get a simultaneous return. True, there are hazards and idealizations in this proposal, and I may have to wait a very long time, but these difficulties are hardly of a different sort from those encountered in the accepted "thought experiment", although they are now compounded and thereby aggravated.

Clearly, it is *not* impossible to make measurements of canonically conjugate quantities as nearly simultaneously as we please if measurement means putting a question to nature and getting a unique answer. What the uncertainty principle means to assert is that this answer—when interpreted in detailed fashion following the precepts of Newtonian mechanics, in a manner which pretends to follow the course of the interaction between photon and electron in every visual particular—makes no sense. It makes no sense on two accounts: first in that the two numbers comprising the answer contain no reference to any definite instant of time at which both were present, since the measuring process does require a finite time. This in itself is not disturbing because the very essence of quantum mechanics enjoins us from employing classical models, visual interpretations of atomic happenings, and the fact remains that we get two numbers. The unique feature of quantum mechanics, as of the uncertainty principle, lies in the failure of what we have called external convergence. Hence, and this is the second reason by the answer attended to above makes no sense, when the measurement is repeated, even with apparatus of indefinitely increasing refinement, the values obtained remain scattered over a non-shrinking range; they approach no limit, but their variances or probable errors, i.e. $W^s(q)$ for position and $W^s(p)$ for momentum, satisfy the relation $W^s(q) \cdot W^s(p) > a \cdot h$, where $h$ is Planck's constant and $a$ a number of order 1. Somehow, the uncertainty relation adverts to some disposition inherent in the state of the electron which manifests itself in the statistical distribution of the measurements made upon it, provided a sufficient sample of measurements is at hand. We shall see below that this disposition is introduced into the situation, not by the act of measurement, but by a prior procedure to be called the preparation of the electron's state, and that it has its locus not so much in human manipulations as in the very essence of the electron.

Many physicists regard the fine distinctions made above as idle and unprofitable embellishments of what everybody knows, or else they disagree with the analysis for reasons never specified. It seems to me, however, that if the preceding analysis is correct, the philosophic significance of the uncertainty principle, and indeed of quantum mechanics as a whole, is profoundly modified. For if the usual version holds, the principle amounts to a proscription of certain kinds of measurement; it says that certain P-plane experiences are impossible; it limits the field of actual empirical occurrences. Now it is my view that any physical theory which places a ban on possible *observational* experiences mortgages the future of science in an intolerable way. For it is the unconquerable mood of science that it will accept any "historically" valid fact of experience and see what it can do with it within its system of explanation, and if a contradiction arises, it is the theoretical system that is sacrificed.

The situation is quite different with respect to the structure of the concepts employed in physical explanation. Here proscription seems quite in order and is indeed practiced at every turn. We agree to use causal theories in preference to non-causal one; we subject equations to covariance with respect to the Lorentz group; we rejected an unobservable elastic ether although, as Poincare pointed out, it could be made to satisfy all observations. The interpretation of uncertainty advocated here places that important principle squarely among the methodological devices in terms of which we agree to describe observational experience. It is properly silent with respect to what can possibly be measured but speaks with eloquence and convincing force of the manner in which the measurements relate themselves to theoretical constructs. In the terms of my own earlier publications, it generates a rule of correspondence, and not a black-out on the P-plane. Uncertainty implies no ban on measurements; it prescribes the structure of certain theories. Nor does it throw a particularly revealing light on the philosophical nature of measurement.

6. There is a mathematical fiction which has tended in some respects to preserve, in others further to confound the naive metaphysical conception that a measurement disturbs a physical system in a predeterminable way. It was used persuasively by non Neumann and later by others who were able to derive from this fiction the correct formalism of quantum mechanics, thus adding another example to the vast array of scientific instances in which correct conclusions were deduced from insupportable premises.

Specifically, the story is this. Quantum mechanics associates operators or matrices with measurable physical quantities. We know, for example, what matrices correspond to the position, the momentum, the energy etc. of a so-called particle. One of the simplest and mathematically most interesting matrices is the so-called statistical matrix $\rho$ which satisfies the equation $\rho^2 = \rho$. For reasons to be given below this is also called the projection matrix, and it can be constructed quite easily in the following way. Suppose we are given a complex column vector **a** of unit length, so that its components behave in accordance with the normalizing relation $\Sigma_i a_i a_i^* = 1$. From every such **a** one can construct a $\rho$. To form the elements $\rho_{ij}$ of the projection matrix all one needs to do is to multiply together two of the components $a_\lambda$; precisely, $\rho_{ij} = a_i a_j^*$. The defining equation is then satisfied, since $(\rho^2)_{ij} = \Sigma_\lambda\, a_i a_\lambda {}^* a_\lambda a_j{}^* = a_i a_j^* = \rho_{ij}$. For this reason the eigen-values of $\rho$ are easily seen to be 1 and 0 suggesting that the matrix ought to correspond to some physical quantity which is characterized by presence or absence, yes or no, success or failure, or some other two-valued aspect. Could it refer to measurement, in the sense that measurement asks whether a specified value is present or not?

The temptation to connect $\rho$ with measurement is further strengthened by another remarkable coincidence, which we present first in mathematical terms. Suppose that x is a column vector. Then $(\rho x)_i = \Sigma_\lambda\, a_i a_\lambda{}^* x_\lambda = (\Sigma_\lambda\, a_\lambda{}^* x_\lambda) a_i = (\mathbf{a}^+ \cdot \mathbf{x}) a_i$, $\mathbf{a}^+$ being the adjoint of the vector $\mathbf{a}$. Thus the result of operating on a vector x with $\rho$ yields $\rho \mathbf{x} = (\mathbf{a}^+ \cdot \mathbf{x})\mathbf{a}$. But $\mathbf{a}^+ \cdot \mathbf{x}$ is the scalar product of a unit vector $\mathbf{a}^+$ and the initial vector x; while $\mathbf{a}$ is another unit vector. Hence $\rho$, when acting on x, changes the direction of x into that of the unit vector $\mathbf{a}$ from which $\rho$ was constructed, and it diminishes the magnitude of x to that of its component along $\mathbf{a}^+$. In somewhat simpler language, $\rho$ "projects" a vector on which it acts upon a specified direction.

Is this not exactly what measurement does to the state of a physical system? If before the measurement the state is given by a vector x (in Hilbert space), then after the measurement, x has been converted into a state characteristic of the measured value, namely $\mathbf{a}$, but multiplied by a coefficient $(\mathbf{a}^+ \cdot \mathbf{x})$ indicating the probability that this will happen. The suggestion is very strong that the interesting matrix $\rho$ be taken as the counterpart of the physical process called measurement.

In plainer language, this assignment entails the following conclusions. If a physical system is in a quantum state which is not an eigenstate of the observable to be measured, then a measurement of that observable causes the system to be suddenly transformed into some eigenstate of the observable. The plausibility of this correspondence between $\rho$ and a measurement is further attested to by the fact that a second measurement following upon the heels of the first can cause no further change in the state of the system, a fact which is mirrored by the property of $\rho$: its iteration has no further effect, $\rho^2 \mathbf{s} = \rho \mathbf{x}$. In the sequel I shall speak of the postulate here outlined in connection with the mathematics which suggested it, as the *projection postulate*. It claims that a measurement converts an arbitrary quantum state into an eigenstate of the measured observable.

7. The physical case in favor of the projection postulate has been argued most strongly and succinctly by Einstein who, curiously, did not believe that the present form of the quantum theory is satisfactory. In 1935 he, in

collaboration with B. Podolski and N. Rosen, attempted to show that quantum theory cannot describe reality. As a sequel to this well known publication I wrote a small article[2] pointing out that Einstein's difficulties, and his socalled paradox, at once vanish when the projection postulate is dropped, whereas the power of quantum mechanics remains unchanged. In a personal answer to my paper Einstein wrote in part: [3]

"The present form of quantum mechanics is adjusted to the following postulate, which seems inevitable in view of the facts of experience:

If a measurement performed upon a system yields a value $m$, then the same measurement performed immediately afterwards yields again the value $m$ with certainty.

Example: If a quantum of light has passed a polarizer $P_1$, then I know with certainty that it will also pass a second polarizer $P_2$ which has its orientation parallel to the first.

This is true independently of the way in which the quantum is produced, hence also in the case in which prior to the passage of the polarizer ($P_1$) the proability for the polarization direction perpendicular to that of $P_1$ was not zero (for instance the case in which the quantum of light comes from a polarizer $P_0$ whose polarization direction forms an acute angle with that of $P_1$).

For these reasons, the assumption is in my opinion inevitable that a measurement modifies the probability amplitudes of a state, that is, produces in the sense of quantum mechanics a *new* state which is an eigenstate with respect to the variables to which the measurement refers."

8. Here is the physical argument in a nutshell, simple and beguiling. *If* the photon passes through $P_1$ then it will surely pass through $P_2$, $P_3$ and any number of other polarizers if they are set parallel to $P_1$. But is the passage through $P_1$ a measurement? Whatever the meaning of this operation, it must provide a positive answer and not merely a hypothetical one. Now to remove the *if* from Einstein's proposition the observer must see whether the photon did in fact pass through $P_1$. For this purpose he may use his eye, a photocell or some other device that will register the photon's presence. In other words, $P_1$ plus photocell constitute a measuring instrument; $P_1$ alone merely *prepares a state*. The example shows the need for a very clear distinction between 1) the preparation of a state and 2) a measurement. In classical physics the two are ordinarily the same, but in quantum mechanics they often differ.

A careful study of the situation considered by Einstein will doubtless lead to an account such as this. To the left of $P_1$ (assuming for definiteness that the photon is known to be on the left of the polarizer) the photon is in a state of known or unknown character, a state which is supposedly not an eigenstate of its spin (polarization). Whether or not that state has been prepared by human intervention is of no interest; it is *un*prepared with respect to the inquiry concerning its spin which is about to be conducted. To the right of $P_1$ the state *is* prepared; it is an eigenstate of the spin. Thus $P_1$ prepares the state, but it does not perform a measurement, since $P_1$ does not tell me whether a photon passed through $P_1$. This is the important character of the act called preparation of state in quantum mechanics: that *it determines the state of a physical system but leaves us in ignorance as to the incumbency of that state after preparation*; it may be a state without a system; i.e. no photon may be present on the right of $P_1$.

To perform a measurement, a photocell must be placed to the right of $P_1$, and the combination, $P_1$ plus photocell, is a measuring instrument, a device which says categorically that a photon with definite and known spin did in fact exist. But this measurement did *not* produce an eigenstate of the spin; indeed it destroyed that state—more than that, it destroyed the photon! Yet it was a good measurement despite its violation of the projection postulate. In constradistinction to the preparation of a state, a measurement certifies that *some system responded to a process, even though we are left in ignorance as as to the state of the system after the response.*

These are the bare requirements of 1) preparation and 2) measurement, requirements which in some sense complement each other. However, there are numerous physical operations which combine the two requirements and may therefore be regarded as both, preparation and measurement. This contingency is very common in macroscopic affairs (and in classical physics) where a machine which turns out nuts or bolts according to specifications may indiscriminately be said to prepare or to measure them. For we know of the finished product that 1) if it is present it is in a certain state and 2) that it is in fact present.

In atomic physics there are likewise instances in which a single operation prepares a state and measures. To be sure for the measurement of photon spins I have not been able to find such an example, as I see no practical way in which the photon can register its presence and retain its polarization. A photon's position, however, can easily be measured in two ways, one effecting a measurement only, and another effecting both a measurement and a preparation of state. The first occurs when the position is determined by means of a photographic plate, where a blackened grain is at once position record and tomb stone of the photon, and where projection into an eigenstate has certainly not taken place. The second is a measurement through the Compton recoil of a charged particle, where the photon is preserved and the state at the moment of recoil is a definite eigenstate of the position ($\delta$-function). Here it is possible that another charged particle, situated near the first immediately after the measurement, might suffer a collision and thereby signify the persistence of the state produced by the measurement. Similar preparation-measurement operations can be made upon the position of a charged particle itself (instead of the photon); indeed, a visible cloud chamber track is nothing but an extended series of such dual events. It would appear, then, as if in this latter class of operations the projection postulate stands aright, as if it characterized some, though not all measurements.

But there are complications even here. While quantum mechanics permits the preparation and the measurement of a position eigenstate, it takes back with one hand what it has given with the other, since it requires that such an eigenstate can not persist for any finite time; according to Schrödinger's equation the state function diffuses with infinite speed. Only for a sufficiently indefinite position measurement do we have an opportunity of testing what is not truly an eigenstate!

Thoughts of this kind, when properly entertained against the seductive surface plausibility of the projection postulate, indict it severely and raise the hope that one might get along without it. Such hope, strange to say, is not frustrated when a positive effort is made to build the foundation of quantum mechanics without the postulate; indeed it becomes perfectly clear on very little consideration that the postulate is *never needed at all*. Suppose we drop it and assign to the individual measuring act no power beyond yielding a number. Instead of making it produce a state, we let it terminate our inquiry concerning the state in question, i.e. the state existing prior to the measurement. With this minimal function, measurement still satisfies its purpose in quantum mechanics. Alone, a single measurement is devoid of significance, as it should be. Performed on an ensemble, however, it generates the distribution discussed in section 3 and permits the collective treatment necessary for the theoretical interpretation of the measured observable. Commitments with respect to any subsequent effect of the measurement on the system are superfluous.

The ensemble which enters the discussion at this point is either a physical assemblage of copresent systems, all similarly prepared, which respond simultaneously to the measuring act, or it is a temporal sequence of identical state preparations upon an individual, each preparation being terminated by a measurement.

Considerations such as these suggest the desirability of an unbiased, careful and exhaustive survey of all classes of physical measurements which does not prejudge their nature in favor of some mathematical conviction.

9. Current disbelief[4] in the correctness of the present formulation of

quantum mechanics has its source at least partly in the grotesque claims of the projection postulate. De Broglie, for example, bases one objection upon the improbability of the "reduction of a wave packet" occurring on measurement. The phrase, reduction of a wave packet, adverts to the projection attending the position measurement of an electron. Suppose the energy of this entity is known exactly, not necessarily by any measurement that has actually been made upon it but by the manner in which it was produced (e.g. photoelectric effect). Its state is then represented by a wavefunction which extends with equal amplitude throughout all space. If a position measurement now succeeds in determining its actual place, the wave will have been "reduced" or, to put the matter more graphically, will have collapsed upon the measured locus, having taken on the value zero everywhere except at one point—provided we accept the projection postulate. This sudden transformation, for which there is no precedent in all of physics, has raised many eyebrows and has led men like De Broglie to assert that the state function cannot represent any physical reality. For if it carries information, the instantaneous collapse violates relativity theory; on the other hand, it might be said to confirm the claims of the advocates of telepathy.

To save the quantum theory in view of these infelicities it has been customary to deny real status to the electron's state function and to regard it as a measure of knowledge which can, in fact, do peculiar things. This avenue is unquestionably open. It leads, however, to the equally unpleasant consequence that physics has seriously begun to describe human knowledge, a subjective aspect of the mind, in terms of differential equations involving physical constants. The point I wish to make is that we are not forced to this conclusion. A removal of the projection postulate removes De Broglie's difficulty, as it eliminates Einstein's. The state function then refers to an objectively real probability like the probability of tossing a head with a penny, a quantity which retains the value ½ even when a throw has yielded a head.

10. The last item to be discussed under the heading of physical measurement and its philosophic interpretation is not directly related to the projection postulate; it has to do with another paradox which measurement has been illicitly called upon to resolve. The second law of thermodynamics asserts that every isolated physical system, such as a gas contained in an absolutely rigid container, increases its entropy. The unusual case in which the entropy remains constant is not of interest here. This means that the system changes its internal state in a certain way, the change leading to conditions of greater and greater probability.

But in quantum mechanics, an isolated system, which can not exchange energy with its surroundings, reaches very rapidly a state in which its energy is definite, if it has not been left in an eigenstate of the energy to begin with. Unfortunately, such a state is a stationary one, i.e. a state in which the system will continue indefinitely. How it can possibly satisfy the second law thus becomes problematic.

Our pragmatists resolve the difficulty in this way. The state of a truly isolated system, they say, is uninteresting because it can not be known. To become known, the state must undergo a measurement. But a measurement "opens" the state, interferes with it, and raises the entropy every time it occurs. The second law does not refer to truly isolated systems, but to systems repeatedly subjected to measurements. The latter act becomes the *deus ex machina* which saves the second law from being trivial or false.

This solution is highly unsatisfactory to me, for I like to think of the second law of thermodynamics as a pronouncement valid independently of intervention. That is to say, measurement should not again be given sacramental unction and expected to perform a redemptive act. Band[5] has pointed out a better way out of the dilemma: it is simple, obvious and devoid of mysticism. To assume a perfectly rigid enclosure, he shows, is a classical falsification of the quantum situation. Such an assumption violates the uncertainty principle, which requires a connection between momentum and position of the walls just sufficient to supply the mechanism that drives the system to more probable states. Perhaps the process which "opens" the system is not measurement, but the inevitable character of nature which is present even in the absence of an observer.

In concluding, I wish particularly to call attention to one other line of investigation designed to eliminate the logical difficulties here uncovered. It is Landé's approach[6], which, though different from the present sketch and more analytic in detail, clearly and sensitively moves to a similar end. I owe Professor Landé gratitude for much inspiration.

## NOTES

[1] For a careful statement of the usual argument see A. March, *Quantum Mechanics of Particles and Wave Fields*. Wiley, N.Y. 1951.
[2] *Phys. Rev.* 49, 240, 1936.
[3] Private letter from which an except is published here (in translation) with permission of the literary executives of Professor Einstein's estate.
[4] See an account by W. Heisenberg in *Niels Bohr and the development of Physics*. Ed. Paul, Rosenfeld and Weisskopf McGraw Hill, 1955.
[5] 'A New Look at von-Neumann's Operator and The Definition of Entropy'. Unpublished.
[6] *Foundations of Quantum Theory*, Yale University Press, 1955, and numerous later articles.

CHAPTER 13

# BACON AND MODERN PHYSICS: A CONFRONTATION

The customary way to honor the memory of a great man is to single out his most meritorious contributions, the accomplishments that have made him famous, and to trace out their importance for our time. This is not the course I wish to follow, for it has always seemed to me that the errors of genius, the lack of foresight on the part of great men, are at least as interesting and often more revealing than their positive achievements. In Bacon's case, it seems curiously true that the power of modern science and its tremendous pragmatic successes spring from its allegiance with Bacon's teachings, and what we deplore in it, in its indomitability and in its apparent lack of relevance for the ethical aspirations of mankind, may owe their existence to an omission in Bacon's understanding of scientific method which subsequent generations of scientists have been slow to discover.

The comments to follow are therefore not in the nature of a eulogy but of a confrontation of Bacon with our present scene. I wish to sketch what is Baconian and what is non-Baconian in our present appraisal of the method of science; at the center of my concerns will be the problems of recent theoretical physics. Our successful understanding of them may be credited in large part to the effective use of Baconian induction; our failure to interpret their solutions in the framework of philosophy and of the totality of human experience may, it is suggested, be the result of a bedazzlement by the very successes of empirical induction. Whatever the conclusion, Bacon's figure dominates it, for he is either the beneficent progenitor of what is good in present science or else the persuasive propagandist who stamped upon it its excessively pragmatic traits.

Let me begin by stating what seem to me to be the salient characteristics of Bacon's unusual mind and of his philosophy of science. As to the former, Bacon says of himself: "I was gifted by nature with desire to seek, patience to doubt, fondness to meditate, slowness to assert, readiness to consider, carefulness to dispose and set in order." Beyond this set of attitudes, which every scientist will doubtless regard as admirable, there lies a positive view of nature, and a doctrine of how she can best be explored, which comes to the fore in Bacon's theory of forms. These forms, which are conceived somewhat in the manner of Platonic ideas, are hidden among the facts of scientific experience. Hence we must analyze experience in order to arrive at valid knowledge of the two forms with *certainty* and with *mechanical ease*. This is achieved by establishing three tables of comparative instances: (1) instances in which a given nature (form) is present; (2) instances in which a given nature is absent; (3) instances in which a given nature differs in amount. As an example, he considers the form called heat. To isolate and comprehend it, the scientist must first survey all properties which heated bodies have in common, then all properties which cold bodies have in common, and finally he must study bodies which are neither hot nor cold. The process of induction thus outlined now goes by the simpler name, the process of elimination. Underlying this view is the conviction that the number of discernible forms is small and finite, that care in investigation will isolate them, and that nature is never so complex as to resist exploration in terms of a discernible set of forms. The tables are therefore regarded by Bacon as possessing a finite number of entries. Philosophers have at times been uneasy with the simplicity of this approach, their misgivings being voiced in the old anecdote concerning the man who wished to isolate the intoxicating element of alcohol. Mixtures of gin and water, whiskey and water, vodka and water, all make a person drunk. It is concluded, therefore, that according to the Baconian directions, one must necessarily infer that the common element of these mixtures, namely water, is the cause of intoxication. There is, of course, no power in such structures, except perhaps to show that it is not always easy to employ the method properly and that nature can perhaps be sufficiently complex to make its application cumbersome. The process of elimination in science certainly remains as necessary and important as Bacon conceived it to be.

The structure of knowledge is reflected in Bacon's famous edifice of the pyramid of disciplines. At the bottom of all knowledge there lies the *prima philosophia*, a body of general truths. These truths are indubitable and self-evident, and their recognition is open to all unbiased men, to men who shun the idols. Upon this basis rests the discipline called natural science, the congeries of subjects often called descriptive science. Next comes physics, the science in which observation enters into combination with reasoning and calculation. The fourth and highest story of the structure of knowledge is metaphysics, the place where reason soars above observations and provides unity for the entire framework. Two things are to be noted about this arrangement: First, that the body of general truths is held to be perennial and inflexible.[1] Second is the affirmation of the validity of metaphysics, an element in Baconian thought which his disciples have frequently ignored or minimized.

The remainder of my discourse falls into three parts. The first is to deal with those aspects of modern science which are clearly in harmony with Bacon's precepts. The second will outline recent findings which transcend his proposed structure of scientific knowledge. In the third part an attempt is made to illustrate the non-Baconian elements of modern physics by reference to a well-known example, the speical theory of relativity.

## 1. BACONIAN ASPECTS OF MODERN SCIENCE

Whenever an important scientific discovery is made and new facts are recognized, these facts are at once marshalled and assembled in accordance with the rules laid down by our philosopher. Novel understanding is brought to bear upon concrete situations; application yields all the technological results which the non-scientist often regards as the sole purpose of the scientific enterprise. Thus the obvious effect of science is technological. It is the visible and impressive movement from discovery through engineering development, the building of small-scale models, the production of new goods and devices, advertising and sale, in short, the establishment of greater comforts of life or, perchance, of greater means of destruction. This chain of events, which is established when a sufficiently significant discovery is made, in the end enhances or at any rate modifies our standard of living, our external circumstances. It is the trend which is clear and open for all to see and which has earned the applause of people in our society today. A little later I shall also speak of another trend which is launched when a deeply challenging discovery is made, a trend much less apparent and more subtle in its progression from phase to phase through scientific discovery and philosophy into human culture. To appraise the efficacy of the technological movement, however, let us consider it a little further and review some chronological incidents which reveal the speed and hence the power which the techological inductive movement has attained.

The following sketch is not intended to be a well-documented, historical account. No more can be claimed for it than that it may convey a general idea of a phenomenon visible in our society today, a phenomenon not unrelated to the scene of this discussion.

It is difficult to define a discovery or even to say when a group of discoveries can be regarded as a single major breakthrough. Nevertheless, it is often possible to assign single dates to the emergence of new ideas, and I shall deal chiefly with instances that allow this. Likewise, the assessment of a year in which a discovery was first utilized in some form of appliance may not be unique. In some examples, however, one can determine the date of a patent.

Among the nebulous discoveries of the past is that of gunpowder. Though anonymous, there is enough evidence to indicate that the discovery was made during the twelfth century as a feat of the alchemists. It is also known that gunpowder was first used in battle during the fourteenth century. A period of about two centuries intervened between discovery and utilization.

As we proceed to modern times this period is greatly contracted. The chemical process leading to the technique of photography was discovered in 1727; Daguerre's invention was patented in 1839. The telephone, which began with a discovery of Oersted in 1820, became a useful instrument in 1876. The period of maturation had been reduced to half a century. Radio began with Maxwell's equations in 1867 and led to the development of wireless devices in 1902. Television was developed between 1922 and 1934, radar between 1925 and 1940. Fission was discovered in 1939 and the first atomic bomb exploded in 1945. The transistor began with a fundamental discovery in 1949 and was patented in 1953. If a curve were drawn in which the time of technological development is plotted against the historical date of the initial discovery, it would be seen to decrease exponentially; and if extrapolation into the future were permitted, a discovery would require only a microsecond to be patented and used in the year 2500. The very absurdity of this conjecture highlights a trend which, if it goes unchecked, may spell disaster for the human race.

I have spoken of this trend as inductive, factual, and technological, and I have ascribed its origin, by implication at least, to our philosopher. This is certainly not wholly fair. The following, however, can safely be asserted. The technological exploitation of scientific ideas which in our time is so vastly accelerated relies on the careful sifting of facts, on a comparison of effects and of the synthesizing of qualities in a manner visualized by Bacon. Statistical analysis, which is the modern counterpart of Bacon's tables of comparative instances, plays an important role in these procedures. And a minor function is assigned to basic essences, abstract meanings, logical consistency and philosophical implications which are rather irrelevant to the Baconian method.

But it cannot be denied that scientific discovery also has obscure effects in the realm of philosophy, ethics, religion, and politics. For every discovery, acting as a fact in initiating the technological trend, also becomes the leaven of an idea. As such it clamors to be understood, and the scientist must provide some sort of pervasive theory in terms of which the discovery takes on significance and organizing power. The new theory usually has features which contradict what was previously regarded as true, and, because of this apostasy, the discovery induces a rearrangement of thought in adjacent and even more distant fields of human concern. Sooner or later the internal structural consistency of the science that was disturbed by discovery is restored, and in this process some cherished beliefs, some tenets of common sense, will have to be surrendered. We note here merely the effect of the Copernican astronomical discovery upon the philosophy and, through Bruno, upon the religion of its age; the manner in which relativity ejected the time-honored concept of absolute motion; the incredible cogency with which modern quantum mechanics has annihilated the concept of solid stuff which formerly constituted matter. Results as challenging as these always have profound effects upon philosophic speculation. They are bound to alter our theory of knowledge, our understanding of the universe and of ourselves. Changes in one branch of experience will entail changes elsewhere, and even though the sequence of changes is uncontrolled and erratic, because we seem less concerned with it than we are with the technological sequence, it nevertheless tends to embrace in the end the entire structure of philosophic, moral, and even religious thought. Scientific discovery is never philosophically neutral.

The speed with which this philosophic movement reaches its culmination can be illustrated by means of a few examples drawn from the more recent past. The discovery of the laws of mechanics, which is the glorious achievement of Galileo and Newton, represents one of the deepest insights into the workings of nature. After a period of about two centuries it gave rise to the mechanical age which we must regard as the technological adjustment of the world to this new discovery. But men also sought for a philosophical transcription of the laws of mechanics. Here maximum success was obtained by Kant some 150 years after Galileo. It still seems strange to the beginning student of philosophy that a system of thought as abstract and artificial as that proposed in Kant's *Critique of Pure Reason* should have had enough persuasive power to sweep over most of the Western World. This historical miracle can be understood only when proper account is taken of the scientific antecedents of Kant's philosophy. These antecedents are found in the physics of Galileo and Newton, a branch of science unique for its singleness of claim, its universality, the categoricity with which it proclaimed its truths. When this science says that bodies attract in accordance with an inverse square law of force, as in the principle of universal gravitation, it means to say that this happens not only here but everywhere else in the universe, that it happens not only today but at all times before and after our epoch. The discovery represents a mathematically simple scheme implying that the world behaves in accordance with elegant formulas and familiar functions, suggesting in effect that the creator's mind was governed by syllogisms and mathematical equations. It is this feature which is reflected with high appropriateness in Kant's philosophy: the regularity found in the world of sense mirrors the pure forms of reason and the categories of the human understanding.

The point we made regarding the diffusion of scientific discovery into wider fields is also well illustrated by this example. After the *Critique of Pure Reason*, Kant wrote the *Critique of Practical Reason*, and formulated the categorical imperative. It is doubtless true that the Prussian concept of duty, which on occasion embroiled the world in wars, is an outgrowth of Kant's ethics which, as we have seen, were formulated as a response to Galilean science. The time between Galileo's discovery and Kant's publications amounts to a century or two.

It is interesting to see whether this period has been shortened in our day. Let us, therefore, review two later discoveries and their philosophical interpretations. At the beginning of the nineteenth century physicists and chemists turned their attention to the laws of heat and discovered the principles of thermodynamics. These laws, in contradistinction to those of mechanics, spoke in a less certain voice; their predictions were statistical, and whatever universality they avowed arose from the laws of large numbers. Thermo-dynamic laws, as we now understand them, permit of exceptions although such exceptions are extremely rare. The laws caution us, because of their probability tenor, to take rare cases seriously, to reckon with the unlikely contingencies which slip through the net of science, but may have profound effects on our lives. A proper philosophical exegesis of this attitude is given by positivism, more particularly by that branch of it called logical positivism, and even more by some forms of existentialism current today. Logical positivism and existentialism seem to have reached maximum appeal in our time; hence the conclusion that the period required for understanding science in a basic philosophical way is still a century or two.

At the beginning of our era a set of challenging discoveries was made. They became the foundation of the subject now known as quantum mechanics. During the intervening fifty years we have learned to solve the equations involved, to apply them to concrete situations and to make practically all the technological predictions which the subject can possibly entail. But do we know its larger implications? Anyone who reads the current literature on the philosophy of quantum mechanics, which is very extensive, is impressed by an amazing fact: the very founders of the theory of quantum mechanics, men like Bohr, Born, Heisenberg, DeBroglie, Einstein and Schrödinger wrote searching articles on the basic meaning of their work, emphasizing the importance of philosophical interpretation; but their views did not and do not agree. Apparently the period of maturation has not yet come to an end; the contraction of the time scale we noted in connection with the technological development has not occurred.

There are cultural conclusions, important in our struggle for existence in a hostile world, to be drawn from these observations, but I shall omit them here. The relation to our theme appears to be this: Bacon has given us no recipe for establishing basic meanings, hence his program could not be carried out in the cultural-philosophic domain. Where science was able to follow him, it succeeded admirably and at an accelerated pace; where it received no directives it floundered and staggered along.

## 2. NON-BACONIAN ASPECTS OF MODERN SCIENCE

It behooves us to ask: Why did science fail in this latter respect; what are the features developed since Bacon's time over which his program lost control? I shall list them under three headings and give a brief description of each, as follows:

*a.* There is no *prima philosophia*, no ultimate, perennial, unchanging truth, at least none to be discivered by science. This recognition became general when non-Euclidean geometries were shown to be possible, consistent and useful; it speaks through the developments in modern logic and has left its imprint on the current scientific conception of axioms. These are not, according to the present version, indubitable and undeniable verities conveyed to us by some *lumen divinum* or *naturale*; they are postulates, basic but tentative hypotheses which further insight and observation can and probably will alter. The scientist's attitude toward fundamental truth is not one of final acceptance but one of commitment, not one of knowledge but of reasonable faith.

*b.* Science is not a set of static discoveries, completable in time, but an unending dynamism which generates new problems as it evolves. There was a time when scientific activity was likened to the solution of a picture puzzle, where pieces are placed in proper locations until a final pattern results. The pieces are the facts to be discovered and arranged, the pattern is the final solution of a scientific problem. The faults of this simile are obvious; they lie in the supposition that science is a two-dimensional affair which exhausts itself in the arrangements of facts. We now know that understanding cannot proceed in two dimensions, that the facts which lie in the plane of the picture puzzle have roots which reach below that plane, into a domain of reasoning, of ideas, of principles which integrate the surface facts and give to them their meaning and their importance. A fact or set of facts alone can never satisfy a scientist; his concern is with the substratum of ideas at least as much as with the surface texture of the facts. Another common fallacy is the supposition that a scientific problem can ever be solved completely, a fact which is suggested by the analogy in question. On the contrary, every solution of a scientific problem presents further challenges, further problems, indeed problems in greater number than were seen before the solution was attempted.

A better analogue to the scientific process, it seems to me, is the growth of a crystal. Here we witness, first of all, a liquid matrix, amorphous in its structure, unpredictable in the chaotic motion of its constituents. Then suddenly, as the temperature of the liquid sinks below its melting point, a solid crystal begins to form. Its growth extends in all directions, although it is difficult to predict in which direction it will proceed with greatest speed. The growth is self-corrective, for occasionally the crystal will leave faults in its wake, but as it progresses it will heal its flaws. Nor does the establishment of a crystal domain pervading the volume of the former liquid change the nature of the stuff which it informs; there is no adulteration of substance but merely the generation of form, of an organized pattern, the achievement of predictablity. This seems to be what the development of science does to the raw experience in its pre-scientific form.

It is clear that this dynamic process cannot be caught in static tables of forms and of comparative instances. Hence I feel that the dynamism of modern scientific enterprise is a non-Baconian element of science.

*c.* Science is not mere discovery, it involves human creativity; it is not inductively extracted from the facts, but constructed in large measure by man's own ingenuity. I am not at all sure that Bacon would have denied such claims; indeed had science at his time been more fully developed, he might have been the first to emphasize this point. But there was hardly the occasion for him to do so, and his disciples at later times have concentrated attention on tables of facts more or less independent of the constructs introduced by man himself. I suppose Bacon himself would have included these human contributions under his metaphysics which forms the top of his pyramid of knowledge.

Metaphysics has fallen into disrepute in some quarters. The reminder I wish to give here is that the scientist cannot get along without it in the sense that he must use principles which are not dictated or defined by observational experience alone. Reference at this point is to such matters as logical utility, verifiability, extensibility, simplicity of concepts and elegance in their use—all concerns which are regulative in scientific experience without being required by the facts alone. Even aesthetic elements are included in this list, as I hope to show in the following section.

To put things simply, one might say that man is no longer a passive, careful, and impartial spectator of nature, as Bacon would picture him, but a being who as an integral and active component of the universe can alter his own experiences and alter the objective world. In the present context, I fear that this kryptic and dogmatically assertive allusion must suffice; more is said about this point in my book *Open Vistas* (Yale University Press, 1961).

## 3. AESTHETICS AND RELATIVITY

As an illustration of the use of non-factual elements in modern science, permit me to consider with you the theory of relativity. Unfortunately, its name implies a relativism, a disposition to state laws conditionally; in short, to avoid absolutes. There is some validity in these implications when they are correctly understood, but the custom among non-scientists is to regard relativity as the doctrine that all statements are relative, indeed to hold with Protagoras that "man is the measure of all things."

In actual fact the nucleus of the theory of relativity is the idea of *invariance*, the idea that the basic laws of science do *not* change under differing circumstances. Invariance finds extensive application in the field of arts, particularly in music, painting, and in decorative design. Rhythm is, strictly speaking, invariance with respect to time displacement; repetition of ornamental pattern is invariance with respect to space displacement; the beauty of a snowflake is attributable to invariance with respect to rotation, and so on. Let me explain the meaning of these remarks.

If you look at the world through a cylindrical telescope tube of circular cross section, you see objects whose location can be specified with respect to the edges of the tube. In particular, lines are seen to make angles with the circumference, and these angles vary between 0 and 180 degrees. Now let the tube be rotated about its axis. Nothing changes as a result of this operation, the angles remain the same as before, and indeed there is no new feature discernible from which the occurrence of a rotation could possibly be inferred. Everything remains invariant or, in more technical language, the circle which forms the cross section of the tube is invariant with respect to all rotations. For this reason, the circle is often called the perfect figure of geometry.

Imagine now that the telescope tube has a square cross section and that again a rotation through an angle differing from 90 degrees or multiples thereof is performed. The world will now look different. Lines which previously made specified angles with one of the edges of the telescope tube will now make different angles, and the fact that a rotation has occurred can be clearly discerned. Thus a square is not invariant with respect to all rotations, but only with respect to rotations through angles of 90 degrees and multiples of 90 degrees. The square is a less symmetric geometric figure than the circle, and for that reason it might be regarded as less beautiful. One aspect of the idea of symmetry is tied to the operation of rotation and draws its meaning from invariance with respect to rotation and the range of this invariance.

The meaning of relativity now becomes clearer. When the circular tube was rotated about its axis the effect of this rotation could not be detected. Absolute position, in a rotational sense, of the telescope edge was meaningless. It is thus seen that the negation of absolutes, in this case absolute orientation, is implied by the existence of an invariance. In the case of the square tube different orientations of the circumference do make sense because they can be detected operationally by comparing the views of the world which are presented under different orientations. Only orientations differing by 90 degrees and mutliples of 90 degrees remain equivalent.

All regular polygons exhibit invariances with respect to rotation. Snowflakes owe their appeal to their six-fold symmetry, that is to say, their patterns are invariant with respect to rotations by sixty degrees. Invariance is thus primarily an aesthetic notion, but we shall now see how effectively it is employed in modern physics as well.

The laws of nature, when written in mathematical form, allow of certain transformations: that is to say, when specific substitutions are made for the variables which occur in them, the laws do not change their forms. All laws of physics, for example, are invariant with respect to space and time displacement, this property being in essence what is often expressed by saying that the laws are causal laws.

Earlier in this century another kind of invariance became interesting and important. It was posed in the question whether or not the laws of nature change their form when they are transformed from one frame of reference to another which is moving at constant velocity with respect to the former. Such systems are called inertial systems. It appeared from novel observations that the law of light propagation should indeed be invariant against transformations from one inertial system to another, but the accepted formula of ordinary mechanics did not bear out this expection. Einstein thus saw himself confronted with this kind of a situation:

He beheld the simplicity and the mathematical appeal of the fundamental law of light propagation. Over against it stood common sense, a reminder coming from classical mechanics according to which certain obvious changes must be made in the variables (Galilean transformations) when passage to a new frame of reference is contemplated. Einstein had to choose between two alternatives: either to reject the aesthetic appeal of simplicity inherent in the basic law and adhere to the dictates of what seemed to have been "proved"; or else to reject common sense and retain the simple elegance of the basic law. As to the "truth" in the situation, the consequences of either step could have been justified before the facts, simply in one, more elaborately in the other alternative.

Einstein decided in favor of simplicity and elegance at the expense of common sense. He thus established the theory of relativity, and everybody knows how fruitful and illuminating his choice has been. In principle, he followed an aesthetic lead, a metaphysical conjecture, and the pay-off even in pragmatic terms was enormous. This motive, which spurns classifications and tables of comparison, is something which Bacon apparently did not foresee. Yet it is increasingly evident to the student of theoretical physics that it needs to be added to his established directives in order to provide comprehension for current problems and researches.

## NOTES

[1] This statement, which at any rate reflects a common view on Bacon, may be at odds with some of his writings. For a critical account see the article by C. J. Ducasse in *Theories of scientific method,* Blake, Ducasse and Madden, Univ. of Washington Press, 1960.

# PART III

## SCIENCE AND HUMAN AFFAIRS

# PART III

Part III attempts to apply conclusions reached in the previous chapters to more general philosophic problems arising in fields other than physics. It asks the question: can the basic method, the epistemology of exact science, be applied to human interactions as they appear in sociology, ethics and history? The answer given is on the whole affirmative. Chapter 17 was given as a rather informed lecture, as its style may suggest.

CHAPTER 14

# WESTERN CULTURE, SCIENTIFIC METHOD AND THE PROBLEM OF ETHICS

The endeavor in this paper is not to solve problems but to state a point of view. For emphasis, this has been done fervently, somewhat dogmatically, and without great circumspection. The analysis of Western culture is certainly incomplete and probably exposes traits common to all cultures. Many of the points here made have recently been presented more forcefully and with better documentation by Professor F. S. C. Northrop.[1] The last part of this paper, however, departs markedly from Northrop's thesis. In it I take a position which will, I fear, incur the criticism of many readers, since it affirms the belief that ethics is a verifiable discipline and that its formal structure should be that of a science. But my intensity of conviction on this point has grown rather than diminished as a result of discussions with fellow scientists and others.

1.1. Among the points of view from which a culture can be analyzed, we select a rather unpopular one, a view that places scientific method in the center of interest, leaving philosophy, esthetics and religion in the background of the cultural scene. The selection is made because the scientific element is believed to be the characteristic *par excellence* of Western culture. In the following I hope to make clear this assertion and to provide evidence for it.

Certainly, its truth is not obvious, for Western culture existed apart from others before the advent of modern science. Hence what must be shown is that the significant traits both of our culture and of modern science spring from similar roots. For we do not wish the assertion to be understood as merely saying that technological achievements have molded the pattern of our external lives. This is a commonplace and would be of no significance in the present discussion if our science were but an incidental factor, powerful but inessential, as it is in some civilizations. Rather, the claim that our culture is typically scientific is here made together with an admission that a way of life wholly dependent on the applicances of science may not possess this character. The Japanese were an example of this apparent paradox.

Endless debates can be argued about whether our culture is this or that, and the results, while edifying, are often fruitless. Such formal debate is not intended here. If no fundamental insight leading to possible action could be gained from an inspection of the present thesis, our interest in convincing the reader would be very small indeed. But there are practical issues of great magnitude which seem to demand considerations of the sort to be presented.

Even at the risk of giving offense, it seems proper to say that history and anthropology have hitherto been disciplines largely marked by creative impotence, except in so far as they give satisfaction to scholars. Perhaps there is no reason why they should be creative in the sense we wish to convey, a sense that has not reference to creation of new knowledge (this is indeed achieved by history and anthropology) but to the establishment of principles for prediction (where they fail). Among the reasons for this deficiency is the desire on the part of the historian to remain in the realm of *facts*; indeed he deems it *unscientific* to go behind facts and to delve into theory. But at this level the essence of our culture cannot be caught: *analysis*, guided by rules of tried effectiveness and grounded in fact, needs serious consideration in the solution of the cultural problem. And the possible gain is considerable. Recognition and study of the main-springs of our culture, made possible by analytic procedures familiar from science and philosophy, carry the promise of converting history into a fertile enterprise, capable of warning and predicting. The least we may expect from the approach to be sketched in this paper is the discovery of conditions foreign to the basic attitudes of Western life, conditions that inaugurate the breakdown of our culture. The most we can hope for, having gained this requisite understanding, is the wisdom to avoid impending disaster.

To illustrate the magnitude of what may be lost by the use of ordinary historical perspective, and what may be gained by the type of analysis here attempted, let us face the terrifying problem of our day. Those who know the scientific facts concerning atomic energy seem convinced that only a form of world government can save life on this planet. Historians and political scientists discourage and belittle the utopian suggestions that physicists have made. Their verdict, based on a study of cultural excrescences of the past, makes world government appear impossible in the near future. The questions we would raise are these: Whence does history suddenly gain predictive power? What is the logic of the argument which concludes from the absence of world federation in the past its impossibility in the future? The urgent query is clearly: What are the permanent, ideal, methodological tenets of our culture, and is a world state compatible with them? Thus the kind of analysis to which we direct attention moves into focus significantly.

1.2. Inferences cannot be drawn from facts, nor are facts ever pregnant with prediction. Inference and prediction require first an *interpretation* of facts: Only a theory about facts renders inference possible. In the following this will be made clear with respect to scientific method. A realization of this crucial condition is necessary at the very beginning of our discussion. To understand culture it is necessary to see what general view of the world and of life is most compatible with the data it presents, and to ponder over its implications. Strangely enough, the layman sometimes regards this excursion beyond data as unscientific and dangerous. Scientists find it difficult to understand that attitude, for they know that the more highly developed a science, the more freely, and the more successfully, does it practice ultra-factual interpretation. We begin, then, by *interpreting* Western culture.

I doubt whether an impartial observer of our civilization could fail to be impressed by its strong emphasis on *stable patterns*; indeed it would seem that, if there be any striking difference between East and West, it expresses itself in such emphasis. We wish for ourselves a certain permanent form of government and tend to consider it as the optimum political arrangement for any nation. We want law and order; we want our jobs secure. We save, and make arrangements that will allow us to control our lives, even in old age, in a manner likely to produce a minimum departure from preceding stages. Our annoyance at some political and social events, and at the weather, arises from the circumstance that they are not completely predictable. We insist on the same sort of justice for all, and our mores tend toward great uniformity. Our various languages have been frozen into crystalline states around lattices of grammar and standards of usage, and we enforce mass education to level off differences of knowledge. When we come to express our religious convictions they tend to form creeds, and in philosophy we announce systems and general *Weltanschauungen*. In our daily lives the phenomena of industrial standardization take on increasingly greater force, and even our tastes for enjoyment, proverbially so divergent, approach a norm.

Others, while accepting this general diagnosis, might see the prevailing feature of Western culture not so much in a search for stability as in emphasis on rational form. Indeed, all the descriptive pronouncements of the preceding paragraph can be explained by assuming that our public life confers attention upon the rational elements of existence, much to the exclusion of the emotive or the esthetic. Doubtless this interpretation goes quite as deeply into the situation as does the other; if causal reasoning were *a propos* in this analysis one might prefer to say that rational form prevails because it is the agency which invests observable patterns with stability and permanence.

Nowhere does search for permanent pattern attain higher intensity, or reach greater success, than in theoretical science. It is for this reason that the term *scientific* so clearly characterizes our culture. It is meant to label its roots, not its external make-up. Basically, we believe in freedom for the same reasons that we accept the theory of the atom; both doctrines offer maximum opportunities for comprehending large portions of human experience in a uniform way. Whatever has given Western culture its uniqueness, whatever impels it to move in a specifiable direction, that same force becomes articulate in scientific theory. Hence its basic features are manifest, if anywhere, in the methodology of science, and they can best be studied there.

The term science must not, however, be construed too loosely; it should signify what is more definitely called rational, or theoretical, science. For it is very doubtful whether such descriptive disciplines as the social studies and psychology have developed peculiar methods of reasoning that are significant in a basic sense. No disrespect is here intended, nor disparagement of the fruitfulness of research in these fields. But their present state is more like the condition of geometry in pre-Pythagorean times, when the three-four-five rule (the forerunner of the Pythagorean theorem) was widely known and used by the Egyptians, though a *proof* of it was lacking. Every science, in its early natural-history stage, proceeds by *correlating* facts, and search for pattern exhausts itself in classification. Only when an element of cogency or necessity is introduced into the body of knowledge, as always occurs when the knowledge is subjected to logical manipulation, does it partake of the character of rational science. And it may be said that our culture has the outstanding property of striving to convert all experience into rational scientific knowledge. This is the precise meaning of our initial assertion.

The suggestion that science has generated or fashioned our culture is no part of my thesis. In endeavoring to gain understanding of an organic development like the history of our civilization, or the growth of living things, it is often dangerous to isolate causal factors; in fact it is often impossible to discriminate between cause and effect. What the thesis does expose is the usefulness of an elucidation of scientific method as a condition for understanding our culture. We therefore proceed with a brief exposition of the methodology of all rational science, suggesting that it will shed light on the cultural environment in which that methodology was born.

1.3. The content of experience when caught in its immediacy is flux; it is heterogeneous, colorful, ever changing and it defies classification. The person in the Western milieu has to be reminded of this. Only by conscious effort can he subdue the consequences of his training and become aware of it. The adult who has outgrown the stage of make-believe, for whom the fairy tale has lost its appeal, succumbs to his cultural heritage so completely that he no longer regards fairy tales as descriptive of primitive sensory and emotional experience. He has lost the ability to wonder at the habit of taking the tree in his yard for granted when he is not looking at it. He has in fact yielded to the stabilizing rule which reifies certain phases of his fleeting consciousness. For a true appraisal of the method of science it is important to recognize this situation.

Sense data are never permanent, cleanly defined or sharply separated into entities. They do not permit application of logic, arithmetic or any sort of rational principle. To apply rules, the scientist—and everyone in our culture is a scientist at this stage of perception—passes by means of a constructive act, usually performed without clear awareness, from crude sense experience to the contemplation of ideal structures called external objects. On a higher plane, he resorts to physical entities of a more abstract nature but always endowed with a degree of fixity which the original data did not possess. It is in terms of these and not sense perceptions that he reasons; to them he applies laws and theories; from them he builds his physical world.

Emphasis upon the transition from sensory apprehension to the general entities here in question, together with a clear recognition of the role played by the latter in our understanding of the world, is not merely of epistemological importance. Modern physical theory requires the assumption that laws and theories do not refer solely to sense data as an essential basis for comprehension. Quantum mechanics, for example, is nonsense unless the distinction between sensory experience and physical systems is carefully made and respected. If any doubt arises, the reader should try to confine his attention to the sensory aspects of a piece of matter while crushing it, dividing it more and more finely, and see if he ever arrives at what the physicist means by his imperceptible electrons and their uncertainties. In such instances the difference between *observables* with their chaotic behavior and *physical states* with their regularities becomes the *sine qua non* of comprehension. Its acceptance is also the condition under which scientific experience at all other stages becomes meaningful.

The coordination of theoretical elements[2] with sensory data is not an arbitrary undertaking. Whatever compels the percipient to associate with the complex of sensations brown, hard, rectangular, and so forth, the conception "desk," is not a capricious urge, but a regulatory device of great generality in our culture, though whether it be a necessary constituent of all perception may well be questioned. For present purposes it suffices to state that there are fixed rules of correspondence, fixed in the sense of general acceptance and temporal invariability, which regulate the transition from the fleeting world of sensation to the stable world of public concern.

Not all philosophers will acknowledge the genuineness of this transition; many prefer to regard the public world as also given in sensation. In the early days of science this stand may have been possible, but it is no longer defensible. For in modern physics the very rules of correspondence between data and "constructs" have become matters of serious attention; no longer do scientists argue about their presence, which alone would suffice to force the distinction, but their specific *contents* have taken on increasing importance. For out of the constructs is formed the public world, and this, which is the world described by theory, the world in which predictions are possible, has become exceedingly complex. In the face of this complexity, correspondences with the world of sense are no longer left to instinctive apprehension, are no longer obvious in their immediate givennesss; they require standardization through suitable definitions, called operational. For instance, when the physicist refers to the temperature appearing in the equations of thermodynamics he does not mean sensory coldness or hotness, but a very precise sort of thing *corresponding* to hotness *via* rules laid down in operational definitions. Certainly, to approach an understanding of scientific method is to grant the operation of these rules; their complete analysis is one of the fascinating problems of meta-science. It may be that their action is specific in our culture, or even that such correspondences are absent in others where life does not rise appreciably above the charming but irrational level of sensory perception. What matters in the present context is only the stabilizing effect which they exert on our formulation of science, and through it on our mental environment at large.

1.4. The fact that certain sense data "suggest" (that is, cause us to pass by the afore-mentioned rules of correspondence to) more theoretical essences endowed with greater permanence and pattern, such as external bodies, molecules or genes, does not in itself constitute the latter as true components of what we call reality. For otherwise hallucinations and optical illusions would have to be accepted into the physical world. There are important additional requirements, of which two large classes will here be mentioned. First of all, the entities constructed as counterparts of sense experiences must have relational properties of a logical and mathematical kind so that they can be combined into theories, that is, formalisms which allow modifications of and substitutions amongst the entities. This feature permits the empirical verification, or validation of the constructed entities because it makes possible not only the passage from an initial set of data to a set of "constructs" by means of the rules of correspondence, but also modification of these constructs by theory and final return to *expected* sense data through the same rules of correspondence. If the expectation always meets the data the "constructs" are said to be valid and the theory is verified.

More interesting than this class of requirements looking toward validation is, at least for the purposes at hand, the second. The reader will be spared much mental effort if introduction to the rather abstract problem now confronted is made through an example. Observable phenomena of heat can be "mapped," as it were, upon two scientific universes: one in which heat is a tenuous form of matter, another in which heat is energy. Both of these "constructs" are valid in the sense of all requirements thus far imposed, for both lead to verifiable and verified consequences. Which of the two universes, then, is to be accepted? Or is it to be supposed that certain experiences require the one, the remainder another? Competing theories often arise in the history of science, and frequently both agree with experiment within limited domains. Empirical procedures do not always discriminate between them, especially when no limit is set upon the degree of conceptual elaboration which a given theory is permitted to attain. Choice is then made by reference to regulatory principles which are not always easy to name. In the present instance, the scientist appeals to maxims of *simplicity*, or *economy of thought*, or *logical elegance*, or *generality of application* to settle the dispute, and the result is that the caloric interpretation has been abandoned. It is this group of regulatory principles that is now being considered, to be sure, without intent of careful analysis.

A complete survey of this field in the light of modern science has not been made. Newtonian physics has been penetrated by lucid investigations of this nature and has led to the epistemologies of the last century, none of which is wholly acceptable today. But the fact remains that metaphysical principles which render theories acceptable are still operative in all branches of rational science, and that science cannot function without them. Among them are certainly the maxims that an ideal theory should ultimately comprise all observable phenomena, that its structure be unique in all applications and that it operate in accordance with a postulate of causality. These vague indications do not exhaust the list; they merely hint at larger technical problems and serve to expose the presence of such problems. What is important for the purposes of our inquiry is not the detailed mode of operation of metascientific principles, but their effect on the body of science and therefore on culture.

This effect is uniformizing and stabilizing. Illustrations of the tendency toward uniformization are available in all historical instances where rival theories formed a synthetic union, where two apparently contradictory interpretations of the same set of facts became reconciled on a plane of higher generality: Witness the resolution of the wave-particle dualism in quantum mechanics today. The stabilizing effect is caused in very large measure by the arduous and exacting nature of the principles to be satisfied. A theory is not discarded lightly for it is difficult to put a better one in its place. Thence arises, in any given age, a relative permanence in the views of the world; but not an absolute permanence such as might exist if metaphysical principles were not coupled with procedures of validation, or if there were no principles at all. This, again, is a peculiarity of our culture. Sometimes, the conviction of permanence is carried too far. The psychological force which drives many a scientist toward discovery is proclaimed to be the desire for truth, and he himself may firmly believe that he has found *the* final interpretation of his result. But history and methodology both deny this claim.

The principles — here called metaphysical, metascientific, or methodological — which regulate the functioning of scientific theories and whose closer scrutiny is avoided in this article, have themselves an amazing degree of permanence. They appear to change far more slowly than theories, forming in a sense the mental climate in which theories arise. They are akin to the categories of old. The desire to make them absolutely permanent, to anchor them upon the soil of rational necessity, runs through and is highly characteristic of Western philosophy. In fact, if I were to name one thinker who presented the gist of Western culture in its purest form I would choose Kant. Not only is the *Critique of Pure Reason* typical of the adjustment which our civilization makes to successful scientific progress (in this case, Newton's), but the general acceptance of his attempt to locate the categories in the very mechanism for understanding, where no one could easily eradicate them, bespeaks the indoctrination originating in the desire for stable pattern which our culture had undergone.

Among the principles described there is one that may be called the leitmotiv of Western science; it presents all the facts already summarized in clearest outline and accounts for most that is characteristic in our attitude toward the world. In it scientific method announces its credo and exposes its metaphysics to clearest view. It is the belief in the convergence of scientific researches, the belief that all experience can be subsumed under general nonconflicting theories obeying requirements like those sketched in preceding paragraphs. Sprung from a vague notion of the uniformity of nature, the idea worked itself through several philosophic formulations; it is a common factor in such divergent views as rationalism and ordinary realism. Refusal to believe in miracles is *the* sign of cultural enlightenment in our midst.

Regarded without preconception, this metaphysical thesis is a strange postulate, based entirely on the success of science to date. It represents a vast extrapolation upon a limited amount of knowledge and casts the shadows of our present selves on infinite horizons. To be sure, not every scientist holds this view, for the positive results of science can never substantiate it. Whenever rival theories emerge, there is danger of an essential bifurcation, for it might conceivably happen that certain phenomena in a given field are explicable only on one theory, while the rest require another. Three decades ago physicists were almost resigned to this contingency, because light exhibited the contradictory aspects of particles and waves. They were rescued from the dilemma by fortunate discoveries for which at the time there was no certain expectation. Yet the belief in the convergence of scientific explanation is one of the fundamental tenets of Western culture, and I am convinced that its repudiation would alter radically the general structure of our thought.

Whether this change would be a loss, or indeed a disaster, cannot be decided on the basis of an analytic inquiry like the present and will for that reason not be argued. At all events it is worth knowing under what circumstances it may be expected. Our period is one in which causes tending toward abolishment of the axiom of convergence are abundant; they are rising to dominance within powerful positivistic doctrines and are strengthened in the popular mind by unprincipled emphasis on the Scientific Approach, no matter how incoherent or misunderstood. The perspective of history shows clearly that a tendency to ignore methodological bases appears when science has made significant advances and the problem of synthesis is difficult. Thought is then inclined to take the easier course and renounce synthesis. This situation exists today. It is also true that genius, including scientific genius, is rarely bound by conscious reflection to the methodological principles of the past, though of course the cultural bond is always present and is usually recognized after a discovery has been assimilated. Today the probability that assimilation may not take place is considerable.

It may be well to summarize our thoughts so far. A cursory survey of our intellectual life shows a progressive development of rational patterns, a tendency toward uniform and stable concepts and toward norms. We seized upon these as distinguishing traits of our culture and tried to find their origin. It was discovered within the hidden preconceptions that lie at the basis of scientific procedure. Analysis of this procedure brought to light a number of methodological attitudes or principles, among them rules of correlation between our sense world and our public or theoretical world, as well as logical requirements to be imposed on concepts. All of these were seen to have the effect of stabilizing science, and all of them, by being tacitly adopted, mold our lives and create our cultural pattern.

1.5. Thus far our aim has been to reduce what appear to be the outstanding characteristics of Western culture to relatively few fundamental convictions, and it was shown how these convictions operate in the field that allows them the greatest range of development, namely, science. If the result is on the whole accepted, then we face a peculiar paradox: Western culture, though having produced a distinctive rational pattern, has not provided a basis for ethics. This is superficially apparent in the wide tolerance in regard to what constitutes ethical behavior, but beomes impressively clear when we inquire what our ethics should be if they had developed consistently along with rational views of the world. Only a brief indication will be attempted.

Viewed logically, every system of explanation is a deductive apparatus that starts with postulates, derives theorems and finally tests specific propositions in the realm of data. A suitable example is the physicist's "understanding" of why bodies fall. The postulate is Newton's law of universal gravitation; from it the laws of free fall, or the laws of planetary motion, or the variation of gravitational force on the earth's surface, may be derived by mathematical processes; specific predictions of these theorems are then checked against observation. The postulate itself is never proved in a strict logical sense; it is reflexively validated by confirming its consequences. This is now known to be true even of the most evident of scientific presuppositions, such as the axioms of geometry and arithmetic, since instances are readily found in which they do not hold but require generalization.

There is nothing in the nature of things, or in human nature, that renders this scheme inapplicable to ethics. Consistently, the ethics of Western culture should be a postulational discipline in which a basic code is adopted, developed through its formal consequences and put into practice. The matter of validation, to be sure, raises difficult problems of its own, different in kind from perceptible truth and falsity in the factual realm. But it is hardly correct to say that these difficulties have prevented ethics from crystallizing into a form cognate with the rational structures of our culture. The cause which stunts the growth of ethics in our sphere lies in an irrational insistence upon *a priori* evidence for the very postulates on which a code is founded, a demand never made in science. It is as though we were unwilling to accept the law of universal gravitation, despite its effectiveness as a formal mode of explanation, unless it came to us accredited with unalterable rational necessity or divine sanction. This cultural schism further manifests itself in an imaginary contrast between the domains of "facts" and of "values," the former of which we have been able to handle while the latter proved elusive. To deal with values, artefacts are used which appear rather oriental in the selectivity and specificity of their action, and which bear no resemblance to the uniformity of the methodological principles regulating science.

In concluding this section, I find it difficult to look away from the urgent problem of our time. Does the foregoing analysis add or remove weight from the pessimistic contention that the history of Western culture makes the achievement of world government improbable? To me it seems that the preponderant traits of this culture lend no plausibility to this assertion; that its uniformizing, stabilizing motive force tends in the direction of world unity. But those aspects in which Western culture has not yet come into its own give cause for apprehension.

In Part 2, a short analysis of ethics from this point of view will be attempted.

2.1. A scientist writing on ethics is, and should be, under suspicion. Hence the following remarks have come as the result of conquering serious hesitations arising from the reflection that formalized ethics is not a field with which I have a wide acquaintance. No assurance can thus be given that what is said is at all original. Only a deep concern for Western culture, and a firm conviction of the compatibility of ethics and science, prompt me to argue about matters in which experts may dispute my competence. However, lest this be construed as superficial modesty, I add that I regard the distance between ethics and science as no greater than that between ethics and politics or between eithcs and history.

There is a curious circumstance, almost without relevance to the basic problems at issue, which should be noted briefly. Many people believe that the most effective code of ethics is that of the scientist. To be sure, it operates in a very limited domain and almost exhausts itself in what is called *scientific honesty*; it rarely improves the social qualities of those who profess it except in relation to other scientists. But within its narrow range it is remarkably successful. Study of this code will not be pursued here, but is recommended to psychologists interested in ethics. The significant point for present purposes is in its manner of enforcement, in the reason for its effectiveness; scientific honesty is not practiced because of religious beliefs, nor through fear of punishment or hope for reward, nor because of love for fellow scientists. It is adopted because it is part and parcel of the scientific profession; one accepts it without question as one accepts the postulates of arithmetic. Being a scientist implies being scientifically honest.

The fundamental question to be considered is this: Can the method of science, in particular of theoretical science, be applied in ethics? If so, and if scientific methodology is an important characteristic of Western culture, as the main part of this paper affirms, then it would seem consistent to study this application.

There is a vast difference between this question and the apparently similar one: Can science, that is, the factual content of science, be applied to *generate* an ethical code? Confusion of these is disastrous even if common, for it leads to a myopic vision of the whole of human activity. Ethics has a subject matter distinct from that of science. Its advocates are right in insisting that a positive decision which confronts the moral individual is *sui generis*, that science has nothing to say about whether it is right to kill. Only when the decision has been formulated can science become effective in bringing about or averting what man as an ethical person has decreed. Science is ethically neutral, and he who denies this elementary recognition has merely forgotten *when* his moral choice was made. Usually he has become so accustomed to a specific ethical standard, such as hedonism, that he is not conscious of his allegiance and regards it merely, but falsely, as a scientific attitude. The ethical "right and wrong" is different from the scientific "right and wrong" and cannot be developed out of the latter. It is therefore improper to speak loosely of a "scientific code of ethics."

Yet it is equally unwise for the romanticist to go off the deep end at this point and proclaim that ethics is the field of responsible choice and of freedom, while science effects nothing more than a description of the material universe. For science, too, has responsible decisions to make, as the preceding analysis has shown. But its decisions have other purposes and are judged by other criteria. The fact remains, therefore, that the pre-empirical situation in science, its need for choosing methodological principles for guidance, is not below comparison with the pre-empirical state of ethics. Hence the answer to the first question—Can the method of theoretical science be applied in ethics? —may well be affirmative.

2.2. To gain further evidence, we recall the essential features of rational science. It starts with *postulates*, that is, general propositions which are accepted as bases for inquiry. These postulates are logically independent of prior fact; they are not inductive generalizations of experience. In their selection, esthetic considerations of elegance, simplicity, and symmetry play as large a role as attention to observable facts, for when a postulate is stated it is frequently uncertain what facts it may imply. Excellent illustrations of these remarks will be found in Maxwell's equations or in the restricted theory of relativity. With respect to the latter it may truly be said that even if facts were to contradict its consequences, physicists would be loath to relinquish it because of the sheer beauty of its postulate of invariance (of physical laws in different intertial systems).

When a scientific postulate has been formulated, it is regarded as *true* for the purposes of subsequent analysis. That no valid or even tentatively valid conclusions may be drawn from it unless its integrity is rigorously respected is a commonplace. The feeling for consistency acquired through a peculiarly authoritarian course of training, a training that brooks no doubt regarding matters of logic, prevents the scientist from using the postulate where it is convenient and ignoring it otherwise in the process of drawing inferences from it. (Nor is this a triviality, for on the basis of *fact* the scientist has just as much reason to doubt that 1+1 = 2 as he has for doubting that the human race abhors murder.) The inferences obtained through such integral deduction are usually specific theorems that have something to say about scientific experience.

The first two stages, postulation and deduction, are finally followed by verification. If what the theorems say about experience is found to be true on observation, sufficiently often and under a sufficiently broad variety of circumstances, deductive procedures and postulates are said to form a valid, or true, theory. Thus the logical state of affairs is very simple: the postulates are reflexively validated as premises of propositions which are true. They can never be *proved* in a deductive sense, since there is no assurance that other postulates may not yield equally true propositions, nor are they inductive generalizations of sensory experience.

The process of verification, however, is not as simple as it looks and is often misinterpreted. The nonscientist, and sometimes the scientist, is likely to regard it as a comparison of what is *predicted* with what *is*, and what *is* need not be subjected to further scrutiny, according to this uncritical view. The saying goes that one is dealing here with a realm of indubitable fact which is either correctly or incorrectly described by theory. Astronomers predict the appearance of a spot of light at a definite place in the sky at a certain time; their theory is correct if the spot appears. But what if the seen spot were a mirage? Hallucinations are not predicted by physical science and yet they occur. In fact, there is a great deal in the field of perception that science makes no pretense of predicting. It is customary to divide sensory experience into two categories: *bona fide* perceptions and illusions. The latter are just as vivid, spontaneous, memorable and convincing as the former; indeed, they cannot be separated from them by any criteria involving sense experience alone. Yet physical science does not deal with them directly: it rejects them on the grounds that they are incompatible with its accepted theoretical formalism. Strangely enough, science is selective in its method of verification; *the effect of its formal structure determines to some degree the manner in which it will expose itself to tests.*

Even aside from this there is the difficulty of knowing whether a given observation corroborates or contradicts the results of calculation, since there is never pin-point coincidence between the two. The usual way out of this misfortune is to take a great number of observations, to regard them as a probability aggregate and then to set in motion the machinery of the probability calculus. All of this is far from obvious and simple; its justification has embraced some of the most technical reasoning of modern logic. What matters here is that the process of scientific verification is a very complex one, made possible by constitutive maxims requiring assent in many of its phases; that it far transcends in elaboration the simple-minded look-and-see operation for which it is popularly mistaken.

The astonishing fact is that scientists never doubt the propriety and essential correctness of their complicated procedures. Why this should be is probably for the psychologist and the anthropologist to say. I believe it to be caused by thoroughly effective conditioning by Western culture in the sense of my initial analysis, in particular to an attitude that permits no doubt with respect to logical axioms and yet demands divine sanction for ethical norms. Perhaps the astonishing qualities of this fact fade away when one contemplates the diverse methods by which we teach (or fail to teach) arithmetic on the one hand, ethics on the other, in elementary schools. If a child says 2+2 =3 we tell him: no, this simply does not happen. If he lies we tell him a story showing that lying is quite possible but leads to unpleasant consequences.

2.3. We now turn to ethics. To be scientific in the sense here proposed, ethics need not be a mechanical discipline concerned solely with efficiencies or with anthropological findings. It must conform to the three stages of scientific methodology: postulation, deduction of specific consequences and verification. Each of these will take a form suitable to the subject matter of ethics, namely human behavior, and may well differ in detail from its scientific counterpart. But the logical structure should be the same. Thus, it should not be required that the postulates be generalizations from anthropological experience nor that they possess *a priori* evidence. And it is equally unscientific to demand verification through agreement with all human behavior. But let us first consider the three stages one after another.

The stage of postulation presents relatively minor difficulties of conception; it involves the formulation of a code of ethics. To be sure, the *practical* difficulties besetting this act are at present enormous, and they are likely to remain great so long as the postulational, or scientific view of ethics, such as that here sponsored, is not generally accepted. These difficulties, while clearly recognized, are not the object of the present discussion. They will be lessened, however, when the purely postulational or, in the strict sense of the word, *hypothetical* character of the ethical code is recognized. The status of this code should be the same as that of the axioms of arithmetic, or of Euclidean geometry, no more sacred and no less compelling. Especially should it be noted that the code requires no prior motivation in supernatural beliefs. Hence the amalgamation of ethics with doctrinal historical religions violates a most essential part of our thesis; and I share the conviction of many that this amalgamation has been a major hindrance in the ethical development of our culture, as well as others. Perhaps this creative act of postulation, wherein man dedicates himself to principles, has much that is akin to religious experience; but if this is the case, then the relation between science and religion is equally close.

The next stage involves the explication of the codified material for use. In science this is largely an ivory-tower affair; in ethics it is a matter for action, action consistent with the chosen code. Failure to be consistent is in one field regarded as a *mistake*, in the other as a *vice*. One wonders whether this distinction is not unfortunate, whether it is not the result of an indoctrination which fails to appreciate the similarity of the premises in the two situations. All one cay say is that possibilities of identifying vice with mistake through more enlightened education have certainly not been explored completely.

The problem of verification in ethics requires more extended comment. Basically, verification is a procedure whereby the validity of the ethical code can be made evident, or can be denied. This is a crucial point, for rejection of postulational ethics is usually grounded on the assertion that such procedures do not exist. True, they do not exist if verification is understood to mean agreement with immediate and indiscriminate anthropological observation, for then the results of an ethical culture would have to be compared trivially with themselves. But with that simple understanding, verification does not work in science, either, as we have attempted to show.

Just what it should be, or will be when postulational ethics is practiced, is not for a scientist to say. But there are several interesting possibilities among which ethics as a going enterprise can choose. First, there is the simple but powerful criterion of group survival, which operates even in the present chaotic situation, whether this be recognized or not. Certainly, if a code of ethics leads to the extinction of those who practice it, that code will automatically disappear.

It is easy, however, to pass beyond such primitive proposals. The suggestion made by some anthropologists that ethics, to be regarded as successful, should produce a society in conformity with typical human behavior, is usually rejected because it raises the *is* to an *ought*. Perhaps this is dismissing it too lightly. For the emphasis should here be on the word *typical*. Clearly, if ethical postulates are checked against their results in a society resulting from the postulates, validation becomes a rubber-stamping procedure. That society may, however, be atypical, and comparison should perhaps be made with the essence of what is common to all human societies. The vulnerability of this position is entirely clear to me, for I am vaguely conscious of the possibility that typical human behavior may not be definable. But the decision on this point cannot be made by the offhand statement that ethics molds society and hence cannot be tested against it; it must wait upon a clear empirical verdict as to the existence of constancies in aggregate human behavior. Weight may be added to this admonition by the remark that in the subject of dynamics the formulation of Aristotle, which held sway for many centuries, made typical behavior of moving bodies appear impossible. Only when Galilei seized upon acceleration as the pivotal element in the description of motions did constancies emerge and simplicity result.

The list of possibilities for verification of ethical systems can be extended. There are many variations of the theme involving the greatest happiness for the greatest number of people. Perhaps the term freedom can be defined sufficiently accurately to yield a means for validation. Nor is the concern for physical health and well-being to be excluded from consideration.

Perhaps this variety of opportunities is bewildering; perhaps it is so confusing as to preclude action. Historically, disagreement with regard to methods of verification has indeed been a retarding factor in the development of codified ethics. In view of this, the question may be asked: is it necessary that there be absolute clarity on this point at the very outset? There was no such clarity in science at the beginning of its history. Before Copernicus a theory that failed to prove the earth to be flat was likely to be rejected. Different methods of empirical validation have interplayed, and even now a precise analysis of the meaning of scientific confirmation is beset with problems. As natural science developed, it gradually evolved acceptable standards of validity. Is it too much to hope that a similar process will take place in ethics when this discipline is properly launched? All that can be shown at present is that standards are in fact available and that such hope is not a logical absurdity.

## NOTES

[1] F. S. C. Northrop, *The Meeting of East and West* (Macmillan, 1946).

[2] There appears to be no generally satisfactory name for these entities. I have formerly called them "constructs" of explanation, but have incurred objections on the grounds that they are not constructed like the symbols of mathematics or like rules for playing a game, but have a definite correlation with sense data. Any term applied to them should, however, suggest clearly that they are neither *concepts* in a generic sense, nor *abstractions* from sensory experience.

Professor Ernest Cassirer, to whom I am greatly indebted, thought the term construct was most appropriate in this context and encouraged me to use it while he taught at Yale.

CHAPTER 15

# PHYSICAL VERSUS HISTORICAL REALITY

The quantum theory of modern physics is too rarely made the object of intensive study by the professional philosopher and the social scientist; it is regarded as an *enfant terrible*, difficult to understand completely yet vaguely amusing because of the odd things it says about the world. In some fundamental respects, however, this theory has an importance far beyond its usefulness in physics, for it develops a methodology applicable, though not yet applied, to phenomena typical in many fields far removed from the physical sciences.

One sometimes hears the dogmatic suggestion that problems of biology, psychology, sociology, history, indeed all matters bearing upon collective human affairs, are immune to treatment by natural science because that science presupposes a kind of objectivity and invariability in the nature of its subjects which human intercourse does not exhibit. This doctrine ignores the fact that the quantum theory does deal successfully with just such collective vagaries as social science and biology present. It has a way of handling the "unpredictable interactions" between a fact and the knowledge of that fact, and furthermore it achieves a significant treatment of observations which, in a very real sense, may kill or annihilate the system under observation. There is a fair amount of literature upon these questions. The present article considers the relevance of quantum theory to problems of the social sciences.

I have argued[1] that the use of probabilities as essential tools in the description of nature has brought about a separation of our experience into two domains: one, composed of immediacies (observations, measurements) that are not all predictable in detail; and another, refined and rational, which is the locus of laws and regularities, of permanent substances, of conservation principles and the like. The former was called *historical*, the latter *physical reality*.

At first sight the word "historical" may seem ill-chosen, for it suggests a technical connection with a well-defined discipline. We mean it here in a looser sense which, it is hoped, historians will condone. Historical events, in so far as they differ from physical events at all, are characterized by these qualities above others: they are significant only when observed; they are rather aloof from theories; they usually involve human decisions. They lie close to the plane of immediacy, in the terminology of The Nature of Physical

Reality. Thus historical here means incidental, emergent; it designates that which simply happens.[2] The word is meant to carry a strong flavor of the "existential." To be sure, all these traits often adhere as well to the events which interest physicists, and they do not provide a completely satisfactory definition. But they will indicate the motive for our choice of words which may perhaps justify itself as our discussion proceeds.[3]

The distinction in question does not, it is true, effect a cleavage in our normal responses to the world about us, for physical reality is the best rational account, and the closest possible approach to a uniform description of spontaneous fact. Indeed the main reason for our reluctance to recognize the distinction lies in the circumstance that ordinary experience hardly requires it: in most of our daily life the rational comes to the surface and meets the deliverances of sense with satisfying congruity. Only on examining the stranger phenomena in the atomic world does the disparity in question reveal itself; in that world exact laws disclose their statistical character, and the novel manner in which they regulate observable facts becomes evident.

Hence arise a number of questions which we here state and then attempt to answer. 1. What, precisely, is the character of the lawlessness of the microcosm? 2. What is the meaning of the laws of physics, and how do they bear upon the lawless events? In this connection attention must be given to the seeming paradox arising from the fact that these laws do operate with dynamic precision in the macrocosm, while their functioning on an atomic scale is but statistical; in other words we must take occasion to study the gradual transformation of lawlessness into regularity which occurs as we ascend the scale of sizes, times and masses. 3. Does history belong to the macrocosm, or do its roots reach down into the chaos of the microcosm? 4. If the latter is true—as I believe it is—what opportunity is there for constructing a causal theory of history? In particular, can physical theory form a model after which a theory of history capable of prediction can be fashioned?

In dealing with questions 1 and 2 we shall leave aside as much as possible the refined language of physics; we endeavor to sketch a picture of the atomic world with the broad strokes of a brush, a picture admittedly surrealistic but faithful, it is hoped, to the fundamental issues involved in our problem. Question 2 will force us to consider the meaning of probability. The remaining questions are wholly epistemological; their answers depend on some of the specific information gained from an analysis of the former two.

In discussing atomic phenomena I shall endeavor to dispense with abstractions and to inject some of the vividness which the atomic world actually possesses. I hope I may be forgiven for an account in the following section that may seem like popularization.

## 1. THE LAWLESSNESS OF THE MICROCOSM

Let us transport ourselves to the world of atoms. In doing so we are not considering the use of wholly phantastic artefacts like flying carpets or time machines which nobody can construct; we are merely supposing our sense organs to be replaced by more sensitive devices which the electronic engineer can actually build. These devices will allow us to perceive very small distances, very short intervals of time and extremely light objects. When observed with this equipment, the microcosm turns out to be a strange place indeed.

Perhaps its most striking feature is the failure of visual continuity: the world is not illuminated and filled with moving things; our eyes are now sensitive to the single darts of light (photons) cast off by single luminous atoms, and the effect is a speckled kind of vision with bright patches emerging here and there from utter darkness, the different patches having different durations. Distant objects of large size and mass have a kind of uniform glow, to be sure, and suggest some cohesion in this chaotic scheme of things, but the smaller things nearby give very little indication of uniformity or pattern.

It is doubtful whether an observer whose experience is limited to this microcosm would find it plausible to speak of the "flow of time" rather than the "emergence of sensed intervals." Continuous space would seem a farfetched abstraction, and our observer would be unlikely to claim that objects —if indeed he were to attribute appearances to the presence of objects—had definite positions at definite instants of time. Certainly he would have no occasion to invent the calculus.

Nevertheless, as was already indicated, this world contains some measure of coordination. Patches of light are not completely random but appear in loosely ordered sequences. There are times when nothing can be seen, and then again the visual field is dotted with perceptions. Furthermore, these perceptions often indicate a preferred location in space, though they rarely mark a point. In view of this our micro-physicist would probably be led to postulate the existence of some sort of objects, vaguely localizable in space and somehow progressing from one place to another. But having never seen a completely continuous path he would not regard them as "moving" in the ordinary sense. Indeed the idea of continuous motion might be as difficult to conceive under these circumstances as is the notion of discontinuous emergence in our molar world. But let us accept the existence of external objects with massess, charges, perhaps colors, position-when-seen, momentum-when-encountered and many other measurable properties. None of these could be said to be "*possessed*" by, or to *adhere* uniformly to the objects in the simple way we infer from our familiar experience. There are however, elaborate operational procedures by means of which these properties can be determined. These will now be considered.

To learn in a pleasant way some of the peculiarities of the microcosm we go on a rabbit hunt. Experience has taught us that there are atomic rabbits, and we invent and use an atomic gun. The task of hitting the animal is not as simple as it is in our world since we see neither the rabbit nor the impact of our bullets. But there are ways of telling whether the quarry has been struck: a certain reaction takes place which, for the sake of simplicity, we shall assume to be an elementary squeak emitted by the dying victim. Forearmed with such knowledge and a gun we arrive at the hunting ground and wait until the darts of vision characteristic of a rabbit's presence appear in our field of perception. We aim at the region of their most frequent occurrence, preferably at the last identified rabbit flash, pull the trigger and hope for the squeal. It does not come the first time, of course, but we repeat the performance until the animal is struck. Just how we are going to find the corpse and bag it are difficult questions which, for our purposes, need not be answered. Let it be noted, however, that all the implements and processes we are invoking have possible atomic analogs which the physicist will recognize after a little thought.

With practice certain rules for successful hunting are acquired. A shotgun works better than a rifle. This will occasion no surprise to our macrocosmic minds, except insofar as the advantage of the gun over the rifle is far greater than merely the greater number of projectiles in the shot cartridge would suggest. It turns out, surprisingly, that the *larger bore* of the gun makes its use more expedient. On further experimenting with a rifle we find careful aiming to be useless; a *fine* gauge precisely directed at the target *increases* the scatter of the bullets and does not improve the aim. On noting this curious result we perform experiments; we shoot rifles of different gauges at a stationary target and observe their impacts. Large stationary targets do in fact exist (see below) and impacts can be made observable by the use of bullets which explode, i.e., emit light, on striking. Normal bullets, which we employed in our rabbit hunt, do not have this useful property.

By experimenting in this way we discover the famous uncertainty principle. It states that a relation between the pattern of impacts on the target and the size of the rifle bore does in fact exist, but the relation is the reverse of what macroscopic common sense would lead us to expect: the finer the bore the greater is the dispersion of the bullets.

Closer study of the impact pattern leads to further interesting observations, in particular to the surprising result that the distribution of shots is not the one suggested by the "error law" but by the diffraction of ordinary light passing through a diaphragm. We have thus come to recognize the existence of regularities in the *aggregate* of events despite the apparent lawlessness of individual occurrences. *If theories are to be devised to account for such happenings, they will have to concern themselves with aggregates rather than with single events.* For the present, however, we put theories aside and return to the vagaries of the microcosm.

We note in passing that hunting elephants is a good deal easier than hunting rabbits. These beasts give a more coherent account of their presence. While they do not produce a continuous progression in our field of vision, their seen emergence nevertheless dots a fairly uniform path. Also, use of an elephant gun with a large bore and more massive bullets reduces the scatter of the shots and makes the chase much more worthwhile.

From sports we turn to science. Galileo would have found it difficult to demonstrate to the inhabitants of the atomic world the validity of the laws of free fall. In transferring his experiments to the microcosm, we again take note of the fact that there are no continuously visible small bodies. Furthermore, remembering the erratic behavior of the individual rabbit, we consider it wise to deal at once with a large collection of objects. Hence we do what, macroscopically speaking, would be called gathering a handful of stones. True, this figure of speech is hardly descriptive of our exact procedures. For the stones are but intermittent patches of luminosity, and to confine them to the space of our hand is a difficult task. When transferred to the hand they are not densely packed nor do they stay at rest; all indications are that they jostle and bounce about, and there seems to be no way of quieting them down.

This unruly collection of bodies is dropped from a place high above ground, perhaps from some microcosmic leaning tower, and their progress is carefully recorded. It is impossible, as we have seen, to trace a single stone, our record being a multiplicity of individual light emissions. Yet there is some semblance of coherence, especially to an observer far away who is unable to distinguish the light flashes coming from individual stones. To him the falling group has the appearance of a swarm of fireflies which is clustered fairly tightly at the beginning but diffuses into a larger and larger assemblage of luminous dots as it approaches the ground. When all the facts are gathered our observer reports: 1. He was unable to trace unambiguously the path of any individual stone. (This is hardly news to us now, after our experience of hunting rabbits.) 2. The center of the swarm seemed to move with an acceleration of $32.2 \text{ft./sec}^2$, i.e., in perfect accord with the macroscopic law of falling bodies. 3. The swarm grew larger and thinned out as time went on.

At this point we, who are accustomed to the regular behavior of falling rocks, are strongly tempted to read their regularities into the miscrocosm.

Every stone, we are prone to say, obeyed the laws of motion as it fell. The reason why we failed to observe it are not hard to see. First, we only caught glimpes of an individual stone and did not trouble to reconstruct its complete path. Second, it was admitted that the stones were not at rest in the beginning. Hence the initial conditions of motion were different for different individuals and their spreading during the fall should have been expected. And thus we go on to assert with apparent safety: *If* we had known the velocity and the position of every particle at the moment of its release we could have calculated its path, and the whole firefly phenomenon could have been predicted.

The argument just stated attempts a *mechanistic denial of atomic uncertainties*. It considers motions as essentially continuous, unique and determinate, and it blames other physical agencies for the vagaries manifest in observations. Before turning to the specific forms of the argument, I wish to examine it on general philosophic grounds and appraise its value as a methodolgocial directive.

The argument is an extrapolation of familiar molar experience. It is highly questionable, however, whether an observer familiar with the ways of the microcosm would have need for an hypothesis of detailed predictability, and consequently whether an investigator who regards the atomic world as primary should properly be disposed to make it. Perhaps these are matters of methodological preference. Let us therefore see what the argument nets us in the way of advantages, or of simplicity of description.

Practical advantages, aside from a certain satisfaction to our intuition, do not result from the hypothesis that atomic uncertainties are due to mechanical agencies. No theory has yet been proposed to render the vagaries understandable in detail, none is able to predict them. Indeed Heisenberg's principle, when disengaged from the "explanations" with which physicists so liberally suffuse it, says precisely that such predictions are impossible.

With respect to simplicity, the case against the mechanistic argument is even stronger. There is a parallel to it in the old theory of the ether, which we here briefly recall. This theory made, in effect, two assertions. One was that space is a medium capable of transmitting electrical and gravitational influences; the second was more particular and insisted in assigning the familiar properties of *material* media to the ether field. Every student of the history of science knows the difficulties and the controversies which ensued because of the tenacious belief on the part of physicists in what at the time appeared to be common sense, the materiality of the ether. The hypothesis was finally discarded because the models invented to account for the behavior of the medium violated every canon of conceptual simplicity. Quite analogously, the facts of the microcosm are now being explained, first by the postulate that *there are, indeed, permanent entities called electrons, protons and so forth*; to this, the scientist who argues for the reduction of atomic uncertainties in terms of insufficient knowledge, unforeseen interactions and the like, adds the further thesis that *the postulated entities have the familiar mechanistic properties of our more primitive experience*. This latter thesis complicates matters needlessly; unless it is eliminated we may waste time in wrestling with problems whose very artificiality, like the structure of an ether molecule half a century ago, belies their importance.

Neither of the last two italicized propositions is obvious in the sense of strict empiricism. The imputation of permanent existence to an atomic stone that reveals its presence only in the manner of the firefly, indeed with lesser consistency than a firefly, is already a posit not wholly forced by the observations. This assumption, however, has proved helpful and is incorporated in all valid theories of nature. Assignment of *exact position at all times*, which is demanded by the mechanist, is another, in fact a more risky hypothesis, as is shown by the circumstance that no theory embracing it has as yet succeeded. This later hypothesis must be abandoned.

The position taken here can be supported to some extent on purely operational grounds: the position-at-all-times of an atomic particle cannot be found directly by observation, no operation for tracing the detailed trajectory of an electron being known. Thus far an appeal to experimental feasibility can be trusted. But operationalism, as always, is not the whole criterion of scientific validity. In this instance it would indict likewise the first postulate concerning the continued identity of electrons, which cannot be observed directly either. The point here is that validity in science relies also on principles beyond those contained in the doctrine of operationalism.[4]

A particularized version of the argument we are criticising is the following. Granted that the luminous spots accompanying what is called the motion of a microsmic stone (e.g., an atom) do not mark a simple continuous path. The reason—says the argument—is to be seen in the unavoidable recoil momentum imparted to the stone by the photon it emits. This causes it to zigzag in a peculiar manner, the corners of its path being made luminous by emissions. If the explanation stops at this point it is innocuous, for it adds nothing of scientific or cognitive value to the patent unpredictability of the stone's behavior. If the argument goes on to give directions for computing the trajectory, it fails.

The idle theory which endeavors to restore continuity by cryptic supplementation of observable facts can take other forms. For instance, if the stone is not self-luminous but perceived by reflected light, the theory can say that the reflected photon disturbed its motion unpredictably; if the stone manifests its presence by collision with another object, the latter can be blamed. Nature's perversity seems forever to prevent the theory from becoming specific or, as I should prefer to put it, seems always to grant it a hiding place.

Man's inability to trace the path of atomic objects is grounded in something far more serious than ignorance; its roots lie in actual indetermination of perceptions. In the case of an ordinary firefly, observed as moving from its scintillations in the dark, ignorance of intermediate positions does not prevent their interpolation and hence a construction of its path. The situation with respect to the atom is completely different; interpolation will not work however cleverly it be conceived.

What, then, has happened to our argument which ran: If we knew the velocity and position of every particle at the beginning of its fall...? It is useless, for according to our best present understanding we never shall know these things. To say that under these wholly impossible circumstances the particle would obey the laws of free fall is quite the same as asserting: If the world were populated by angels there would be no wars. And finally, it should be pointed out that the argument, even when accepted, will not account correctly for the spreading of the swarm of stones as it "falls." This involves what physicists call "diffusion of a wave packet" and has no macrocosmic counterpart. It is thus necessary to take the lawlessness of the microcosm as it affects individual objects at its face value, to desist from trying to embellish it.

But it is quite essential that we recognize the regularity-in-the-mean exhibited by the cneter of the swarm of stones as it fell. Yet, how are these facts to be reconciled? What is it in the individual particle that makes it obey laws of the aggregate? Does it know what the other particles are doing and behave in relative conformity with them? Is an attracting force holding all particles together while some sort of individualistic repulsion keeps them apart? Science holds neither of these specific suppositions to be adequate. It proceeds on the most neutral plane that will join regularity-in-the-mean with individual caprice; it assigns neither purposes nor forces, but colorless *probabilities* as innate qualities to the microcosmic stones. This, it may be shown, is the least committal and the most effective thing to do in such circumstances. However, before studying probabilities, as we do in the next section, we return for another look at this elementary atomic world which we now inhabit.

Astronomical objects far away appear and behave like the bodies of our molar experience, as indeed they should, since they are the things we ordinarily encounter. Quantum theory accounts for this by showing that probabilities congeal to certainties when the masses in question become large. Large masses, therefore, do possess positions at all times and continuous paths to a very high degree of approximation. Nor is there a contradiction in this, or a mystery in the transformation of probabilities into near-certainties. An illustration of such a transformation which can be visualized is a loaded die. As the load is increased and the weight placed nearer and nearer one surface, the die's behavior becomes increasingly regular and predictable.

The laws of *optics* in the microcosm—to take one final example—are as peculiar as those of mechanics. On heating a large body it is found to glow, not with uniform incandescence, but with pointlike luminous spurts of different colors. As the temperature is raised, the bluer ones begin to predominate at the expense of the redder scintillations. No law suggests what color will appear at a given point at any specified time, but in the aggregate the color distribution agrees with Planck's law at every instant. A small object (a single atom), too, may be subjected to heat treatment and become self-luminous. But the light it emits is not continuous; it reminds even more strongly of a firefly in the dark, except that now the emissions are in different colors. No rule governs the details of the color sequence, yet in the long run the frequency of the individual colors obeys the laws of quantum theory. It may surprise us, however, that every elementary kind of body has its own assortment of colors by which it can be identified, the irregularity of sequence notwithstanding.

These examples may be regarded as fairly typical of the microcosm. They defy lawful description when attention is focussed on emergent, incidental, perceptual detail, yielding orderly pattern only when treated *en masse*.

## 2. REALITY IN THE MICROCOSM

Having learned this lesson we return to our accustomed sphere. Our bewildering experience will be summed up in the simple question: *What things are real in the atomic world?* On the molar scale of magnitudes a similar reflection hardly arose. What we saw was describable by continuous processes and by accurate laws. Perception was predictable, the emergent and the incidental were merely unrecognized features on the one regular face of nature. Atomic nature presents two faces and therewith a dilemma.

Or is there no dilemma? Suppose you saw an airplane in the sky and identified it as a B29. You look away, and after a moment you observe it again, noting to your astonishment that it is a Thunderbolt. After a while you observe it once more to be a B29. Would you not conclude that you were mistaken at least once about the identity of the plane? Nevertheless, as sense impressions, all three observations may have been real enough; so real that each might have been the occasion for a definite course of action under critical circumstances. If you discredit any of them it is not because it is less real as an observation, but because it is incompatible with the detailed lawfulness of nature, which does not allow a B29 to transform itself into a Thunderbolt spontaneously. In the microcosm there is no such lawfulness in detail, and the criterion for rejecting an individual observation as incompatible with others does not exist. Hence we repeat the question: What is real, the individual darts of the microcosmic fireflies, or whatever inheres in them to make the aggregate of darts conform with laws?

Perhaps both are real. Admitting this without qualification, however, is fatal to the proper understanding of large branches of modern physics, and it obscures whatever significance already formulated deductive science may have for other disciplines. In short, failure to make a distinction is an invitation to ignore a problem. Nor has the problem in fact been overlooked by theoretical physicists. Their voices, however, are discordant, often reflecting deep esthetic convictions, sometimes philosophic preconceptions, and occasionally disregard of basic philosophic issues. Three different responses are on record: the answer of the rationalist who favors the coherent aspects of our experience and regards *them* as primarily accented with reality; the plea of the positivist who recognizes the schism but limits his reality to observations, relegating to theory a secondary importance commensurate with its auxiliary function as handmaid to fact; finally, there is the council of the skeptic who, acknowledging the schism, sees in it an indication of error in our fundamental theories of nature. Representatives of the last-mentioned attitude are in the minority. Clearly, we cannot deal with their position because it will not grant the validity of modern science, which we are accepting (without prejudice, of course, to future improvements of its structure).

The other two answers bear responsible messages; partial, to be sure, but complementary. We wish to analyze them against the background of procedures actually used in science. This is best done, perhaps, by considering specific examples.

Every observation, every measurement, indeed every perception introduces errors. A measurement without error is an absurdity. Let the measurement (and measurement is after all only refined perception) be of a star's position in the sky, of the length of this table, of an automobile's speed, or of an electric current; its outcome is never to be believed exactly. This is apparent in the circumstance that the same number will not ordinarily result when the measurement is repeated, regardless of the care taken in performing it. Characteristically, only the careless experimenter and the ignorant observer believe raw nature to be unambiguous. To be sure, the different numbers found by the astronomer for the latitude (and longitude) of a star in successive observations lie within a reasonable interval, and this convinces him that he is somewhere near a "true" value of the quantity he seeks, that nature is not fooling him with hallucinations. Yet in a very fundamental sense he is witnessing a behavior not unlike the lawless emissions of the microscopic firefly; we thus see that the macrocosm is not wholly without its vagaries, but that it confines them sufficiently so that the observer with some credulity can feign their absence: he can blame himself for nature's equivocality and call the departures from a true value "errors."

But what is the *true* value? Let us look into scientific practice. If ten measurements of a physical quantity yield ten slightly different values, not one of them is necessarily regarded as true. Their *arithmetic mean* is singled out for this distinction, even though it may not have occurred among the measurements.[8] The justice of this choice is not provided by the ten measurements, nor by any finite number of observations; it comes from a belief in, or rather the postulation of, a certain uniformity of nature. Thus the very determination of a true value, and in the end the selection of whatever is believed to be true perceptory fact, involves a reference to law and order not immediately presented in the sensory structure of that fact. Here we find the clearest expression of the attitude which has led to the development of deductive science: it relies upon rational elements to straighten out erratic data. It does not ignore the latter's presence, nor does it accept them unrefined. It distills from them an essence and *calls* it *true*. But the nature of the essence is partly determined by the process of distillation.

Now it is this true value which science takes to be characteristic of its reality. If an electron be real, its charge and mass are assumed to be true values in the outlined sense, whether or not they have occurred in any measurement.

The real iceberg is not the exposed portion which the sailor sees; it is a largely unseen object compounded of the truths extracted from former observations and joined by postulations. The farther something is removed from immediate perception, as in the case of atomic entities or the facts of ancient history, the more dependent is its real character upon the lawfulness of the content in which it appears; it is real if that content is true. And here again, true does not necessarily mean "observed," anymore than it does in the process of measurement. It means "inferred from observations," and the nature of an inference, a word too often carelessly used, far transcends observations. What is physically real is rather close to the ideal.

Truth does not imply finality. The term is not to be taken in an extravagant metaphysical sense but signifies simply the best available. Truth may change. The scientist readily admits that he never knows a true value with infinite precision. This right to maneuver gives him the advantage he needs in rationalizing his observations, in making the best of an equivocal nature pitted against himself, an agent with rational propensities which force him to construct reality in accordance with rules.

A similar lesson can be drawn from many other instances. Real entities have often been inferred from lacunae in natural order before their existence could be certified by the standards of empirical science. Elements were predicted from gaps in the periodic table, planets because of irregular movements of known heavenly bodies; radio waves owed their conception as real constitutents of nature to the simplicity of the equations of electromagnetism which implied their existence. The most significant advances of modern physics were motivated or at least anticipated, by conjectures based upon the neatness of our universal laws; cases in point are the discoveries of the positive electron, the neutron and several types of meson. The whole case of the neutrino rests upon the empirically slender premise of valid conservation laws: this particle simply *has* to exist if the principles of energy, momemtum and spin are to be retained. Yet it has never been seen in the sense that other elementary particles have been observed. There is good evidence that it can never be seen, and even this consideration does not count against its admissibility as a component of real nature. How can such generosity be countenanced except by granting that the real draws its sustenance in large measure from a belief in the lawfulness of the cosmos?

The same sentiment finds its expression in the philosophic view which identifies the real with the elements of our experience that are causally connected in time and space. Current doctrines of materialism, relativity, much of causality are reared upon this rationalistic credo. Hence it is safe to say— and this is one of the theses supporting the remainder of the present discussion—that physical science would lose its hold on reality if an appeal to law and order were interdicted as a major claim.

We have seen, however, that lawfulness is at a premium in the perceptory realm of the microcosm. There, regularity is found primarily in aggregates, or, when assigned back to individual events, in the *probabilities* which inhere in these events. Laws govern these probabilities, they do not govern single occurrences. To be in harmony with the spirit of physical science we must therefore accept a conclusion unpalatable to many thinkers of the past, the conclusion that *probabilities are endowed with a measure of reality*. What this means in detail and what pitfalls must here be avoided will be analyzed in the following section.

Now we do not claim that it is fair to put these arguments in reverse, i.e.

to pronounce events resistant to lawful description, *unreal*. The perception of a single dart of light certainly happens, and the lawful multitude is made up of them. The fact of an hallucination is real and may be of great historical importance. Non-predictability hardly lessens the practical significance of certain unique occurrences both in the microcosm and the macrocosm. But the point is that they arrange themselves within the structure of our cognitive experience in a manner different from the simple order envisioned by traditional laws of nature.

To reconcile these disparities, and to accentuate their presence, I have advocated in *The Nature of Physical Reality* a distinction between *physical* and *historical* reality.[4] It seems to me that the need for this illuminating division is very great indeed, not only for the sake of terminology but also as a means for stating clearly what science can and what it cannot do. The data of immediate experience always belong to the latter sphere, the enduring entities of physics always to the former. But the spheres often overlap. In the macroscopic world they are nearly coincident, for the seen trajectory of a molar missile is also describable by the laws of mechanics (to take the simplest example). This accounts for the unimportance of our distinction in classical physics. In the microscopic realm, however, the two spheres break apart and science becomes obscure unless this break is noticed.

## 3. PROBABILITIES [5]

Single events, as we have seen by studying the world of atoms, have in general only probabilities for occurring. In special cases, particularly in cases involving objects of ordinary size, these probabilities take on values very close to 1 and thus reduce to certainties. But let us consider an elementary particle (our former rabbit) for which experience indicates that the probability of its being in a given place cannot be 1 for any finite interval of time. Physical laws predict its *mean* position in a number of observations; for any given position they indicate a probability only.

*Now suppose an observation is made and the particle is seen at a definite place. Must we not conclude that at the moment of observation the object was surely at the place where it was seen? And if we grant this much, are we not driven to admit certainty for its position at one instant? Expressing our suspicion more formally, we seem constrained to say: the act of observing the particle has caused the probability of its position, which was less than 1 prior to the observation, to jump suddenly to the value 1 during the act of observation.*

This consequence expresses an orthodox view, widely accepted and emphasized in a number of textbooks on quantum mechanics. If correct, it raises difficulties with some of the remarks made in the earlier parts of the present paper, and indeed with some of the basic axioms of quantum theory. For instance, it would be preposterous on this view to subject probabilities to physical laws—the discontinuous change during observation being precisely the feature that defies these laws. The advantage of using probabilities as regular, and as real, entities arose from their immunity to such erratic changes, and the result now tentatively reached once more injects lawlessness into them. The gain sought is thus destroyed.

This is not the only defect of the view under consideration. If it is valid, there is no sense in talking about probabilities at all. For it is then obvious that a single observation can determine a supposedly erratic property exactly. Hence the initial conditions in any problem of motion *are* ascertainable (remember the example of the falling particle!) and experiment will always prove our theory to be in error.[6] Instead of providing a valid theory, the notion at issue can only demonstrate that none is attainable.

It also implies that, whenever an observation is made, physical reality suddenly transforms itself into historical reality. There is no a priori reason why such a transformation should not occur; however, the mathematical features of this conversion are most perplexing and, we feel, objectionable. Suppose that we have given a charged particle an exact momentum by sending it through an accelerating chamber equipped with diaphragms. We then know precisely in what direction and with what speed it is going, but we cannot say at all where it is. The probability of position for this particle has the same small value everywhere. Now let a measurement of the particle's position be made, perhaps by noting the point at which a silver grain is blackened (after development) on a photographic plate. According to the thesis under criticism we must then say: the probability after the measurement is zero everywhere except near the position of the blackened grain.

Thus as physical reality, the probability extended through all of space, like a continuous medium devoid of matter. It formed in fact a field. At the instant of observation this field proceeds to vanish everywhere in space and concentrates itself as historical reality upon a point, where it takes on an infinite density. And all this because some human being chose to make an observation! Aside from the miraculous features of this theory, one wonders whether it was designed to deal with the physics of particles or the psychology of perception.

The confusion and the welter of contradictions accompanying the thesis stated in the earlier italicised paragraph of this section disappear when we avoid a simple error in our understanding of the term probability. To discover the error we consider a familiar example.

The physical condition of a regular die may be specified in many ways, some more complete than others. The die might be described as having a certain mass, shape and size. It might be said to have six equal black faces with white dots on them. Another perfectly good way is to assign the probabilities for the appearance of the numbers from 1 to 6 when the die is thrown at random. These are well known to have the value 1/6 each. We repeat, knowledge of these six equal numbers is just as significant with respect to the intrinsic nature of the physical object, the die, as the knowledge that it has six equal faces or that it has a certain mass and size. These parcels of information are not equivalent, to be sure; but each can serve as a basis for the prediction of certain physical occurrences. The psychological stigma of incomplete knowledge which we habitually attach to probabilities must be erased. Let us fix our attention on these probabilities.

Suppose the die is thrown and a five appears. According to the reasoning employed in the italicized paragraph above, the former probabilities (1/6, 1/6, $\cdots$ 1/6), for the numbers 1, 2 $\cdots$ 6 have now suddenly changed to (0, 0, 0, 0, 1, 0). Still it is obvious that the physical characteristics of the die have not been atlered by the incident of the throw, and the reader doubtless has an uneasy feeling that the meaning of the word probability has shifted during this discourse. Clearly, here is what happened. Initially we meant by probability the quality of the die by virtue of which the results of a long series of throws, say $n$ in number, will contain $n/6$ ones, $n/6$ twos, and so forth. In the end we meant by probability the degree of certainty of *our knowledge* with respect to the outcome of a particular throw. These two are not the same logically, and the confusion incurred earlier was occasioned by our mistaking them to be identical. Their difference is well understood in the theory of probability, where the distinction between the frequency interpretation and its counterparts (Laplacian, a priori definition and others) is fully recognized. If we stick consistently to the frequency interpretation, a single throw, or any number of throws, alter nothing so long as the physical character of the die remains unchanged, and the conclusion reached above is fallacious. On the certainty-of-knowledge interpretation the conclusion follows.

Now the certainty-of-knowledge interpretation in the present form is not tenable because it is hopelessly indefinite. To be acceptable it requires a statement of the *evidence* to which the knowledge relates. Probability becomes a function of two variables: the event considered and the evidence available. If the evidence is confined to the normal properties of the die, the probability for throwing a five is 1/6; if it includes knowledge that the thrower habitually cheats and has a way of getting sixes, the probability is less than 1/6; if the throw has already occurred and has yielded a five, and if its outcome is included in the evidence, all other evidence becomes irrelevant and the probability is clearly 1. Strictly speaking, all these probabilities are different physical entities and must not be confused. Hence, if the degree-of-knowledge interpretation is to be employed the evidence variable must be held constant during its use; in our example, evidence must be restricted to knowledge of the normal properties of the die, nothing smuggled in *en route*. But then the two interpretations agree, the probability does not change when the die is thrown and the dilemma is avoided. Henceforth, we shall employ the frequency definition of probability, as is customary in most scientific work. It will be called the objective interpretation.

Description of physical states in terms of probabilities need not have the trivial character exhibited by the ordinary die, whose properties may be specified by writing: $P_1 = P_2 = \cdots = P_6 = 1/6$. Here the probabilities $P_i$ for throwing the numbers $i$ are constant in time. The die can, however, be imagined to have an internal structure which changes in time. Suppose for definiteness that it is hollow and contains a sprocket with a small weight at its end. The other end of the sprocket is fixed to an axle extending parallel to an edge through the center of the cube, and this axle is driven by a small clockwork. We now have a strangely loaded die, but one in which the load revolves when the motor goes. The probabilities $P_i$ are functions of the time. If the mechanism is known these probabilities can be calculated; the reverse, however, is not true. Quantum mechanics asserts that there is no mechanism, that the probabilistic behavior is in the nature of the physical object and is ultimate.

Yielding nonetheless to our propensity for mechanical models we could invest our die with further appliances. Let us assume that the sprocket can be set in one definite position by some manipulation from the outside which does not interfere with its being thrown. The same operation also starts the clockwork. We shall speak of this operation as "activating the die." Normally the die is in a stationary state, its probabilities are constants in time. After activation they become functions of the time.

How can we determine the variable probabilities by measurement? If the die is known to be in a stationary state it may be thrown a sufficient number of times and the relative frequencies can be computed. Otherwise, more elaborate procedures are required. Merely repeating throws will not do when the die is activated, because different throws then catch it in different internal states, and a computation of relative frequencies is meaningless. Two correct methods for determining the probabilities in their time dependence are at hand. One is to activate the die, wait a time $t$ and throw it, repeat this procedure many times, always observing the interval $t$ before a throw. The results can then be used to calculate $P_i(t)$. Another method is applicable when many dice with the same internal structure are available. It consists in activating all of them at once, waiting a time $t$, and then throwing them all.

Such are the typical features of the quantum theory of measurements. The die corresponds to an atomic system, e.g., a hydrogen atom. This can be in a stationary state, as it will be for example when it has been left alone for a sufficient period of time. In that condition we are unable to say where the electron is relative to its proton, but we can perform measurements (illumination by short x-rays) each of which will locate the electron at some point $r$, not of course at the same point. From a sufficient number of such measurements we construct $P_r$.

The situation is different when the atom is "activated." There are many forms of activation, called "preparation of state" in quantum theory. Perhaps the simplest is exposure to a light wave, which causes the atom to be in a time-dependent state. The probabilities are determined as in the case of the activated die: either by repeating many times the act of switching on the light, waiting a period $t$ and measuring, using a single atom; or by exposing many atoms simultaneously to the light, waiting a time $t$ and then measuring all of them at once. The result is found to be a $P_r$ which is a function of $t$, the same for both methods. The latter method is the one which the physicist most frequently uses.

It is not our intention to discuss fully the quantum theory of measurements, which presents complexities for which the activated die provides no analogs. For instance, to make the story more realistic the die should often break after it falls, so that another one must be used when the throw is to be repeated. A fall can also change the setting of the sprocket and thus produce further difficulties which have here been ignored. The intent of our discussion was primarily to illustrate the sense in which probabilities can be *objective* physical quantities.

## 4. HISTORICAL REALITY

The lawful world of physical existences contains all external objects, from stars down to our own bodies and to electrons. It contains the states in terms of which the objects are described, the fields they generate, the time and space in which they are embedded. Notably, too, it contains the causally evolving *probabilities*, the states of the quantum theory which modern physics has taught us to regard as functionally ultimate. Man has, of course, experience of this world, not experience in the narrow sense of empiricism but in the wider one of constructing concepts and of creating rules of correspondence. This is the universe of strictly physical reality.

Over against it stands the multitude of immediacies over which, as we have seen, physical causation has lost its direct control. A sensation, a measurement, an observation, a will, an action, and certainly a psychological introspection, belong to this class. I do not argue that it is always possible to tell whether a given element of experience is to be assigned to this class with certainty—as in many other instances, here, also, experience shows no sharp boundaries. The failure of a sharp logical distinction is never serious when recognized. At any rate the items last enumerated, and others which partake similarly of the character of immediacy (spontaneity, coerciveness), in our experience will be said to compose *historical reality*.

Epistemologically, the two worlds are related by rules of correspondence (1). My sensation of an object is the historical component of the event in question, reification is the rule of correspondence, and the postulated external entity (desk, tree, lamp) is the physical component of the experience. The distinction made is admittedly useless and grotesque in the ordinary instances of regular cognition, where lawfulness extends into the historical realm and thus annexes it to the domain of physical reality. Classical physics was the formalization of this all-pervasive causal doctrine. Recent discoveries, described in the earlier portions of this paper, force the distinction upon us. The microcosm obviously fails to convey sense if its lawful and its historical phases are confused, and to what extent the distinction can be ignored in the large-scale world of action requires to be investigated. (See below.)

An electron, if it moves in accordance with classical laws of mechanics, describes a physical path, a trajectory. It has no history. The actual electron, subject to the laws of the quantum theory, appears unpredictably here or there; it has no path but a history. To be sure, it also has a determinate physical state associated with its "motion," a wave function in this case. This wave function, however, hovers abstractly over its history, guiding it by enforcing a sort of disposition without concretely assigning its fate. Historicity involves knowing, it implies observation; it arises through a union of a knower and his object of knowledge.

The inveterate mechanist tries to explain historicity as an aberration from path-like behavior through an appeal to "unpredictable interactions," as we have seen in section 3. So long as this mode of reasoning is forced to fall back on "unpredictable" matters it fails to achieve its mechanistic end and becomes what we have called an idle theory.

An equally idle, but no more idle theory, is one herewith proposed: the electron itself, as an individual, *decides* what value of a physical observable it will exhibit in the act of measurement. While nothing of scientific importance depends on whether we accept this dogma—and I believe the meaning of decision in the context is not very clear—physics can not refute it any more than it can invalidate the mechanist's assertion. What is true is this: to account for experience in its fullness even in the atomic realm, physics requires supplementation by aspects of actuality, incidence, decision—in short, historicity.

All that precedes seems to show that the accent on emergence is particularly strong in the atomic world, that the atom is the prime actor in the drama of history. When many atoms cooperate, when masses and distances become large, classical physics with its unhistorical lawfulness results. Our elephant hunt was far less exciting than our rabbit hunt. Quantum theory "reduces" to classical physics in the molar world.

While this is generally true, there are important exceptions to the rule. Nature permits arrangements in which the randomizing effects of large numbers does not occur, instruments through which the caprice of the microcosm can be projected into the world of ordinary experience. Every amplifier is such a device. A Geiger counter amplifies the passage of a single elementary particle by precipitating an avalanche of ions when a few are initially produced, and the current thus generated can be further increased by the use of electron tubes. As is well known, much of modern physical research employs such arrangements. Feedback mechanisms achieve the same purpose of amplification, and the biophysicist is apparently discovering their widespread occurrence in organized nature. There is indeed an increasing mass of evidence to indicate that the delicate balance of metabolism and self-maintenance called life depends for its establishment on precisely those mechanisms which are able to amplify a random atomic impulse into historically significant proportions.

An example often cited in this connection is the mutation of a gene. A single X-ray photon can bring this about. Suppose now that the frequency of this photon has been observed and is known. According to the uncertainty principle the position of the photon is then entirely random. Indulging in an anthropmorphism, we might say it is wholly "up to the photon" where it will interact with its surroundings. If it chooses to invade the neighborhood of a gene, the latter undergoes a change which may, under certain circumstances, spell the doom of an individual.

In the inorganic world similar processes of random triggering are easily found. A somewhat unrealistic but impressive one is the release of a uranium bomb by a single neutron. Place a sufficiently large block of $U^{235}$ in a space free of neutrons, then allow a single neutron at some distance from the block to move toward it. Quantum mechanically, its wave function is known, and along with it we know the probability that it will impinge upon the bomb and cause disaster. The historic fact whether it will or not is left open by the physics of the situation.

These instances demonstrate the ingression of atomic historicity into the generally lawful macrocosm. The merger between physical and historical reality of which we spoke, and which takes place to some degree in the molar world, is not complete. Our distinction carries its validity far beyond the atomic domain and must be reckoned with everywhere. Astronomy is about the only science which is relatively immune to it.

Yet all this, while true, seems to have very little bearing upon the problems of history as this subject is usually understood. History is the arena of *human action*, and action has not yet entered our discourse. What, then, is action? When analyzed, I believe it is seen to be a composite of arrangements and processes in accord with physical laws, *plus* here and there an element of voluntaristic *decision*. It is the decision which transfers action from the physical to the historical universe, or, rather, makes it an inhabitant of both realms.

We used the word decision, albeit in a loose and tentative manner, when the discussion was about electrons in the act of manifesting their presence to the perceiver. There we were unable to invest the term with accurate meaning. In human action there is a similar element of decision, similar in the sense that it also transcends physical lawfulness (as did the manifestation of the electron's position); but in human action we can study it by introspection. And it is recognized as conscious, active, voluntaristic intervention, a true component of historicity.

The problem of human freedom might seem to enter here. Its traditional features, however, are peculiarly foreign to the present line of thought. What is obviously, introspectively true, is the occurrence of voluntary decisions, the existence of what earlier philosophers called the human will. Physics has nothing to say about the freedom of that will, about its dependence on motivation, habit and so on. The fact of *conscious decision* is clear for all to see, and the latitude needed for action consequent upon decision is guaranteed by the probabilistic features of physical reality. This is as far as we need to go.

Indeed it is risky to go farther. The physicist who tries to prove freedom on the basis of quantum theory invariably meets a misfortune, whether he recognizes it or not. For if he makes $\psi$-functions govern human behavior he can prove *randomness* of action, but never freedom. He can show that man will act in accordance with ethical precepts a certain percentage of the time, that he will act immorally in another percentage of instances. On this theory man's behavior whould be a set of random doings, some good, some bad, without a clue indicating which are good and which are bad. This is not the kind of thing philosophers call freedom. What the argument needs to make its case is again the element of decision, of historicity.

Since we left the atomic world our discourse has increasingly taken on the character of speculation, at least to the extent that conscious decision, and action, have assumed a dominant role. We now return to safer matters and show in what manner the possibility of action, regardless of its psychological essence, arises from the modifications of physical lawfulness.

Consider again the neutron on its way to the uranium bomb. If classical physics were true, a single set of observations on the position and velocity of the neutron at a suitable time would decide whether an explosion will occur. It would leave room for action only to the very limited extent that, if the neutron is found headed for the bomb, we can try to intercept and deflect it before the impact. Usually this is impossible because the speed even of a thermal neutron is greater than ours, and we are forced to resign ourselves to fate instead of being agents in the course of events.

Notice, too, that the very decision to intercept the neutron must be taken as a physiologically determined consequence of physical reality, whose historicity can be but an illusion. Decision as such is indeed an impossibility on the basis of classical mechanics.

The new physics, with its concession of autonomy to historical reality, leaves greater room for action and avoids this difficulty. Even if a set of observations[7] reveals the neutron to be headed for the bomb we can still hope and pray for our survival, because what is now dynamically determined is a probability of collision, not a necessity. There is no cause for fatalism, but accentuated need for action. This is true even if, classically speaking, the neutron is seemingly winning the race for collision with the bomb. In this instance, as in all others, the physical situation leaves alternatives which action can seize in numerous ways. Decision fits neatly into the spaces presented by the semi-deterministic honeycomb of historical reality. But we do not pretend to have shown that it actually resides there, nor why. This is an illuminating conjecture made reasonable by the probabilistic nature of physical reality.

To illustrate this point minutely, let us analyze a decision. The suggestion is always strong that we should reduce the psychological act of deciding to physiological bases. In other words, when explaining the outcome of a so-called choice we advert to physiological processes taking place in the brain, to reverberating neuron currents, firing synapses and the like, and we assume, following traditional doctrine, that in the ultimate analysis molecular events determine the outcome of our choice. *This avenue is now blocked*, for clearly such a process of reduction will land us in the realm of atomic uncertainties. There is no unique road from the event of voluntary decision to the laws of physical reality. Again, we are forced to take decision as an irreducible act, a component of historical reality which stands aloof from physical lawfulness.

A summary of conclusions now reached might run as follows. Nineteenth Century natural science conceived of man as a detached spectator of an objective universe. It held the spectator—spectacle polarity to be genuine and fundamental. During the present century, discoveries concerning the nature of atoms rendered this doctrine untenable. The nucleus of a new philosophy of nature emerged with Heisenberg's principle of uncertainty, whose basic meaning implies a fusion of the knower with the known. The theory grew with amazing speed and sucess; it led to a mathematical formalism which, in order to attain its purpose, namely lawful description of experience, has to speak of probabilities rather than unique events. Individual events are no longer related in causal fashion, although in the domain of probabilities causality still reigns. Thus has been introduced another, more significant principle of division than the old spectator—spectacle bifurcation: the distinction between physical and historical reality has appeared.

Along with these devlopments, man has been transformed from a spectator to an active participant in the drama of becoming. Room has been made for decision and choice, which had no place in the older scheme of things. What was formerly fate has become history.

Before laying away his pen the philosopher, though disclaiming all right to speak as an historian, desires to raise a thought or two for historians to ponder. Physical science has yielded autonomy to the historical process. The detachment of the latter from physical lawfulness is the more pronounced the greater the abundance of momentous, unique events having a potency to release an avalanche of history. We live in an era charged with such potencies. The distinction which this paper set out to describe may have a sinister chance of becoming fatal before it is universally recognized. Hence its grave importance.

The other thought concerns the possibility of a science of history. Let no one deny this possibility on the grounds that history has too many variables to be susceptible of scientific treatment, or that it deals with human situations in which inquiry has a profound effect upon what is sought to be known. Natural science has solved both of these difficulties in its long course of evolution, the first by judiciously eliminating irrelevant variables and searching for significant ones, the second by injecting probabilities into the last constituents of its universe. It may be supposed, therefore, that history can take on the structure of science by adopting the pattern of physics. If this plausible thesis is accepted it follows that history, like physics, can predict only mass phenomena such as economic cycles, large-scale migrations, periods of cultural activity, and the like. But it will be unable to address itself to those peak events, to the emergence of powerful personalities, which have so decisive an influence on the course of human affairs.

A science of history that wishes to bring these critical phenomena within its sphere of prediction must not follow slavishly the pattern of physical science; such a science must strike out on its own along paths hitherto uncharted by existing disciplines.

## NOTES

[1] *The Nature of Physical Reality.*
[2] Support for this terminology may be found in the writings of many historians. "The problem of the historian is to tell what actually happened. The more clearly and completely he succeeds the more perfectly he has solved that problem. Straightforward description is the very first and essential requirement of his calling and the highest thing he can achieve" (v. Humboldt, W., 'Uber die Aufgabe des Geschichtsschreibers', *Werke*, ed. A. Leitzmann, IV, 1905, p. 35).
[3] A most illuminating review of the various conceptions of what constitutes historical knowledge may be found in E. Cassirer., *The Problem of Knowledge,* tr. W. H. Waglum and C. W. Hendel, New Haven: Yale University Press, 1950. Cassirer's analysis contains many suggestions to which the present article, in part, owes its origin.
[4] As show *in extenso* in my *The Nature of Physical Reality.*
[5] This section deals with slightly more technical matters arising from an assignment of physical reality to probabilities. They must be faced if the view here presented is to be acceptable. Readers interested only in the main argument may well omit this part.
[6] Textbooks obviate this conclusion by admitting that *some* initial properties are indeed ascertainable exactly, but not the full complement needed for a determination of the motion. This is prevented by the uncertainty principle. In the problem of the falling body, for instance, the initial position can be measured with accuracy, but then the momentum can be determined with no precision at all, according to the textbook version. But there is a misunderstanding here. The uncertainty principle says this: If the position of the particle at a given time is known with certainty, i.e., if its condition is such that a series of position measurements may be presumed to give nearly the same answer, then the error in the momentum measurement is infinite, i.e., a series of momentum measurements would yield extremely erratic answers. The principle says nothing about happenings in single observations; in particular it does not assert the impossibility of simultaneous position and momentum measurements. Nothing prevents such measurements from succeeding in the sense of yielding numbers, which is what measurements are meant to do. The uncertainty principle warns that these numbers have no significance in determining the dynamic state of the particle. The view we are about to present leads to exactly this conclusion and avoids the annoying self-contradictory issues sometimes encountered.

[7] It is often asserted that a measurement of its position and a simultaneous measurement of its velocity, which are necessary to certify that the neutron is headed for the bomb, cannot be made. This is incorrect. Such measurements, when made, have no predictive value, which is, of course, the point here at issue.

CHAPTER 16

# THE NEW VIEW OF MAN IN HIS PHYSICAL ENVIRONMENT

> Man is not merely made for science, but science is made for man. It expresses his deepest intellectual needs, as well as his careful observations. It is an effort to bring internal meanings into harmony with external vertifications.
>
> Josiah Royce, Introduction to Poincaré's *Foundations of Science.*

## 1. WAYS OF STUDYING MAN

When dealing with a subject as large, engaging, and obscure as man, an author is obliged to say forthwith how he intends to approach his problem. For there are many roads leading to partial knowledge of the nature of man; their courses need to be sketched and their goals must be correlated.

A person who wants an estimate of the conditon of an automobile before buying it ordinarily does two things. One, he examines its appearance, its parts, and inspects its mechanism; two, he gives it a road test. Man, too, can be studied in these two ways: first by painstaking inspection of his external and internal make-up and then by watching him in action. Anatomy and psychology perform the task of inspection; they are in a limited sense the "sciences of man" and provide the kind of specific, relevant, but partial knowledge which the mechanic gains when he opens the hood. This approach to the nature of man is not the one I am able or expected to conduct.

It is a road test that will be made in this study. The ride could go through many different countrysides, exposing many of the extensive achievements of man, from art and religion down to technology. However, not knowing my way in most of these areas, I am forced to take you through the field of physical science, endeavoring to appraise *its* bearing on the nature of man. True, this road will lead only to partial knowledge; still it seems proper to couple this protestation of modesty with what may seem an unwarranted challenge; I affirm that the results of modern physical science are of extraordinary importance to an understanding of the nature of man; and furthermore, that this peculiar relevance, hitherto largely ignored, is greatly in need of exposition and emphasis today.

Science as a human activity bespeaks the capabilities of man; as an accomplishment it reveals the nature of its agent. But far more important than this causative relation is the reaction of scientific achievement on man himself, is the manner in which accepted scientific doctrines fashion and modify the cultural essence of man. For man is primarily what he conceives himself to be, and this conception, this self-appraisal is formed vis-a-vis and in relation to the changing picture of the universe around him and to the philosophy which that picture suggests.[1]

To trace in some detail the reaction of science upon man himself I shall consider briefly some of the effects of every major scientific discovery. The obvious effect is technological, it is the visible and impressive movement from discovery through commercial development, production of new goods and devices, advertising, and sale, toward the establishment of greater comforts of life. This causal chain, which in the end enhances or at any rate modifies our so-called standard of living, our external circumstances, will here for brevity be called the "obvious" movement.

But every truly great scientific discovery launches also another trend, much less apparent and more subtle in its progression from phase to phase through human culture. The discovery, acting as a *fact* in initiating the obvious movement, becomes the lever of an *idea* in the other. It clamors to be understood, and the scientist must provide some sort of theory in terms of which the discovery takes on significance and organizing power. The theory contains novel features, features contradicting what was previously regarded as true; and by virtue of this apostasy the discovery induces a rearrangement of thought in adjacent fields. Sooner or later the internal structural consistency of the science that was disturbed by discovery is restored, but in the process some cherished beliefs, some aspects of common sense have had to be surrendered. Thus relativity theory has repudiated the notion of universal simultaneity and, to some extent, the simple intuitions of geometry; quantum mechanics denies the continuity of motion; and a good deal of time will probably elapse before men cease to feel that such theoretical consequences of discovery violate common sense.

Results as challenging as these cannot fail to have a profound effect on philosophic speculations. Indeed philosophy, in time, must and does take account of the ideological consequences of scientific knowledge, first by changing its cosmological beliefs and perhaps its theory of knowledge. Changes in one branch will entail changes elsewhere, and even though the sequence of alterations is uncontrolled and haphazard (chiefly because we are less conscious of them than we are of the technological sequence), they nevertheless tend to embrace the entire structure of philosophic thought before their course is run. Thus this movement ends in new views on the nature of the universe, the relation of man to the universe, and the relation of man to man. Ethics, sociology, politics are ultimately subject to infestation by the germ that is born when a discovery in pure science is made. This movement, which terminates in a change of the cultural milieu of man, will be called the "obscure" movement. There is a paradox in the fact that Marxism professes the supremacy of the obvious movement while disparaging the other, whereas the Western democracies reverse the declaration, pay lip service to ideas, but practice what Marxism professes.

It is most reasonable to postulate, and I believe history shows, that human society enjoys maximum stability when the two movements are in balance. This was often true, both in Western and in Eastern cultures, during the ten or twenty centuries that preceded ours. And the balance resulted from the lumbering slowness of both movements. Gun powder was discovered in the 12th century, used in warfare two hundred years later. Galileo and Newton found the laws of mechanics in the 17th century, the machine age arose in the 18th and 19th Oersted discovered the magnetic field of electric currents in 1820, electric motors became industrially important nearly one hundred years later. Thus the technological gestation period, the time required for the obvious movement to be completed, was of the order of a century, and there is clear evidence already in the examples cited that it is progressively decreasing.

Considering now the obscure movement, one finds no signs of speed-up. Kant, the philosopher who more than any other developed the philosophic framework for Newtonian physics, published approximately one hundred years after Newton. Modern empiricism, which is clearly a philosophic version of the great discoveries in thermodynamics and statistical mechanics of a century ago, has reached its zenith in our time. And the scientific revolutions that occurred in the atomic field at the beginning of the century, the unprecedented, galvanizing pronouncements of the quantum theory, have not found satisfactory philosophic consolidation to this very day—but look at the swiftness of technology; the discovery of fission resulted in the development of the most perilous of weapons in less than ten years.

Not only in the West, but everywhere do we see evidence of such unbalance between the two movements that start from scientific discovery. The obvious one has been vastly accelerated in all parts of the globe, the obscure has not found its goal. It flounders and gropes without rational guidance on both sides of the iron curtain, and the ideological cleavage between East and West is, in part at least, symptomatic of the failure of the hidden movement to have completed its run. For it stands to reason that a common science will engender a measure of agreement in philosophic outlook across all artificial curtains when equilibrium between the two movements is finally established, when our intellectual atmosphere is congenial with our applied science.

Hence arises the suggestion, vague perhaps and insecure at this point, that two problems should be of very serious concern to the thoughtful student: how to speed up the obscure cultural movement so as to bring it into step with the other, how to make the obvious movement less obvious, the obscure one more evident. Both ends can be achieved by a shift in emphasis from technical science to the philosophical problems surrounding and pervading science, by consciously taking stock of the need for philosophic digestion of discovery and of our patent failure to achieve it. Fortunately, this need is being recognized by universities and research foundations, though not very clearly as yet by government agencies and industry. But it is not the primary purpose of this paper to note these generalities and to urge appropriate consideration; in what follows, I desire chiefly to show how in fact the obscure movement has proceeded in the past and, if possible, to project a few of the features of the coming philosophy which is in harmony with present physical science. And through all this, I shall outline the views of man and of human destiny implied by outmoded and finally by modern physics.

## 2. OLD VIEWS OF MAN

### 1. *Mechanistic materialism*

The vastness of such a task forces one to make selections, and there is no guarantee that these will be regarded by everyone as appropriate. The claim I make is that the three old doctrines here presented for consideration—materialism, logical empiricism, and existentialism—are indeed responses to specificfic groups of scientific discoveries succeeding each other in time, that they do spell out different, incompatible concepts of man, and that—because of the slowness of the obscure movement—they continue to dominate the modern cultural scene. Even in the limited context of these selections there will be no room for scholarly thoroughness, careful historical analysis, or complete documentation. Nor will the presentation be entirely objective, for it is inevitable that an author's convictions and his dedication to the theme he expresses should incline him to exaggeration. I include this caveat even though I resolve to be fair in my appaisals.

Materialism is a doctrine of many forms, the most plausible and influential of which is what will here be termed *mechanistic materialism*. It is completely characterized by its affirmation of two theses: a) Matter obeys the laws of classical mechanics; b) To be, is to be material, i.e., nothing exists that is not material. The former commits materialism to the hypothesis of universal continuity. All changes occurring in matter must be continuous changes; objects change their sizes and shapes by infinitesmal gradations; they move along continuous paths in three-dimensional space. This character of motion may be said to mean that the position $x$ of any object is a continuous function of the time, $x = f(t)$. To this point we shall return later. If it is felt that the continuity hypothesis is self-evident and logically necessary, let it be recalled that St. Thomas contradicts this allegation by insisting that continuity, far from being necessary, may or may not apply to the motion of his angels. "Motus angeli," he says, "potest esse continuus et discontinuus sicut vult ... Et sic angelus in uno instante potest esse in uno loco, et in alio instante in alio loco, nullo tempore intermedio existente."

Thesis (a), though sometimes questioned in the remoter past (cf. for instance Zeno's paradoxes), became part and parcel of physical science through the successes of Galileo and Newton. Forces are continuous functions of space and time; they are proportional to accelerations, and by these premises there is defined a differential equation with solutions corresponding to continuous trajectories in space and time. Were it not for the success of this analysis is might be questionable whether science would have embraced the continuity of motion so completely; for experience, even when very refined, does not always endorse it. But scientific success and plausibility established the continuity thesis and made it an integral part of the world view which followed Newton's science.

Thesis (b), the identification of existence with materiality, is likewise implied in Newton's work. It became a philosophic conviction at a later time, perhaps as the result of two further developments which enhanced its power and its certainty. Early in the 19th century there arose a sweeping scientific conviction: beginning as a tentative belief in the impossibility of perpetual motion, it developed into the principle of conservation of energy, the certainty that energy can change its form but is basically indestructible. Yet this knowledge was nothing more than an inductive generalization of a multitude of facts, for no one had derived the conservation law from first principles concerning the nature of the universe.

Helmholtz succeeded in doing this. In 1847 he published a famous paper wherein he showed energy conservation to be a consequence of two simple assumptions. One was that nature consists of mass points, i.e., small particles of matter; the other that the force between every pair of mass points is a central force, that is, a force acting along the line joining the two particles. This proof was a tremendous scientific achievement, acclaimed everywhere, and it induced many scientists, because of its dazzling brilliance, to believe the premises of Helmholtz's syllogism along with its conclusion. If conservation can be logically established on the basis of these two simple postulates, they *must* be true. Every logician knows full well, of course, that this reasoning involves the fallacy of "affirming the consequent," that the same conclusion may be deducible from different hypotheses—but few people bothered about the logic of the situation and few withstood the conviction that nature did, in fact, consist of nothing but mass particles known as matter.

A decade after Helmholtz this view was reinforced by a doctrine of quite a different sort. Darwin published his *Origin of Species,* a work which, on the face of it, seems unrelated to Helmholtz's contribution. But again its world-shaking significance led to an acceptance not only of its essential intent but also of the frills and flavors that accompanied its presentation. The emphasis on the survival of the fittest, the tooth-and-claw behavior of all creatures, seemed to harmonize in philosophic pattern with the physical materialism of the time. And toward the end of the century, fed by other tributaries much like these two, there rose the powerful stream of mechanistic materialism. Man believed that there was indeed nothing in the world that was not material, nothing that failed to obey the laws of mechanics.

## 2. *Empiricism, logical positivism*

The strict, precise, and unconditional character of the laws of Newtonian dynamics or celestial mechanics with its suggestion that "reason applied to nature," the consequent rationalism of Kant and others, lost support in some quarters when physics turned its attention to the subjects of heat and thermodynamics. Here was a field in which dynamic regularity was not the norm; its laws resulted, strictly speaking, as rare anomalies from the chaotic interplay of large numbers of molecules. The chief era of these discoveries began in the late 18th and extended through the first half of the 19th century, and the names associated with them are Lavoisier, Black, Count Rumford, Davy, Mayer, Joule, Carnot, and Clausius.

Thermodynamics is the most empirical of the physical sciences. Its theorems are relations between an excessive number of experimental variables; it thrives in a situation spurned by other branches of physics, namely, one in which more variables are used than are actually needed. Because its measured quantities are not logically independent, thermodynamic formulas exhibit that well known disfigurement by subscripts added to partial derivatives, an outward indication of its earthy stature, of its factbound significance. There are no neat and elegant second-order differential equations with solutions representing the unique history of a thermodynamic system; the connection between formula and measurement is always emphasized. Nor are the basic laws very simple. The most embracing "law" (in the sense of mechanics, i.e., an equation connecting variables of state) is the equation of state, it is different for every substance and has extremely complicated forms for all real bodies. The contrast with Newton's law of universal gravitation is remarkable and is philosophically suggestive.

Furthermore, even the greatest generalizations encountered in this branch of science, the so-called laws of thermodynamics (which everywhere else would be named principles), entered the scene as inductive inferences from a large mass of experiments and not as deductive consequences of some simple and pervasive conjecture. Much ingenuity has been lavished on the question whether they are as true as the laws of mechanics, or whether they permit exceptions, and even now textbooks sometimes say that water can freeze on the stove if you wait long enough.

The reason for this wary and circumspect approach to the validity of thermodynamics lies partly in its history and its formal structure, but primarily in the reformulation which its discoveries induced within the science of mechanics itself. For in the process of readjustment enforced by the discrepant new knowledge regarding heat, as described earlier, the ideas of mechanics were enlarged to include the subject of statistical mechanics. This contains all the theorems of Newtonian mechanics plus special postulates concerning the *probabilities* of molecular motions. Only with the use of probabilities can theory account for observed thermodynamic behavior. And the need for probabilities, a novel feature in the explanatory scheme of physics, puts the imprint of looseness and ambiguity, which only actual observation can resolve, upon the theories of heat.

The philosophic implication of all this is perfectly clear. Even if theory says water will boil, one must not trust that prediction without qualification. For experience *may* show that it will freeze. It is all a matter of probabilities. Laws are approximate, and the childish amazement expressed by those who hold that "reason applies to nature" marvels at a fairy tale. Nature fundamentally defies reason; she goes her own erratic way, producing regularity through the law of large numbers, by sheer exhaustion of alternatives for aberration. Law, strictly, is an illusion. And in the midst of this universal play of chance, man is a creature endeavoring successfully to make the best of things, betting on the basis of probabilities.

The preceding account is not an accurate description of the present views of men (Carnap, Feigl, Frank, and others) who call themselves positivists or empiricists. What I intended to sketch is empiricism as it first arose, and as it ought to be if it had remained unmingled with other considerations. Its emergence was inevitable, for it is the terminus of that obscure movement which started from the science of heat and thermodynamics.

### 3. EXISTENTIALISM

There are scholars who deny the claim of existentialism to be a philosophy; few indeed will recognize it as a world view developed in response to science. It is in the first place a working attitude of artists, a pervasive mood which passionately seeks to justify the sordid as well as the magnificent contingencies of existence; it includes Nietzsche's joy over the death of God, and Tillich's quietly pious "courage to be." With Kierkegaard it is the resolution, brought on by irritation at the static concepts of traditional philosophy, to progress from the habit of understanding backward to one of living forward. In Malraux the accent is on the absurdity of life and on the need to endow it with significance through adventure; the heroic gambler is the object of justifiable admiration. The works of Sartre, a "widower of God" according to his own testimony, express and portray the nausea of existence.[2] But all these men, whether they admit it or not, stage a revolt against science. Their attitude is a response to science, albeit a negative one; their philosophy is, so to speak, a result of the obscure movement jumping its track.

Before I attempt to demonstrate this seemingly unsympathetic assertion, I should say in all fairness that it is an over-statement and not wholly true. For there is an element in the attitude of existentialism that reflects a deep insight of very recent science; I refer to the fact that theoretical physics, by its appeal to probability reasoning, has relinquished its hold upon individual events, on single observations. In a sense these are left untreated by most recent doctrine, and so a special appeal for attention to what is existentially given and scientifically fatherless is just. It may be held, however, that this is an *ex-post-facto* conjecture, an artificial regularization of a movement which has fundamentally broken its bond with science.

For existentialism, when it speaks philosophically, declares war upon "essences," saying that existence comes before essence. What it means is that the unregulated contingencies, the bare and given facts and immediacies of our experience, take precedence over the regularities and constancies constructed or found by reason. The essences of existentialism, when freed of poetic disdain, are the constant entities and the laws of science. These are the citadel against which the onslaught of that movement is directed.

The rebellion broke forth openly and under philosophic generalship in the writings of Heidegger. He admits that science is a noteworthy attack upon truth or *Sein* or Being, but one not likely to achieve final and full success.

For Being is, in his representation, an existential something that lives in the solitude of human experience and will not be caught alive in scientific traps. It is like rare game which man can espy, stalk, and observe quietly, but which will flee the noisy scene of science, and when the scientist does finally overpower and capture it, he comes to hold the corpse of truth and not living truth. Such a view, when stated more adroity and without the use of metaphor, tends to assume a measure of persuasive plausibility sufficient to make it the stock in trade of many humanists and artists; it is the central creed of the hard-dying attitude which insists on a basic cleavage between science and the humanities.

The rebellion against science which has taken the name existentialism is aided and abetted by two significant facts. First, modern science has become increasingly and at times forbiddingly abstract, and the artist is repelled by it because in plain truth he cannot understand it. Hence, in curious reversal of that medieval attitude which led the scholar to scorn the craftsman because he spoke too vulgarly and in his native vernacular, the existentialist now spurns the scientist because he uses an esoteric tongue called mathematics which, in the eyes of some, poisons Being before it is apprehended. The second fact giving strength to existentialism is the chaos of modern history which belies reason, order, and essence. Quiet desperation, probing the depths of human tragedy, contemplating death and coming up with the resolve to *be*, to *be* in the face of absurdity—those are understandable attitudes in the modern world of politics, natural to those who have severed their relation to science and to the order which science reveals.

Existentialism concerns itself with the nature of man more directly than any other modern philosophy. Through its commitment to the priority of fact, the doctrine is forced to portray man fundamentally as a creature cast out into a universe devoid of reason. Man engulfed by the abyss of being, finding himself alone, capable of anxiety and sure of death, such is the primary frame of human existence. Relief is sought in diverse ways, by redemption in Kierkegaard and Marcel, by "gambling one's life on a stake higher than oneself" or by "transforming as wide an experience as possible into consciousness" in Malraux. Some, especially the novelists, seek to ameliorate man's prime state of irrational abjection by ethical and political manifestations; they activate, to quote Henri Peyre's excellent summary, "the desire not to remain unmoved by the anguish of other men suffering from the threat of war, by social inequality, or by economic injustice. Several of them have taken sides, usually with the extreme left, in political issues: but they have raised such issues to the height of metaphysical speculation and envisaged evil as a cosmic phenomenon, though one which lies within the power of man to redress in part."

Freedom is a fact and a cornerstone of existential experience; being isolated, unconditioned, and blind, it is at once a burden and a source of anguish. Still it compensates in a significant way for the forlornness of human existence and gives an active concern for the future to a creature whose past is meaningless. Man must make the best of freedom, his singular and most cherished gift. But one gets the impression from reading the literature that its use, its unstinted and enthusiastic use, is also its complete justification. One finds neither a deep concern for the restraint of freedom nor a search for an explanation of this unusual phenomenon in a world that largely lacks it. Existentialism takes freedom as a fact, not as a paradox.

### 3. THE NEW VIEW OF MAN

1. *The decay of materialism and the rise of nonmechanistic science*

In our century science has grown in new directions and has changed to a position that denies the claims of mechanistic materialism and of the extreme forms of logical positivism. And by opening up promising vistas to the gaze of man, it has made existentialism pointless. To the physicist these developments are well known, though even he rarely sees them in their rich philosophic context; I am unable to sketch them here in a manner carrying persuasion for those unacquainted with modern physics, and refer therefore to some of the numerous expositions in the literature. In summary, the salient facts are these.

We have learned to prove the law of conservation of energy, and most other laws, on the basis of axioms far more general than Helmholtz's simple premises, and historical occasion for our belief in these materialistic premises is gone. The knowing reader will recall the proofs of conservation for an electromagnetic field which is based on Maxwell's equation, and the theories involving the matter-energy tensor.

Matter itself was supposed to exist in two forms, continuous and discrete. The quintessence of continuous matter was the luminiferous ether; the atom, first throught of as a small pellet of stuff, symbolized the other. Let me review briefly the unhappy fate of the ether which has been so vividly described by Whittaker[3] in one of the classics of modern science.

The ether was supposedly the tenuous material medium filling space and carrying the light waves as air and water carry sound. Men searched for it through various of its manifestations. If it were material, it ought to have a density, but it was too light to be detected. If it conveyed a wave, it ought to be an elastic medium—indeed from the high value of the speed of light one might caclulate that its rigidity should be enormous—but no evidence of such phantastic properties was found. The ether presented a further anomaly inasmuch as the waves it carried were transverse and not longitudinal. To account for this it had to be equipped with molecules of most remarkable structure, molecules specially designed for the purpose and not encountered anywhere else. In the latter quarter of the 19th century, the great era of materialism—or as I should preferably put it, the era of Rube Goldberg devices—this challenge was met with equanimity and poise, and a successful model was proposed. In Whittaker's words, here is the model of an ether molecule.

Suppose . . . that a structure is formed of spheres, each sphere being in the centre of the tetrahedron formed by its four nearest neighbours. Let each sphere be joined to these four neighbours by rigid bars, which have spherical caps at their ends so as to slide freely on the spheres. Such a structure would, for small deformations, behave like an incompressible perfect fluid. Now attach to each bar a pair of gyroscopically-mounted flywheels, rotating with equal and opposite angular velocities, and having their axes in the line of the bar: a bar thus equipped will require a couple to hold it at rest in any position inclined to its original position, and the structure as a whole will possess that kind of quasi-elasticity which was first imagined by MacCullagh.[4]

Materialism succeeds, but at a price that strains imaginative resources and makes us wonder as to its adequacy.

During the same period attention also turned to the state of motion of the ether. Was it entrained by moving celestial objects, so that each of them carried with it its own private atmosphere of ether? Or was it stagnant, allowing all bodies to move through it freely? Astronomers denied the first alternative; the physicists Michelson and Morley made the second untenable by performing in 1887 their ingenious experiment. The ether became a mere word, the noun for the verb "to wiggle"; it was the grin of Alice's cat after the cat had vanished. Then Einstein taught us, through his special theory of relativity, that even the word is unnecessary, and modern physics is very comfortable, indeed far better off, without a material ether. Theory has become simple again after MacCullagh's nightmare, and the idea of continuous matter has passed beyond the scientific horizon.

Discrete matter, the atom, has of course remained with us. But it is no longer that pellet of stuff, hard and impenetrable; it has transformed itself into something highly abstract and difficult to picture, into a set of mathematical singularities moving in a space pervaded by electric and magnetic fields. The stuff has been pretty much knocked out of it, and the atom has become a rather empty structure.

The thesis of mechanism has had similar reverses. The motions of atomic particles cannot be pictured with that intuitable directness which the idea of continuity conveys. Elements of the Thomistic angels have crept into the situation. An electron in its motion about a nucleus no longer has a well-defined path, it has "probabilities of being observed." And it is these probabilities, not its actual positions, that are being calculated and predicted by physical theory. Mechanistic description, as we have seen, presented the position $x$ as a function of the time, $x = x(t)$; quantum theory, though still called quantum mechanics, only involves the probability $P(x)$. It confers importance, not on a single observation (such as noting that the electron is at $x$) but on an aggregate of observations (how many times out of a thousand was the electron found at $x$?) endowed with probabilities. Through these changes the physcist has above all learned two lessons: (a) Reliance on mechanical models suggested by common sense is sometimes dangerous and misleading; (b) *Formal* principles of mathematics and of logic must often replace pictorial intuition.

In might seem as if the very tide which swept away the foundations of mechanistic materialism carried in the substance on which logical positivism is built. This, however, is true only in a very limited sense. For with the new emphasis on probabilities, which is so easily but wrongly interpreted as a loosening up of nature's laws, there came increasing *formalization* of thought rather than a surrender of the precision of science. Quantum mechanics is not like thermody-namics; its probabilities are not the result of human defects, not the kind of lesser evil that makes us try for profit by betting—they are facts of nature cast in a new role, inexorable in their own new meaning. The mere suggestion that we can never know exactly, that empirical knowledge is necessarily fragmentary, is quite abortive, for quantum theory still has that assertive splendor which tends to impose reason upon nature, and it lacks the disposition of *nil admirari* which is typical of the true empiricist.

## 2. The coming philosophy

That hidden movement launched by the discoveries I have outlined is, unfortunately, still somewhere in its middle course, for it has lacked the drive and the resources which propel the other movement at its unprecedented rate. One cannot emphasize this point too strongly before teachers and scholars, whose responsibility it is to integrate our learning and to restore the sanity of our culture: It is not enough for us to fear or to admire science, the greatest challenge of our day is to *humanize* science, that is, to speed the evolution that will set it into an organic relation with philosophy, with culture and with life. Until this is achieved a sketch like the present remains a fallible prophecy. Nevertheless it seems that some features of the coming philosophy can be discerned today without great risk of error, and these I will now undertake to sketch.

a) We are witnessing at present and shall continue to see an enormous widening of the horizon of knowledge and of scientific tolerance. That man will learn more facts is trite and hardly worth recording. But there are two ways of expanding knowledge; one is the accretion of data in the manner of the physicist at the end of the last century, of the busy-body who understood his subject and looked for the next decimal place in the numerical constants of nature. The other is the openminded reception of novel truths will full cognizance of their heresy to past convictions. It is the latter disposition that characterizes the progress of recent science. The smug and complacent attitude of the 19th-century physicist who, to choose a trivial example, denied the existence of a human soul because he could not weigh it or locate its whereabouts, is gone forever. The modern scientist may still deny the soul, but on better grounds. Whitehead's impressive reference to the fallacy of simple location, the reminder that existence need not be tied to location, was startling in its day but is almost a commonplace to modern physics. The rise of interest in objective and dispassionate investigations of phenomena once called occult is in harmony with the new spirit of the time.

There is another way to put the story: science has lost its dogmatism. Present developments show it to be an ever-unfolding human enterprise, a self-corrective dynamism continually engaged in modifying and improving its theoretical structure, a Heraclitic flow of facts and ideas. Science disavows static and final truths, giving in every age a different answer to man's eternal questions, approaching certainty as an ideal limit. It knows that its principles as well as its facts are changing, and it has renounced the error of believing in the possibility of explaining all human experiences, past and future, in terms of those principles and laws which are *now* called science. Such tolerance, and such modesty, will surely be traits of the coming philosophy and the rising man.

b) Rarely in the past has physical science addressed itself to philosophy's agonizing question: What is reality? But it does so now. And the answer, still tentative and timidly proposed, may well be its greatest and profoundest benediction. For it integrates traditional views and allows them, purged and widened, to exist in harmony.

The theory of knowledge and of man has from the very beginning suffered from a dualism which philosophers have not been able to eradicate. It appeared in the antithesis between Thales and Heraclitus, Anselm and

Roscellinus, Kant and Berkeley, and, to name a current manifestation, between Einstein and Bohr.[5] When stripped of all external complexity the problem is simple indeed. All these men are asking what is real. One group answers: we take the real to be the invariant, rational aspect of experience. The other says, in the words of Berkeley, "esse est percipi." One group singles out as credible, interesting, and valid the elements of knowledge that are permanent, lawful, and therefore rational; the other seizes upon immediate experiences, perceptions, actions, scientific observations, in short the "positive" phases of knowledge. Hence the present conflict between rationalism and positivism.

Many have said that this conflict is an idle and a verbal one, that the choice of position is a private matter of taste. This was true until the advent of quantum mechanics, for the so-called crisis in physics is nothing but the termination of an unphilosophic period in science, a roll call on the meaning of reality. The theoretical physicst has always been at heart a rationalist, for his laws, his constructs, his equations deal clearly and exclusively with permanent and lawful gleanings from a larger experience, while the experimenter often, though not always, was a positivist because he endowed direct observation, manipulatory operations, and the like with special and ultimate significance. But in classical physics this did not matter, because theory predicted in every detail what was observed; every P-plane fact had a unique construct as its counterpart[6] and in the presence of this universal correlation a choice between rationalism and positivism was unimportant. The positivist had the P-plane and could infer from it the theoretical notions if he so desired; he made a defensible point when he argued that theories were *renderings* of facts made in answer to our desire for economy of thought. The rationalist had his constructs, his essences, his universals and could in principle deduce the concrete world; he lost nothing except the respect of his adversaries when he slighted sensory experience and dubbed it a mere manifestation of rational essences. Each picture was complete, each mirrored the other, and it was hard to tell which was the original and which the reflection. The mirror, in a sense, was man, and his function was trivial.

Now, as we have seen, the immediate facts of experience are no longer predictable in detail from theory, reason has lost its absolute hold on fact, essences no longer bind existences completely. It is as though in a fundamental way rationalism had been split apart from positivism. To use a phrase of Bohr's, they have become complementary to each other. This complementarity, however, is not an idle reduplication of one realm by another, as some still hold; on the contrary, it forces upon us the need for recognizing a distinction between (at least) two kinds of reals: that which enters into our reasoning about nature, the rational entities, and the laws of physics—atoms, electrons, fields, and probabilities; and that which assails us coercively through our senses or results from our active participation in the world. I can do no better than to call them *physical* and *historical* reality. They are no longer completely isomorphic, even though they are coincident in a very large domain of experience. They do break apart, indeed they are most widely separated in those individual human acts that have historic significance, acts which by their uniqueness and their singleness are not parts of a collective whole to which the probabilities of the quantum theory must be referred.

Historical reality, thus conceived, is not far from the existentialist's world. And modern science documents to some extent the validity of his views. But it shows them to be part of a more embracive picture, for the world of physical reality is still there, its laws regulate events in the aggregate even if they have lost the stringency to govern every detail. This conclusion is reached by an analysis of modern science, and it seems to restore to life that richness which rationalism and existentialism individually tend to destroy.

c) As we have seen, the mechanistic philosophy of the 19th century presented human freedom either as a paradox or an illusion. Kant, who preceded this movement and who found a more idealistic interpretation of Newton's mechanics, suggested a more delicate treatment of the problem of freedom but had to leave it in the limbo of his antinomies. Quantum theory throws an entirely new light on freedom.

It says that the law of cause and effect still holds with respect to the constructs and essences that constitute physical reality, to which since Kant it had always been applied. The basic differential equations describing the changes of physical systems are as precise as ever, and the entities they govern—atoms, nuclei, and electrons—continue to have properties like charge and mass which are measurable with indefinite precision. In this realm causality prevails.

But among the qualities governed by those causal laws are now to be found probabilities, and these in a sense make the causal scheme rather academic. For they refer to what is observed or, more generally, what is directly experienced. Probabilities of events can be predicted with accuracy while individual events elude all forecasts. The situation in atomic physics is much the same as in the actuarial field, where the number of deaths in a large group can be predicted despite complete ignorance as to who will die. There is lawfulness in the realm of essences, but much chance in the field of observable facts. Let us see what this means with respect to freedom.

No one knows how to apply quantum theory to a human being or, for that matter, to a complex molecule. Hence we have to pitch our arguments on a simpler plane. Assume an electron to be moving toward a target, for example a television screen, and suppose all knowledge concerning its state of motion to be available. This knowledge must and does conform to strict laws of nature, to the essences of the rationalist. Despite this, however, it does not specify the point of impact; it predicts only the mean position at which a million electrons, when similarly projected, will hit the screen. Exactly where this one will fall, all available knowledge, all imaginable experimentation cannot make sure. Causal analysis leaves lacunae of uncertainty in the behavior of this particular electron, and it does not say how they are filled. Furthermore, it is the well-considered belief of most physicists that physical theory, by its very nature, will not fill them.

Now the presence of these lacunae in physical causation does not spell freedom. If this concept could be applied to an electron, mere uncertainty as to the spot it is going to hit does not make it free; rather, it makes it unpredictable or randomly behaved. If the same kind of uncertainty were present in man, his actions would be erratic, and one simply could not tell whether he was free, i.e., capable of responsible decisions, in the sense of the moralist.

If, on the other hand, principles not known to present science but quite different from the laws of physics—and I should say principles of a *psychophysical* nature, principles that involve the presence of consciousness and that are as inapplicable to unreflective matter as the laws of thermody-namics are to single particles—if such principles could be shown to fill the lacunae of physical causation, something like freedom might result. To refer back to our hypothetical example: If the electron itself decided where, within the penumbra of physical determination, it would impinge on the screen, then it could truly be said to be free.

It is clear that physics is a long way from explaining that human freedom which declares its presence so forcefully in our introspective consciousness. But it does make room for it, it takes freedom from the wastebasket of paradoxes, illusions, and irrelevancies, and lifts it to the shelf of challenging problems to be solved. The scientist is no longer a fool when he talks about freedom, and man is no longer restrained from professing it. Even the form of a possible solution of the problem is visible in outline. Attaining it will probably involve two steps. First, it must be shown that the uncertaintiy surrounding the physical behavior of small atomic entities persists uncancelled through the composition of atoms into organisms. There is already some evidence for this. Secondly, it must be made clear how, in the process of material organization, a self-reflective and self-determinative principle becomes active. And there is some precedence for this in the Pauli exclusion principle, as I have indicated elsewhere.[7]

d) Man's new understanding of nature has resulted in a dispersal of scientific dogmatism, in a sense for the distinction between physical and historical or existential reality as well as in their conciliation; it has removed the paradox of human freedom. Finally, almost as the fulfillment of a promise vaguely implied by this series of advances, modern science releases man from his restrictive role as a spectator in a universe that is quite complete and objective without him. He has become an active participant in the drama of existence; no longer a passive watcher of a stage preset, he has come to know that the stage is different for his presence.

I do not mean this in the trivial sense that human beings change the nature of the physical world through their actions. There is nothing new in that, and it is as proper now as heretofore to contemplate this fact for moral comfort. Yes, it was always agreed that man can change the world through action. But what about knowledge, what about the search for truth? Put in a simple way, the new fact is that search for truth modifies truth, that there is an effect of the knower on the known, that *knowledge, too, is action.* Four decades ago the typical observation of science was the measurement of the position of a star, an act wholly detached from the celestial object far away and insignificant to its further motion. The astronomer took pride in being able to make such measurements without disturbing the star, and envious physicists considered this kind of non-disturbing measurement as their ideal. Today, with our principal concern about the atom, we regard such observations as atypical, as limiting cases never realized, and we consider as normal the measurement of the position of an electron, wherein the fact that an observation has been made is crucial to the fate of the electron. Even the process of acquiring knowledge, we now learn, has the determinative quality of action upon a previously stolid but now pliable universe.

When the genius of Heisenberg first confronted the physicist with this interpretation of the measuring process, he evinced a shocked reaction, for his whole concept of objectivity was shaken and his neat distinction between the spectator and the spectacle broke down. Had he been less conventional (I crave your indulgence for this truism voiced so long after the fact) he would have noted that his was the only science, indeed the only branch of learning which thus far ignored the interaction between knowledge and the known; the new discovery made physics more like biology, social science, history, and economics, where the occurence of what is now called "feedback" between the subjective measurement-knowledge-prediction process and the external system subjected to the process, was already a commonplace. The biologist knew that certain of his observations inevitably kill an organism; the physicist was alarmed to realize he could not measure a photon's position without destroying the photon. The alarm has subsided, biology and physics face similar problems, and another of the passive features of the old mechanical universe had disappeared.

Beside the uncertainty principle, to which the last remarks have been confined, there are other indications leading to the conclusion that the curtain between the knower and the known is disappearing, and suggesting a necessary involvement of man in nature not only through action but through knowledge as well. There remains no abode for detached contemplation which makes no difference to the world. Man, freed of earlier prejudices, with a humbler approach to ultimates and absolutes even in science, with a new appreciation of the existential qualities of experience and a glimmer of freedom that is neither paradox nor illusion, has finally seen his *facts* turn into *acts*. It would seem that the evolving philosophy commensurate with such fragments of present vision must portray man in a measure of wholeness and dignity he did not hitherto possess.

## NOTES

[1] This point has been made with greatest cogency by F. S. C. Northrop in his well known books *The Meeting of East and West* (New York: Macmillan, 1946) and *The Taming of the Nations* (New York: Macmillan, 1953).

[2] Henri Peyre, "Literature and Philosophy in Contemporary France," in F.S.C. Northrop, ed., *Ideological Differences and World Order* (New Haven: Yale University Press, 1949).

[3] E.T. Whittaker, *A History of the Theories of Aether and Electricity* (London, 1910), p. 157.

[4] *Ibid.*

[5] H. Margenau, "Advantages and Disadvantages of Various Interpretations of the Quantum Theory," *Physics Today* 7, No. 10, 6-13 (October, 1954).

[6] Roughly speaking (in the terminology of *The Nature of Physical Reality*), the "P-plane" is the locus of all immediate (perceptory or "protocol") experiences, while "constructs" denote the rational, conceptual experiences.

[7] *The Nature of Physical Reality*.

CHAPTER 17

## SCIENCE AND HUMAN AFFAIRS

1. The question of whether the methods of science can be applied to human affairs is a difficult one, and a question upon which men are rather evenly divided. Some think it ought to be answered affirmatively, some think the answer should be negative. In order to arrive at a disciplined and reasonable answer it is necessary for us to think first of all about the meaning of science, and there are various misconceptions in that respect.

Many people seem to feel that science is merely a catalog of verifiable facts. In this view, everything that we know with certainty belongs in the book of science. The things about which we are in doubt, but concerning which we hold not completely well-founded beliefs, are not scientific. Factualness is the criterion of science.

People holding this view frequently employ a simile to describe or allegorize the scientific enterprise. This simile is the picture puzzle. One sees the scientist making discoveries, uncovering a fact here and another there, and these facts, if they are proved, and if they belong to science, must in the end form a pattern. There must be a sort of congruity about the elements of the puzzle which allows them to be put together into a satisfying pattern which one can recognize as the solution of a scientific problem. Science in this sense is a two-dimensional game—the fitting of fact to fact, resulting in the solution of a problem.

It seems to me that this view leaves out of account something intrinsic in the method of science—something that is characteristic of the scientific pursuit. It describes scientific knowledge correctly, but hardly scientific understanding.

The pattern is wrong because it is only two-dimensional, remaining on the surface of the facts. It does not convey understanding in the full sense. A third dimension is involved in the full practice of scientific research. This third dimension is that of the ideas, the concepts, the laws, the principles which underlie the facts, and within which they have their roots. One cannot be a scientist without giving disciplined attention to the meaning of the facts —indeed a fact means nothing unless it makes reference to an idea, to a context of reason from which it receives relevance and significance.

A simile which seems to me to portray far more adequately the true concerns of the scientist is a growing crystal. Many of us have seen this simple lecture demonstration in physics or chemistry. One takes a vessel having two parallel glass sides and fills it with a liquid just above its melting point. He then arranges a polaroid in front and a polaroid behind and shines light through the arrangement, projecting light on a screen. At first one sees streaming a liquid, forming a moving pattern which is interesting because of the irregularity of the motion. It is interesting and beautiful in that sense, but it does not convey the facility for prediction of order. Then suddenly as the temperature of the substance reaches its melting point, something occurs which one could not have foreseen. At some spot (chance only decrees where that spot will be) a seed of a crystal forms. This seed becomes the origin of a pattern of regularity and geometric beauty. What was previously an amorphous liquid now takes on shape, form, organization. The substance remains unchanged, but is simply endowed with pattern and organization.

This, I think, is what science does to our experience. As we survey the sum total of our experiences, by and large amorphous, we feel that they are incapable of being used for the prediction of order. But sooner or later, within this amorphous mass of experience, the crystal of science begins to form, and the domain of experience which it engulfs becomes regular; through it and in it, we can understand; the light of reason shines through it; whereas the rest of our experience remains in a sense opaque.

Now I do not mean to convey that by virtue of its crystallization into a scientific domain our experiences becomes more important or valuable. It is a fact, and I do not think an unfortunate fact, that most of our concerns today still lie in the amorphous part of our experience. Very few of the important decisions we make in our daily lives are based on scientific considerations. Science at the present time encompasses only a very small portion of our human interest. It does work in physics and chemistry, it is beginning to launch a growth in such fields as economics, psychology, and biology, but by and large it is still restricted to a relatively small domain of our thoughts and interests.

Science has built in it a tendency to grow. We cannot restrain a crystal from growing, neither can we know precisely where it will grow. So it is with scientific progress. No one knows where the arm of science will stretch, or what part of our experience it will organize next. But it is possible to influence this growth. One can plant "seeds" in the crystal, and then a formation about this center will occur. Similarly, one can inject science artificially into a certain amorphous body of fact and expect it to grow. But this is after all a rather arbitrary procedure and has not very often been successfully tried.

There is another consequence which I should like to draw from this picture of the growing crystal as a simile of science. This growth is not malignant.

Nothing is adulterated by the conversion from the amorphous state to the crystalline state. The substance remains unchanged, the chemical nature of the crystal is unaltered. Thus it is with our experience. What happens to our experience is simply that a new kind of light shines throught it. We see its relation to other kinds of experience, and that is all. The facts remain the same. But they have taken on a depth of meaning previously invisible.

One can see this every time a new science is born. The celebrated Pythagorean theorem was at one time a non-scientific fact, a mere matter of general observation. The ancient Egyptians knew how to survey the land along the Nile by means of a rule which they called the 3-4-5 rule. They would take a rope and put eleven knots in it, so that the rope would be divided into twelve sections. They then bent the rope—first after three sections, then after four—thus making the rope close on itself, but keeping the lines straight. They knew they then had a right triangle. In other words they knew the factual content of the Pythagorean theorem. Why then do we celebrate Pythagoras? Why do we assign his name to this theorem, when in fact the ancient Egyptians knew it? For the reason that Pythagoras converted a piece of knowledge into a scientific theorem. He asked the question, what is the rationale of the fact? What is the texture of reason within which this theorem—this fact—takes on significance? And this conversion from a factual, amorphous set of circumstances into a scientific theory is what made a man like Pythagoras famous. (Whether Pythagoras did in fact discover the theorem is rather doubtful. There is evidence that someone else proved it before him, but that is beside the point.)

There are certain consequences which flow from this. Science is equipped with facilities for self-correction. This means of scientific pursuit also corrects itself as it proceeds. Science always makes errors. There is no set of scientific theorems, no theory of science, which is free of error, or which we should not expect to be corrected as we go on. New experiences continually arise. The amorphous domain is unlimited, and therefore I think it is unreasonable for us to expect that science will ever cover everything that can occur in human experience. We never need to be afraid of that. On the other hand, if the analogy of the crystal is true, if this continual growth takes place, then we must never expect that great millennium of perfect knowledge to be at hand. We have discussed the "picture puzzle" concept according to which a problem can be solved completely, so that when it is solved the knowledge is complete. According to this, when we know all the facts and have put them all together into neat little patterns, then the golden era will be at hand and we can regulate human affairs in a manner controlled by science. But according to my version of science, this will never happen. There will always be more and more amorphous experiences which have not yet been converted into scientific domains. Science always tries to achieve this conversion and the process will continue without end.

2. I now turn to the question, "Can this method of science be applied to human affairs?" It has been proven tremendously successful in the inorganic world, in part even in the biological world, but what about human affairs? What is it that distinguishes the social domain—human behavior—or, if you please, the whole organic world, from the inorganic world?

It is clear that in the first place the social world manifests far greater complexity than those which are studied in physics, chemistry, and mathematics. And what is meant by complexity? Nothing more than the number of variables present. If we wish to describe the motion of a star, or a falling stone, we need in principle only two variables. To know for example what an earth satellite is going to do, all we need to determine is where it is and with what velocity it is moving. These two variables permit the prediction of the future state of that object for all time. Even if certain small perturbations are present, as we know they are, the state can be predicted quite well within a reasonable period of months at least. Now consider a human being. A baby is born. We want to determine its state in such a way that we can predict the manner of the baby's growth even for only a few days. What are the variables we must observe, and how many are there? The problem is really a double one. It is doubly confused because in the first place there are undoubtedly many variables, and in the second place we do not even know what they are. This confusion, this proliferation of variables, is visible everywhere in the organic field and certainly in human behavior. Therefore, it is claimed, and rightly claimed, I suppose, that because of the complexity, because of the great number of variables present in the social field, scientific method must, if it can be applied at all, be greatly modified.

Secondly, I should like to call attention to the problem of the control of variables. It is said that in science we can arrive at simple laws because it is possible for scientists to control the conditions of observation. For example, suppose we investigate the behavior of a real gas, hoping to arrive at what is called the equation of state of that gas, namely a relation between its pressure, volume, and temperature. We find that a literal transcription of the actual observations made on the gas leads to a very complicated equation, so complicated that we do not even know the functional form that expresses it completely and accurately. The general equation of state of a real gas is exceedingly complex. We can make it simple, however, provided we make the pressure small, or the volume big—attenuating the gas, making it have fewer molecules per unit volume. Under these circumstances the pressure-volume relationship is in accordance with a simple law called the ideal gas law (PV = RT). The simplicity of that law relies upon an idealization which is in a sense a distortion of the facts. We subject the experimental object to extreme conditions and find a simple law which holds quite accurately only then. Only in this limit can we describe the behavior of the substance by a simple formula.

Now in the social world this approach will not work. If we want to investigate the laws of a society we cannot subject the society to arbitrary extreme conditions—we cannot refine it (in the sense of making it very thin and tenuous), or take people out of the society and then manipulate it—and hope to arrive at laws describing human behavior. This makes no sense. We cannot control the variables; therefore a grave question arises with respect to the applicability of scientific method to human affairs.

A third point, which is rather interesting and has come to the fore in this complex of arguments quite recently, has to do with feedback. The phenomenon of feedback involves an interplay between what is being measured and the result of the measurement I might call it epistemic feedback because it couples knowledge with being, and epistemology is the branch of philosophy in which the relation between knowledge and being is studied. Economists and psychologists have known of the occurrence of epistemic feedback throughout the history of their sciences. It is commonplace for economists to remark that an economic prediction has an effect upon what is being predicted. Thus, there are self-fulfilling predictions, like the prediction of a run on a bank. There are also self-defeating predictions. If I were to predict that a certain stock will go up by, let us say, 20 points next week, and if everybody thought I was right in this prediction, then of course this result would not occur at all. The stock would not rise 20 points but probably 50, because everybody would go out in a rush, buy it, and push its price up. The same kind of feedback occurs in psychology, where, for example, by asking a person if he is angry we can actually make him angry. When feedback occurs it appears clear, at first sight at least, that the ordinary method of science is not available for unravelling the difficulties of the situation.

Finally there is the point, very often made, that men are able to make decisions. The possibility of decision adds a new element to the situation which is not present in the inanimate world. When men can make decisions it seems impossible to make a valid prediction of their behavior. Decision in human behavior amounts behavioristically to a departure from mechanical expectation.

There are four particularized points, then, which are made against the thesis that the method of the natural sciences can be applied to human affairs.

First of all, the complexity of the phenomena in terms of the number of variables; secondly, the lack of control of the variables; thirdly, the existence of feedback in the social sphere; and finally, the occurrence of decisions on the part of human beings. We shall return to these points later.

3. There are some very much more sweeping and incisive objections launched against the method of science, by philosophies which I shall very briefly survey. A sweeping indictment, which is made principally by men like Heidegger, takes this form. It is asserted that the scientific method is an artificial one. In order to ascertain being and truth, one has to proceed with caution; being and truth are like elusive game, which one must stalk carefully, creep up upon, listen to. One must never set traps. The scientist does set traps, and what he finally gets hold of is not the living body of truth, but merely its corpse. So this view distinguishes between two methods of apprehending facts and truths: the one forcible, violating the body of truth, and the other gentle, naturally apprehending living truth.

On analyzing this argument, one cannot help but feel that the distinction between the two kinds of approaches to truth is quite artificial. What is a forced scientific approach to truth in contradistinction to the natural approach? If I look at a table without my glasses, then, if I were to take this argument seriously, I would be apprehending the table in its natural way without violating. If I look at it with my glasses I would violate it somehow. Of course I already have some lenses in my eyes and they do not violate! This is not a very convincing sort of argument.

But suppose there were a way in which the scientist could come to know the atom, besides the one he chooses. Clearly if there were two ways, one gentle and the other violating, the scientist would prefer the former. The point is that in many branches of science there is only one way and the supposed distinction between violating and non-violating approaches is wholly out of place. About all, I think, that one can say about this point of view is that it is mistaken, being based on ignorance of science.

4. Now we seem to have reached the conclusion that there are grave difficulties attached to any attempt to apply scientific method to the social world. On the other hand it is perfectly clear that attempts in this direction have on some occasions been successful. These actual circumstances stand against the conclusion we seem to have reached. Let me remind you of a few of the successful social scientific theories—theories based upon and patterned after certain theories in the natural sciences. One or two are quite striking.

The city rank-size law is an interesting example. We write the names of the cities of the United States in the order of size. Next, we assign a number to each city, according to its rank—New York 1, Chicago 2, and so on. Then we put down the actual populations of the cities, and multiply each rank number by the population of the corresponding city. We get always the same result, approximately, for about the first hundred cities—a rather amazing fact. There are certain troubles with this rank-size law, and I will not affirm that it is truly valid, but it has been advanced as a scientific law. It works not only for cities, but also for other things—for instance for the volume of sales of the department stores in the country—and if we are willing to generalize the law a bit by means of a slight mathematical complication, we can make it fit even a larger set of facts.

There appears to be something in this law, although it seems very strange. As a matter of fact, I am reminded immediately of certain chemical and physical laws. There is a particular resemblance, here, to the law relating the abundances of the chemical elements. If we write the elements—hydrogen, helium, lithium, and so on—in chemical order according to atomic number, and then write their relative abundances, we find a certain law relating the abundance to the atomic number. This law was a puzzle for a long time, but has now been unravelled. There is a theory expressed in terms of an original temperature and density of all the matter in the primordial state of the universe, from which, by applying the laws of thermodynamics, one can derive this law for the distribution of the abundances of the elements. Perhaps the city rank-size law will prove to have some analogous explanation.

Another rather interesting thing is the law for the propagation of rumors. This was discovered by sociologists during the war. The Air Force was interested in spreading rumors; hence it wanted an investigation of the manner in which rumors spread. Sociologists therefore loaded airplanes with leaflets carrying rumors. These leaflets were dropped on a certain city; then after a week or two, researchers would go around to the various cities in the neighborhood and inquire by door-to-door canvassing how many people had heard the rumors. The result was very interesting. It turned out that if the rumors were dropped at a certain city (called the origin), then the number ($N$) of people in a city a distance $R$ from the origin was proportional to the strength ($S$) of the rumor, that is to say the number of leaflets dropped, and inversely proportional to $R$. Symbolically, $N = KS/R$, where $K$ is some constant.

Now every physicist looking at that sees immediately that it is like Coulomb's law for the field produced by an electric charge. Of course as usually given in textbooks of physics, Coulomb's law contains the factor $1/R^2$ rather than $1/R$; but that is because it is applied to three-dimensional situations. When we apply Coulomb's law to two-dimensional situations, it becomes a $1/R$ law. Conversely, if rumors ever spread in three dimensions, I am sure they will follow the $1/R^2$ law. Perhaps this will happen when space travel is established.

A very effective contribution to social science along the lines of natural science has been made in the last ten years by the operations research people. Von Neumann and Morgenstern wrote a very impressive book on the theory of games and economic behavior. Perhaps for the first time, they solved in straight mathematical terms certain baffling questions which are typical of competitive human behavior. Two duelists who want to kill each other start out, each with a gun in his hand. The guns are of equal power, and the men are equally good shots. Each has only one shot. Now as they advance toward each other the question arises, where should one begin to shoot? If I shoot too early, my shot goes astray and I am sure to be killed, because the other fellow can now advance at will. In the theory of games, this problem receives a simple mathematical solution.

Another example from the theory of games is the queuing problem—a situation which we encounter often in shops. The shopkeeper does not know how many clerks to employ, because in the morning there are many customers and at noon there are none, so the question is how many should he employ to maximize his profits while holding his customers. This problem was solved mathematically and the results were applied to airports during the war by the Air Force.

Thus we can give many examples of problems in social science which have received satisfying solutions. To cut a long story short, it seems that there are cases of a satisfactory application of scientific method, as we know it in the natural sciences, to human affairs.

5. Let us then review the four points named in Section 2, to see whether they are really valid.

Is the number of variables a crucial item? It is true that in the present sciences which have become exact we often get along with a small or at least smallish number of variables. I cited the example of the moving satellite or the falling stone. To describe the motion of the stone, we need only two facts, as previously given; this is an example of why physics is simple. But in Aristotle's time, in the infancy of physics, things were not so neat. There was a theory of motion which required the following information in order to become predictive. Aristotle had to ask first of all where the body started from—the dynamic cause. Then he had to know where the body was going—the efficient cause. Then he had to know the constitution of the body—how much of it was earth, how much of it was water, how much fire, how much air—because fire would drive the stone upward, air would keep it there, earth and water would make it go downward. There were other things involved also.

The point is that when physics was young, one had to have many more variables in order to be able to predict. Apparently the number of variables, or the simplicity of the science, depends on its age. It took the genius of men like Newton and Galileo to make mechanics simple. They succeeded in introducing a new variable, something that could not have been seen immediately, something rather far-fetched at least as it appeared at the time. They introduced the concept of acceleration, and showed that in terms of acceleration the motion of bodies became simple, whereas in terms of the Aristotelian concepts surrounding the problem of motion it remained most complex.

Perhaps in the social sciences we have not yet gone far enough to simplify the situation. I am not suggesting that this is true, but merely propose this idea for your consideration. If it means anything at all, it urges caution with respect to the proposition that because at the present time the social sciences are complex, we must never try to apply the scientific method to them.

The second point, I think, can be demolished in a similar way. It concerns the control of variables. Is the control of variables really necessary in science? The oldest and most perfect of all sciences is astronomy. But the astronomer cannot control his variables, he cannot push the stars around, he must take potluck in his observations. And look how he has succeeded. Clearly, then, the possibility of control may be an aid to the advancement of a science, but it is certainly not a necessary condition.

The other two points require perhaps slightly more extensive consideration.

Let us consider the third one—the occurrence of feedback. It was formerly believed by the natural scientists that the ideal measurement was one which allowed the world to be separated into a spectacle and a spectator and in which the presence of the spectator made no difference to the spectacle. There was, it was hoped, a transparent curtain between the spectacle out there and the spectator here. I could see what was happening and my seeing it made no difference on the other side of the curtain. This conviction of the possibility of non-disturbing measurements arose from the early science of astronomy, and crept into physics. The physicist felt that if he was really good and made careful observations, there was no ingression of the method of measurement into that which was to be measured.

But quantum mechanics, which changed so many things, has changed this too. We know now that when one makes a measurement he disturbs the system in an uncontrollable way. Thus what we find in nature depends in part on what we do. In this sense the facts of science have become acts on the part of man. Man has entered into the history of nature. He cannot shut himself out. There is, even in inanimate nature, some feedback between the knower and the known. It cannot be excluded. The physicist at the end of the last century who believed that he could, in principle, make measurements which do not disturb systems, was simply wrong. He thought that only in his domain of physics—in the realm dominated by the queen of all sciences—such measurements were possible. But he had to find out that he was no so privileged.

Now did this destroy his ability to predict? No, but it did something to his method. Scientific method had to change from a conception of state in terms of clear dynamical variables, of individual events, to the acceptance and the study of probability aggregates—of aggregates of events or ensembles.

Finally, there is the occurrence of decision, and this is doubtless something which sets the living world apart from inanimate nature. What effects does it have? Because of the possibility of decision, the social scientist cannot make accurate predictions with respect to individuals. He has to turn statistician. The possibility of human decision makes it necessary for the social scientist to study the problem of suicide, for example, in terms of large numbers. He can predict how many people in a given society will commit suicide, as exemplified by the theories of Durkheim which were so successful in predicting suicide and other behavior. But he cannot say who is going to commit suicide. He has to take the attitude of the life insurance researcher or the actuarian who speaks in terms of probability.

Now precisely the same thing has happened in physics, and in all the sciences depending on physics. Why? Because decision, which has been recognized as being possible in the social world, in human beings, has now been found to be present, in a certain sense, even on the part of electrons, neutrons, and all the other little things making up our universe. In a way, we can say that they make "decisions" too.

Let us return to a previous example. Suppose I have a large block of plutonium and a neutron. This neutron is now headed for the block of plutonium. When it gets there, the plutonium will become an exploding atomic bomb. Of course, you say, this means we are doomed. We might run, but it will do us no good because the neutron will certainly get to the plutonium before we can run far. This is the fatalistic prediction of classical physics. In reality it is by no means true. If we measure the position and velocity of this neutron and find that it is headed towards the block of plutonium, the neutron may not get there at all. It might violate the expectation based upon the measurement we have made upon it.

Now we may say, in a way, that the neutron made a "decision" in the behavioristic sense, which is the only sense of the word we can employ in this connection because we cannot endow the neutron with consciousness.

What do you mean when you say that I am making a decision? You might, and probably do, mean this. Suppose I am headed for the door. You say,

"Thank goodness he's leaving—he's concluding his lecture." But I might violate your expectation by stopping in the middle of the room, coming back, and continuing to talk. In this case you would say, "Oh well, he has made a decision contrary to my expectation."

What is a decision, watched from the outside, a decision as it would be described by a behaviorist? Simply a departure from an expectation. But that is what the neutron performed too, in not hitting the plutonium bomb.

From this point of view, therefore, the situation in physical science, or quantum theory at any rate, is not greatly different from the corresponding situation in social science. And the moral—the lesson we can learn here—is that presumably the social scientist, if he wishes to follow the pattern of the natural scientist, must turn statistical in his methodology, but will certainly not be constrained to give up the use of scientific method.

6.  Now all this sounds pretty; but it seems quite academic, I know, and does not carry conviction. Why not? I think because we all have a strange psychological block. We feel that the application of science to human thoughts, interests, and concerns, and to the values which establish the relations between human beings, is inappropriate. We think of the scientific approach as being somehow degrading. It does not seem to belong in human affairs. Cold calculation and prediction of what men are going to do is somehow out of accord with morals and with humanist taste.

What is this argument good for? What does it mean to say that the application of scientific methods is degrading? Of course no scientist ever means that one ought to analyze a painting in the sense of subjecting it to chemical analysis, or that one should dissect a poem. The scientist wants to make things quantitative and measurable. And there is a popular feeling against this —an aversion which simply deems the application of scientific method misplaced and says nothing more about it. I think that this attitude—this rather widespread conviction—is nothing but a ghost, a hangover from ancient eras and still inhabiting the ivory towers of some professors in the humanities and perhaps the studios of some artists. But that is all; because on closer inspection, the ghost vanishes. There is not a shred of truth in the assertion that the application of scientific method is degrading. There is not a shred of truth in the assertion that values cannot be quantified, that values are somehow elusive because of their very nature.

This was an attitude taken by many artists about twenty or thirty years ago, but there is an interesting story about it. The artists would say, "Color has a subjective value, something which you will never catch in scientific traps." The physicist would say, "Let's see. I can describe color in terms of two attributes, each of which I can measure. One is the wave length, the other is the intensity. If these two numbers are fixed, the colors are determined." The artist would then paint two canvases of blue, each with the same hue, or wave length, and intensity, and find that the two canvases would look wholly different to the observing eye. There was indeed a subjective quality, an essence, which this scientific description failed to catch. And so the artist was right in saying, "There you are; you cannot capture the essence of color by your scientific method." But then something happened. Schrödinger and others discovered that a third variable should be introduced—a variable now called saturation. Color should be described in terms of wave length, intensity, and saturation—three numbers, not two. When this was done and when two canvases of blue were produced, each with the same values of the three variables, they looked exactly alike. The elusive quality of color was gone. It could now be quantified, it could now be measured, and it could now be reproduced.

Has pictorial art suffered from this scientific discovery? Ask the artist. In all honesty he will admit that quite the opposite is true. As a result of this discovery, this encroachment, if you please, of the scientific method into art, the artist himself has been given better tools for creative work. And the men who enjoy art have been given greater facilities for enjoyment.

The same story could be told about music, or about any of the arts. An application of scientific method has never done them any harm. And as we survey the whole, and as I conclude the story, it does not seem as though there are any essential obstacles in the way of our attempting, at least, to apply the method which has been found so successful in the natural sciences even to human affairs.

CHAPTER 18

## THE NEW STYLE OF SCIENCE

One of the oldest legends of our culture dates back to the era before the Libyan dynasties of Egypt, many centuries before the Christian era. It relates to the town of Sais, in the delta of the Nile, where a great temple had been dedicated to Osiris, the god of the underworld. The ruins of that temple are still visible today.

It is said that this temple contained a mysterious picture, covered with a veil and inscribed by the tantalizing words: "The Truth." Mortal man was forbidden to lift the veil, and the priests of Osiris enforced this statute with severest rigor.

A youth, dedicated to the discovery of truth, perhaps a person we would now call a scientist, once entered the temple and saw the covered image. He asked his guide whether he knew what was hidden under the veil, but he received a horrified denial and an official account of the ancient law. Thoughtfully, the youth left the temple that day but an irresistible thirst for knowledge of truth forced him to return at night with intent at sacrilege. In the ghostly light of the moon, he entered the hall of Osiris and lifted the veil from the image. What he saw, nobody knows, but the legend insists that he was found near death, lying at the foot of the picture, by the attendants of the temple the next morning. Revived, he would not speak of his experience except to regret it. His life, thereafter, was spiritless, his actions were undistinguished, and he sank into an early grave.

There the legend stands at the very beginning of our history, non-committal like the Sphinx, foreboding human agony over truth, symbolizing one of the great and noble passions of man. The legend has not lacked interpretations. Some writers have made it imply the finiteness of the human mind which cannot comprehend absolute truth; the German poet Schiller has given the story moral content, claiming that truth is fatal to a sinful conscience: "Wer dem, der zu der Wahrheit geht durch Schuld, sie wird ihm nimmermehr erfreulich sein." Others have said that only God can reveal truth, and He will not be forced to it by human impetuosity.

I should like to suggest another resolution of this ancient riddle, a solution peculiarly significant in our day of crises, in which science stands accused of having perverted man's spirit and eroded his life. But my suggestions will be empty or incomprehensible unless I sketch for you some of the changes which our understanding of science has undergone, unless I portray it as the challenging adventure, the never-ending enterprise which it has come to be. Let me depart, therefore, from the story of Sais and attempt to resolve its mystery when the foundation for the solution has been laid.

A large part of my talk will deal with the contrast between the science of the 19th century and the science of today. The former was factual, its business was the discovery of ever more accurate and reliable data and the determination of the constants of nature to an increasing number of decimal places; ours is a human adventure, alive with challenges and ideas, hopes and frustrations; its concepts vastly transcend the domain of measurable fact. The older view regarded theory as important only insofar as it aids and promotes the discovery of facts; the power of concepts in themselves, the force of abstract logic were not appreciated as fully as they are today. The philosopher who caught the spirit of the older science in his writings was Ernest Mach; he characterized the task of science as the careful recording and systematization of sensory fact. In a famous passage he said that science requires theory for the sake of economy of thought; theory is not an essential ingredient of science, and its role is temporary; theories are leaves which, though they are essential for a time, the living tree of science sheds when it is no longer in need of them.

The chief aim of science, in this older conception, is to discover new elements, new substances, new planets, new stars, new species of plants and animals. Most prizes honoring achievements in physics in the last century were awarded for discoveries of new elements and elementary particles and other unsuspected facts about nature. This has aided and abetted the belief that the only function of science is to turn up rocks to see what is under them.

Discovery brought forth not only facts but principles as well. These, like the principle for conservation of energy, of mass, the tacit belief that space and time are infinite and the universe is eternal, were likewise held to be indubitably true, and the suggestion that they could possibly be questioned would have been rejected by most thinkers as wholly unscientific. In those days the contrast between science and philosophy, and even more significantly the contrast between science and religion, was extreme.

The starting point of science in that bygone era was in axiomatic, infallible and certain truth. Logic was based on indubitable propositions, the so-called laws of identity, contradiction and the excluded middle; geometry started with the axioms of Euclid, and these permitted no doubt. Physics operated with all these absolutes and added a few of its own. The scientist thought of himself as possessing most, if not all, of the static verities of knowledge, and he was proud in this knowledge, inflexible in his appraisal of what was scientific and what was not. The words, natural and supernatural, date back to this period of certainty; they suggest that science knows the principles that control this universe and can render a clear and irrefutable judgment as to what is possible and what is not. Thus arose the conflict between science and religion, the conflict between one set of static certainties and another. Today, as you will see, all this has changed. Facts are as important as ever, but only in conjunction with ideas. Facts in isolation have lost the glamor that surrounded them.

Only the *popular* understanding of science perpetuates the 19th century bias. To the man in the street discoverable facts still exhaust the meaning of science; he regards them in the style of the 19th century as interesting in themselves, as though they were sacred in their givenness, no matter how trivial they are. Thus, when we read how an ancient tribe that never got anywhere and is now extinct, raised its cattle or tilled its fields or educated its young, we are supposed to be tremendously impressed; a study of the Kinsey report on the sexual habits of unspecified human males and females is supposed to leave us speechless, and the headline which I recently saw: "Scienttists discover that the reading of crime stories by teen-agers contributes to delinquency," is taken not as a platitude but as suggesting a solemn solution to vexing problems. Science, then, according to this pristine understanding, is an aggregate of facts, often of trivia, put together with a suffusion of majesty about their mere factualness.

Nor is this interpretation confined to the popular mind. Unfortunately, scholars in the humanities, too, sometimes regard the emphasis on facts, the avoidance and spurning of theory as the sole characteristic of science. Greatly impressed with the fecundity and power of science, they endeavor to convert their own disciplines into branches of science by advocating a restriction to facts.

Such was, for example, the attitude of Ranke, the great historian of the last century, when he claimed that historical research must limit itself to factual reporting, to the testimony of eye witnesses; that it falsifies and adulterates history when presenting theories and interpretations. Contrast this, for example, with the contemporary richness of Toynbee's writings, of a man who avows a theory of history and believes, rightly or wrongly, that historical studies reveal a lawful trend and portray ends in human development much in the manner in which science uncovers the laws of nature.

There is a simile, earlier used to illustrate the aim of science, which renders this factual emphasis of old-style science very accurately. It is the picture puzzle. The scientist is like the man who tries to solve it. He examines every piece of the puzzle with meticulous care and seeks to place it in its proper foreordained position, quite certain that it fits somewhere. The pieces are the observable facts of nature; science forms them into patterns, and a pattern represents the solution of the problem at hand.

This interpretation of science is certainly not wholly in error; indeed it describes the initial stages of the scientific quest with reasonable accuracy. But it implies promises which science cannot fulfill. It suggests completability of the scientific task. It says that when the pattern is finished, the particular problem is solved, and another, perhaps an unrelated one, must be begun. Further pursuit of this image has led to the expectation that some day, when a sufficient number of puzzles has been solved, when perhaps all questions of science have received their final answers, the golden era will have arrived and one envisions a condition of mankind, blissful or utterly distressing, according to one's point of view, in which no factual problem remains. All material affairs can be arranged and man can live happily ever after. Or, in the other version, human life will be degraded to the level of a machine.

The picture puzzle simile has another glaring imperfection, for it represents the concern of science as a most simplistic game in which the facts provide their own integrating unity through their coherence with one another; it neglects the substratum of ideas, the dimension below the surface pattern of facts in which the unifying roots of scientific knowledge are to be found.

The image of science as the ultimate criterion of truth, as the savior of man, as the final instance of appeal in every quandary, has produced its caricature as a sacred cow. *Science As a Sacred Cow* is in fact the title of a book, written in 1950 by Anthony Standen. The title is the best part of the book, which represents a confused but entertaining account of the scientific method. An acknowledgement on the title page dedicates the book to the Long Island Railroad, which, the author says, is responsible for its creation because of the long waits it caused the author who, at the time of writing, was commuting between Long Island and New York City.

I ask your forebearance for having mentioned these irrelevancies; the point I want to make is simply this: If the 19th century appraisal of science with its emphasis on self-significant facts, on inflexible and permanent empirical truth were correct, I should think it deserving of the title, sacred cow.

Furthermore, if this view were correct then clearly what is already known in science permits historians to chart the path on which the sacred cow is moving. Many of them are doing this, and they have become prophets of crises. Let me survey a few of the crises which we are said to be confronting.

First, there is the ever growing danger of an excessive energy release. To see this threat in proper perspective let me recall a few simple facts. The human hand is able to impart to a stone about 50 foot-pounds of energy; the slingshot multiplies this output by 2, the bow gives its arrow 6 times as much energy, T.N.T. introduces a factor 12 over the bow, the thermonuclear reaction a factor of many millions. If science marches on, and extrapolation is permitted, the single shot that destroys the earth will come in our lifetime. But even before then, say the prophets of doom, wars will have terminated most useful life.

In case you are not frightened by *these* prophets, there is the opposite calamity called the population explosion. It has been calculated that at the present acceleration of the birth rate there will only be standing room for man on this globe in the year 2500, provided he manages to kill off all lower forms of competition for space. Moreover, about 98 percent of us will be colored or else we'll all be mulattoes.

At a much earlier date physical laboratories will have been moved away from mother earth because the radius of the cyclotrons to be built in about 100 years is going to be larger, according to a simple extension of the events now taking place, than the radius of our planet.

Research will increase at an enormous rate; the span between discovery and commercial use of new facts will be greatly shortened. Indeed the technological development which required over 100 years in the case of the telephone, 4 years in the case of the transistor, will need only a microsecond for a device conceived in the year 2500, if present indications are to be trusted.

The human brain will not be called upon to carry on the work that leads from discovery to patent; automation will be so nearly complete that man's mind can devote itself to more pleasant things.

Scientific documentation has recently been expanded in unimaginable fashion, and the mathematical prognosis is that in the year 2500 the surface of the earth will be covered 6 feet high with IBM-coded scientific documents.

And worst of all, the rate of production of scientists is increasing in a manner most fearful to behold. A simple computation, based on the facts of the population increase and of the growth rate of scientific manpower, a computation quite as accurate as those which led to the preceding predictions, shows cogently that in the year 2500 there will be eight times as many scientists as people!

Horrid things are being said in the name of science, and they are believed because science is supposed to be the sacred cow. What I hope now to demonstrate is that the course of science is not as inexorable as the fact-finding process, the belief that truth grows by accretion of data and that all its accretions are permanent, make it out to be. Science is as variable as the human spirit, and as unpredictable. Hence the premise of stability and inevitability which underlies the foregoing prognoses is unacceptable; scientific extrapolations based upon the past are unreliable. In fact if science, modern science, is correctly understood in its fullness, it can be counted upon to forestall such crises. Having enabled us to swim the seas like fish, to fly through the air like birds, to rise into space like demigods, it may finally teach us to walk this earth like men.

To see this we must study carefully what has gone on in the basic scientific disciplines during the last several decades. Before turning to this, however, let me dispel the picture-puzzle analogy and put in its place an image which, as the sequel will show, portrays the function and the competence of science in a more faithful way.

Allegories are not to be trusted in detail, yet they are pleasing guides for our imagination and are harmless if their limitations are understood. Making such allowances, I return to my favored previous example, a crystal growing in the amorphous matrix of its own substance. The matrix, which should be regarded as indefinite in extent, is our primary unorganized experience, formless, ever evolving new and unknown features in defiance of reason and of regular expectation. I mean the unpredictable observations, the emotions, the spontaneous ideas, the esthetic reactions, the surprises, the hopes and fears that assail us without being deliberately engendered by our thoughts and actions. These are the raw material for the scientific process. Their lack of order, capricious and challenging, sets for us the task of rationalizing them by "bringing them under concepts," to use a phrase coined by the philosopher Kant. The ingredients of primary experience thus become crystallized, organized by relations which create regularity and predictability; the crystal of science begins to grow. Where there was chaos within our experience there is now order; every item falls in its proper place and causal prognosis of facts becomes possible within its domain.

A crystal in a melt does not differ in chemical composition from the melt itself, except in the regularity and symmetry of its constituent molecules. The same is true of the crystal of science: To convert knowledge into scientific knowledge does not alter its substance, it merely imparts regularity and predictability. And the growth is beneficent, not malignant; it never kills but enhances the potentialities of experience for good or evil—as man decrees.

Our allegory reflects other interesting features of the scientific process: As a crystal grows it heals its flaws; so in science, its advance is self-corrective, errors are discovered in retrospect and a whole region of thought may suddenly change its structure as new facts and ideas emerge.

The most important feature, however, is this. Many of you have doubtless watched the erratic manner in which a crystal expands in its own liquid.

There is a lecture demonstration, often performed in courses in physics and chemistry, which makes the process beautifully visible. The "seed" from which growth begins has a random position; nobody can tell where the organizing trend will start. When a small crystalline cluster has been formed, the direction in which it will extend itself is equally unpredictable; sometimes a series of small regular facets appears, at other times the growth will stretch a long arm toward nothing. Or a new seed may form in some separate location and an isolated growth will start. Such randomness, such incalculability inheres likewise in the development of science, and it is liable to make all extrapolations based upon the scientific movements of the past a mockery. It is for this reason and for some others that the predicted crises do not impress me very much.

The crystal image I have advocated has one serious fault. It suggests that growth will stop when the liquid matrix is used up. The scientific process never stops because the matrix in which science forms is intrinsically unlimited in extent. We have an infinite capacity for primary experience; new facts continually arise from the fertile ground of being, so that the amorphous substance of experience not yet regularized by the scientific process is always vastly larger than the volume occupied by the crystal. Science, it seems, forever strives to capture and convert the domain of unformalized knowledge but will always lag far behind in its dynamic reach.

There will never be a condition in which science has attained full control over all experience; the golden era or, if you please, the phase of machinelike regimentation will never come. By way of pretense at a mathematical characterization of this situation I would be tempted to say that the ratio of what we know scientifically to what we do not understand in scientific terms has been, is and will forever be zero. This is the reason why I believe that science will never do away with politics, poetry and religion.

Science is a very limited force in every individual life. A moment's thought will reveal to you how very few of the important decisions you are called upon to perform are made, or could be made, on the basis of science. The domain of that discipline is unexpectedly small, but within it science has tremendous power.

We have seen also that its growth is erratic, its progress as incalculable as the human spirit. I now want to emphasize that the relation between science and the human spirit is not as remote as is sometimes believed, that science has as large a share in spiritual concerns as any other human activity. To elaborate this last point I shall devote the remainder of this discourse to the theme: *Modern science and man's spiritual or cultural concerns.*

In the English-speaking world—and almost exclusively there—science is invariably coupled with technology; the two words appear in rigid conjunction in practically all official utterances, indicating an almost general belief that the results of science are purely, or at any rate predominantly, technological. The movement from scientific discovery through experimentation, through the construction of new laboratory apparatus, commercial devices, advertisement and finally the sale of new goods which affect our way of living, our external milieu, is obvious and commonplace, needing no elaboration here. That development proceeds under the neon lights of general publicity and is aided by all the resources of our technological society. As I mentioned earlier, its duration has been greatly reduced so that at present the period between discovery and commercial use, which lasted for centuries in the past, involves only a few years.

But we are not sufficiently aware of another, obscurer movement which is launched whenever a fundamental scientific discovery is made. For such a discovery necessarily challenges some former belief, pronounces apostacy to some cherished ideas and therefore clamors for accommodation in a new structure of thought. It calls for fresh theory, enlarged understanding, often occasioning changes in our basic conception of the world and our conception of man. Sooner or later these effects of scientific discovery are felt in such distant areas as the theory of knowledge, in metaphysics, politics and ethics. No area of human concern is altogether immune to the germ that is born when a truly deep scientific discovery is made.

Let me illustrate this obscure development by a few examples. The year 1642 is notable in the history of science because it marks the birth of one and the death of another scientific genius. In that year Newton was born and Galileo died. The great achievement of these men was the discovery of the laws of motion, the foundation of the science of mechanics. I shall pass over the historical details that led from there to the machine age and single out for comment the diffusion of that discovery into the fields which we do not ordinarily associate with science: philosophy, moral theory, politics and ultimately the arena of international history.

The striking features of the laws of mechanics are their simplicity and the stringency of their control upon nature. They have been called categorical and absolute because they seem to tolerate no exceptions. This posed several philosophical questions: Why does nature behave in ways that appear simple to the human spectator, why does the cosmos reverberate in human syllogisms and exhibit the solution of low-order differential equations? Why the causal consistency, the unique regularity in the motion of bodies?

Many philosophers supplied answers to these questions, none succeeding as well, however, as Immanuel Kant. If you study his teachings in isolation, out of their historical context, you find them abstruse, implausible and repulsive. His greatness becomes apparent only when you project them upon the background of Newtonian physics. For he answers precisely the questions I just raised. Regularity and uniqueness, order and categoricity arise, he says, because man invests the universe with the forms of his own reason; he sees features which he himself has inserted in the world. You may recall his theory of time and space as pure forms of intuition, his categories which regularize the "rhapsody of perceptions," all designed for the purpose of making such laws as those discovered by Galileo and Newton understandable.

Science had thus invaded the theory of knowledge, but the invasion did not end there. After the *Critique of Pure Reason*, published in 1787, Kant wrote the *Critiques of Practical Reason and of Judgment*, and here he propounded the view that human actions ought to be and ultimately are controlled by laws as rigid and inflexible as those of the physical world. He formulated the basic law of ethics, the categorical imperative, and evolved that stern concept of duty, of obligation beyond pleasure and exterior ends, which swept over, and dominated the moral fabric of Kant's world. Nor did it remain isolated within ethics. Kantian duty became the backbone of the Prussian state, took on political forms in Germany's civil service system called *Beamtentum*—and the imaginative historian has no difficulty in tracing connections between that rigid ideology and some of the wars fought during the last century.

The effects of science can not be confined to the limited domain we call science in the English language. Science is more than technology; its implications and effects are as wide as man's mind.

The very philosophies that govern current thinking are grounded in scientific discoveries. At the beginning of the 19th century physicists and chemists turned their attention away from the mechanical behavior of bodies and studied the phenomena of heat. They came upon the famous laws of thermodynamics, whose interpretation happens to be quite different from Kant's suggested explanation of mechanics.

Basically, they are statistical laws affirming regularities which arise from the chance motions of large numbers of molecules and therefore permit exceptions. The odds that water will freeze on a hot stove are small but finite despite the law of thermodynamics which interdicts such a happening. Thus, science found that some of its laws can be contradicted; actual events may slip through the net of scientific analysis and defy the laws. A philosophy which takes due cognizance of such occurrences is one that must reckon not only with the laws of science, but also with the discrepant, existential, positive contingencies actualized in human experience. That philosophy is called positivism or, with certain variations and differences of emphasis, existentialism.

While these examples show that scientific accomplishments are often the seeds of new philosophies, others could have been selected to indicate that philosophic views can induce a climate favorable to, and sometimes directly responsible for, scientific discoveries. Hence I do not wish to argue that the relation here illustrated is irreversible. What needs emphasis in the West today is the competence of science to influence philosophy and through philosophy other phases of the human spirit.

Elsewhere, especially in Communist and Eastern cultures, the philosophical significance of science is fully recognized, at times grossly exaggerated. It is to our peril I think, that we counter the Russian argument purporting that all philosophy depends upon the features of material science, with the inane doctrine of the philosophical neutrality of science, alleging that the latter can be interpreted in any manner one wishes, according to one's taste. This claim is neither true nor even expedient propaganda; in fact in renounces wholesome opportunities for exposing errors in the Communist ideology with which the doctrine of neutrality never allows us to come to grips.

If science does have philosophical implications, the interesting question arises as to what messages modern science has to convey. What is humanly significant about the approaches and the substance of modern mathematics, physics and chemistry, astronomy, psychology, sociology and the life sciences? My own limited competence forces me to restrict my comment to the first three of these disciplines, but I am sure that a person versed in the social and the life sciences would have an equally important story to tell. Since I have dealt with this theme at greater length in *Open Vistas*, I shall content myself here with a résumé and mere allusions to details. Five headings will be chosen as clichés which, by themselves, mean little. When filled with substance, and reflected upon at leisure, they will be found to yield the skeleton of a future philosophy, not a philosophy of scientism nor a shallow humanism, but a panorama which raises man's eyes to new heights and gives him courage to meet the future.

## 1. THE DECAY OF MATERIALISM

Materialism, as I understand it, is the systematic elaboration of two basic beliefs: One, that everything existing is composed of matter, and two, that matter behaves in accordance with the ordinary laws of moving objects, e.g., Newton's laws of motion. Like most philosophic terms, the word materialism has a wider variety of meanings, some of which, like "dialectical materialism," have very little to do with the original sense of the word. There is in fact no philosophic doctrine which has changed as rapidly, has proved as flexible, as "diamat" which did indeed spring from the view I am discussing, i.e., the view based on the two theses mentioned, but has now departed notably from them.

The scientific events which contributed to the decay of materialism are themselves a fascinating story, especially insofar as they concern the failure of science to reveal the metaphysical essence of matter. Initially it was taken in the Greek sense of *pleon*, the full, as that which fills space without having internal structure. This idea developed without much change into the chemist's impenetrable stuff which still underlies the thinking of most people who are untrained in physics. That science, however, experienced some baffling disappointments in its quest for the nature of the stuff called matter. To the question, does matter fill *all* space?, it received a negative answer in the various experiments that initiated the theory of relativity, experiments which revealed the non-existence of the all-pervading continuous medium called the luminiferous ether. Matter had thus retreated into the hard globules known as atoms, and these were thought to move in an otherwise empty space.

Further investigation, however, resolved the atom into tiny particles, electrons and nucleons, which according to Bohr's early conception whirl about in planetary fashion, therby installing the harmony of the spheres in the very microcosm. But alas, this lovely picture proved erroneous, for the "onta" (nucleons and electrons) cannot truly be said to move at all; they are afflicted by the renowned dualism which leaves us uncertain as to whether they are particles or waves. Even if they are particles, the last remnants of the original matter, the space they fill within the atom is infinitesimal in comparison with the size of the atom itself. If the atom were magnified until it occupies this room, its nucleus would be no larger than a barely visible speck of dust and the electron the size of a marble. The remainder of the space would be vacant. And still this is not the end of the story: It has become very doubtful whether the speck of dust and the marble are full of stuff; they may indeed be nothing more than mathematical singularities haunting space.

Such has been the fate of matter, the only stuff materialism regards as real. Turning now to the second thesis, according to which matter, presumably in the form of particles (whose existence is, as we have seen, beclouded by doubt), moves in the manner of visible objects, we witness a similar debacle. The laws of mechanics, we now know, do not hold for the objects of the microcosm; they have been replaced by the laws of quantum mechanics, which are so strange as to be almost incomprehensible. What concerns us here is the fact that they do not allow us to picture the progression of an electron through space, the process we ordinarily call motion, as a continuous passage at all. Strictly speaking we cannot assign to the denisons of the microcosm such attributes as position and velocity but are required to treat them in a manner so abstract and shadowy as to destroy every vestige of pictorialization in the representation of their behavior. This likewise destroys the validity of the second platform of our doctrine, leaving materialism not only invalid and outmoded but without residue of scientific substance whatsoever. Materialism was a respectable philosophic view at the end of the 19th century; it has now become an anachronism.

## 2. THE BANKRUPTCY OF "COMMON SENSE"

Without further explanation this cliché, I fear, is meaningless or at least misleading. The trouble is with the definition of common sense. men of affairs claim to have it and to rely on it successfully, but even they have difficulties when asked to say just what it is. Some call it logic—but logic is much more restricted and formal than common sense. Some call it intuition, which is tantamount to replacing one mystery by another. Others equate it to experience, an equation which is trivially true and no definition at all.

Yet when the challenging and unexpected consequences of the theory of relativity, for example the retardation of moving clocks and the shortening of moving objects, the failure of simultaneity and the increase in mass with motion, were first announced, philosophers professed scepticism and disbelief on the grounds that these effects contradicted common sense. Experimental proof has now convinced them that these consequences are true, but many still balk at the so-called "twin paradox," which affirms that a travelling twin will, when returning to meet his brother, be younger than his brother. Again it is said to be common sense which prevents them from accepting this verdict.

The uncertainty principle of Heisenberg alleges that when the velocity of a "particle" is known with precision, its position cannot be known at all. Well, says common sense, this may be true because of the imperfection of our measuring instruments; but when the physicist goes on to claim that precise knowledge of velocity precludes all possibility of knowing a particle's position, or that strictly speaking it has no position at all, that, according to common sense, is incomprehensible and certainly not true. It insists that a particle must at every instant possess both a position and a velocity; quantum mechanics must therefore be wrong.

Common sense has often made similar strictures in the past. When it became apparent that man's mind was not located in his brain, nor in his heart, nor in his stomach, that indeed one could not attribute to it any determinate position at all, this monitor tended to say the mind did not exist. Yet psychology was not disturbed by that suggestion. Common sense was merely transferring what it knew of physical science into the new area of psychology which it did not understand.

Rather than multiply the list of instances in which the facile predictions of common sense have gone wrong, let me distill from them an *ex-post-facto* definition of this versatile agency: It is simply the familiar set of generally accepted facts and ideas which advancing science leaves in its wake, the residue of its accomplishments that has found entrance into the understanding of the "man in the street."

In the past the difference between what went on at the forefront of science and what it leaves behind was not very great; to be sure, the outcome was more refined than the raw material it took in, but the categories of conception were the same. Today, input and output are even qualitatively different; the man in the street is baffled and incredulous, and he makes a last frantic effort to assert himself in appealing to common sense. But science disavows it, discredits it and entrusts itself to the rigorous regime of logic and mathematics, even if they lead into unfamiliar territory. And strangely, this new attitude gives the worker at the forefront of science no concern, for he knows full well that sooner or later his astonishing findings will become part of what is called common sense, which is not a monitor but a laggard who stumbles along behind the scientist's advance into the unknown.

The recognition of the nature of common sense and its disavowal before rigorous principles is, I believe, one of the pregnant philosophic principles of modern science, making it more venturesome, less bound by tradition. Philosophy will doubtless capture this breezy spirit when the time is ripe—its conditioning by the thinking of the past is still too strong.

## 3. RENUNCIATION OF MECHANICAL MODELS IN SCIENTIFIC EXPLANATIONS

The present item is closely related to the preceding one. I have indicated that physics has surrendered the idea of continuous passage through space, the facile concept of visual motion, whenever it applies to microcosmic entities. This, together with other similar developments, has marked the end of a period in which science sought explanations in imageful mechanisms. Some 80 years ago men like Hertz, Helmholtz and Maxwell gave expression to the conviction that no physical explanation can be complete or ultimately satisfying unless it provides a model composed of masses held together by rigid joints or attractive forces, moving in some pictorially plausible fashion—in short, unless it constructed in thought a miniature machine capable of producing precisely the observed phenomena and none that are not observed.

So strong was this tendency that men were willing to accept the model of an ether molecule, invented by MacCullagh for the sake of assuring that the ether will transmit transverse and not longitudinal vibrations, a molecule composed in accordance with directions quoted on p. 275.

The experiment of Michelson and Morley in 1887 then led to the conclusion that there was no ether: space was completely empty. More recently, however, the ghost of the ether has begun to haunt the recesses of physics in uncanny ways. Dirac's electron theory suggests the existence of an infinite number of negative kinetic-energy electrons filling all of space. There are thought to be unobservable. But a single vacancy in their spatial distribution, a so-called hole, may be shown to have the properties of a positron. Perhaps, then, space is not empty; the ether has changed its character, but in a most unmechanical fashion.

It is true that physicists began to feel uneasy at this proposal and began to wonder whether picturesqueness and mechanical conceivability were not being purchased at too high a price, the price being simplicity of conception. Such wonder grew into doubt, and the doubt congealed into what I do not hesitate today to call the certainty that there exist many things in the world, indeed things of interest to physical science, which cannot be pictured in terms of mechanisms. Particles to which one cannot assign position at all times, systems that pass from one energy state to another without ever manifesting themselves in intermediate states, entities without mass, fields whose substance is as tenuous as a probability, have begun to interest the physicist seriously. The freedom he has gained by relinquishing the ancient Rube Goldberg mechanisms has brought him unexpected power and somewhat esoteric satisfaction. Instead of relying on visual clarity, as Descartes once did, he commits himself to the rigor of abstract thinking, using mechanical models when they are available, to be sure, but not trusting them to the limit.

This aspect of the new style of science is bound to enlarge the concept of reality which plays an important role in philosophic thinking. Physical reality has taken on qualities that are rather far removed from direct sensory experience.

### 4. FREEDOM

In spite of the claim made by a few physicists, quantum mechanics has not demonstrated human freedom. It is true that atomic events are uncaused in the usual sense. Only large statistical aggregates of observation conform to deterministic laws; single events do not and are therefore unpredictable, random, and in this sense uncaused. It is also true that in certain known instances the indeterminacy of the microcosm enters the world of large-scale phenomena: A single neutron which is subject to the causelessness of the atomic domain can set up a chain reaction in a block of fissionable material and produce a disastrous explosion; a single X-ray photon whose passage is again outside the causal chain of events can enter the nucleus of a living cell and produce a mutation that destroys or improves the organism; the visible light which falls upon my retina, if sufficiently weak, partakes of the randomness of atoms and yet may be the stimulus for an important action. While it is very doubtful that *all* human acts thus arise in the turbulent world of atomic magnitudes where events are free of causal chains, some certainly do, and these events could never be predicted.

But there is a difference between turbulence and guided motion, between chance and freedom. If the picture I have tentatively drawn, which represents man as an amplifier of atomic uncertainties, were indeed correct, then man's actions should be erratic and unpredictable but not free: they would be determinate in the aggregate, random in their individual occurrences. Chance and freedom are not the same. Freedom requires a measure of chance as its very precondition; it is impossible without the indeterminacy which chance provides. But it is more than chance; it is chance plus choice. Man must be able to choose responsibly between causally indiscriminate alternatives in order to be free. Quantum mechanics demonstrates the existence of chance; it says nothing about choice. A solution of that problem lies at present beyond the competence of physics.

Why, then, was freedom included in the list of philosophically interesting features of modern science? Because its status as a *problem* has been restored even if it remains short of a solution. The materialism of the older science left no room for it. If all events in the world were controlled by the laws of classical mechanics, absolute determination would be a demonstrable fact; fatalism would be the only reasonable attitude toward the future. In principle that science denied the chance which is the precondition for choice and freedom.

I should add that many philosophers will not grant this point. There were theories in the days of classical physics which tried to make a case for freedom. Some of them simply denied the applicability of the laws of physics to the circumstances surrounding the human will, creating as it were a dualism of mind vs. matter. Others explained freedom in terms of ignorance or error, leaving man with a consciousness of choice because he did not know the precise state of affairs which in fact had already predetermined the outcome of his action.

Modern science makes such complex philosophical arguments unnecessary. It shows forth the possibility of chance in the physical world and leaves the debate without further comment. And by this unpretentious move it converts human freedom from a dead issue into a challenging problem whose solution it assigns to other hands.

The effects of this change are already evident on the current philosophic scene. The newer philosophies, particularly the various existentialisms, make much more of freedom than did their predecessors. The connection between this development and the new style of science is almost never recognized: the obscure movement is forging ahead whether we consciously attend to it or not.

## 5. INVOLVEMENT

The last generation thought of nature as a spectacle which is basically distinct from man the spectator. This cleavage was thought proper became science prided itself on being able to view the world without disturbing it. You could and did alter nature by acting upon it but not by merely observing it. There were natural *facts* ascertained by observation, and there were human *acts* designed to effect changes.

In the sphere of animate things, particularly in social and economic matters, the distinction between spectacle and spectator was not so firm: the observation sometimes intrinsically altered what was being observed. Knowledge interfered with what was known. Certain kinds of self-fulfilling knowledge (a bank is going to fail; inflation is sure to come), defeatist attitudes whose mere presence aside from deliberate acts caused changes in the world, had been recognized. These, however, were thought to be peculiar to the living world.

Meanwhile, feedback was discovered as an important regulating principle even among inanimate things. Engineers, physicists and mathematicians seized upon it to construct regulating devices, calculating machines, and mechanical brains. Feedback, however, is a transaction between physical phenomena that stand in a relation of cause and effect.

Heisenberg's uncertainty principle finally made the culminating claim that there exists in the physical world not only feedback between objective processes, but feedback between the knower and the known, a condition which one might call epistemological feedback. The very process of perceiving an atomic entity—and thus in principle all entities—entails a change in what is being perceived and this change, so Heisenberg and others have demonstrated, is as incalculable and unpredictable as it is unavoidable. In contrast with the feedback encountered in engineering, epistemological feedback lacks the element of control, placing the yield as well as the physical effect of every observation among the inchoate, spontaneous contingencies of our experience. There is no longer a field of mere occurrences, separated in principle from a field of actions. Man's involvement in the world is complete and irremediable; to put it briefly I might say that all his facts have been turned into acts.

Now I think the practical consequences of this consideration upon philosophy will be slight, for philosophers have never taken the physicist's claim to be able to observe the world without essential immersion in it very seriously. One may even look upon the uncertainty principle as the physicist's belated concession that he is, after all, no better off than scientists in other areas. Yet this very admission, based as it is upon remarkable new findings, is to me immensely encouraging because it manifests a concurrence, a convergence of philosophic attitudes among the different sciences which suggests that in all probability our knowledge marches toward a unique, coherent goal.

### 6. DISPERSAL OF DOGMATISM

This goal is the object of my last cliché, the dispersal of dogmatism. I had originally intended to call this section "The new faith of science," a phrase which seems to stand in contrast with the one I chose. Let me state at once, therefore, that I regard faith and dogma as antonyms. A dogma is a set of static beliefs which one is not free to doubt; faith, however, presupposes doubt, results from doubt as an act of commitment in circumstances where one can not know for sure. Moreover, faith is a dynamic thing which may alter its allegiance for good reasons when new evidence makes the old commitments absurd. Modern science is inspired by this kind of faith.

To see this it is well to turn back and review the teachings of mathematics and physics in the last century, when, as I hope to show, their attitude was dogmatic. The starting point of each of these disciplines was believed to lie in certain axioms, and these axioms were unalterably, indubitably and ultimately true, as I have already indicated in the beginning of this lecture. Thus, prior to the middle of the last century, occasional suggestions which cast doubt on the axioms of Euclidean geometry were not taken seriously. These propositions concerning points, lines, triangles and parallels were obviously true by inspection; no sane mind could doubt them because they were implanted in man's reason by God or else they shone forth by some *lumen naturale*. These were absolute, perennial truths, and the details of science somehow flowed from them. They, too, could be proved in ways that would make them certain forever. The monumental documentation of this philosophy is found in the system of Thomas Aquinas.

About 100 years ago a remarkable mathematical discovery was made, a discovery which resulted in the invention of non-Euclidean geometries. Experts seriously questioned the necessity, and then the validity of one of Euclid's postulates. They went beyond the stage of formal doubt and constructed systems of geometry actually denying the so-called parallel postulate, and their endeavors, which the dogmatists hoped would lead to nonsense, actually created homogeneous, clear and interesting bodies of theorems quite safe from internal contradictions. Which of these axioms, then, were true, Euclid's or Lobatchevski's? Evidently both made claims to truth, and a discrimination as to their acceptability required recourse to other kinds of evidence, namely empirical evidence.

Axioms could no longer be held to be indubitably true, to carry within themselves the affidavits of their validity. They turned into postulated, basic hypotheses which are maintained as long as the consequences entailed by them agree with the facts of experience. The mathematician does not rest secure in knowing them to be true; he avows them for the purpose of explaining the world, much as the moral person dedicates himself to norms for the purpose of satisfactory living. I do not wish to maintain that these two acts are comparable in all respects, nor that they involve the same sort of emotional dedication; what I do assert is that they are logically the same, that one can find absolute certitude neither in the starting points of scientific knowledge nor—in spite of the absolutist claims made in traditional ethics—in the starting points of responsible human action.

What happened in mathematics also took place in physical science. Space and time, previously regarded as Euclidean and infinite in an axiomatic and unalterable sense, were now conceived as manifolds about which reason was not the sole arbiter. Conservation principles, like those concerning energy and mass, were seriously subjected to doubt. In short, every affirmation of absolute truth in science immediately gathered about itself a dense cloud of suspicion, and the present state of affairs may be described as follows.

Science harbors no absolute, no final truth. It has its beginnings in commitments to postulates whose consequences are continually being exposed to the test of experience. When new facts contradict these consequences the postulates are altered, usually in minor ways, but they sometimes suffer complete rejection. Science is thus an open enterprise, driving ahead toward new knowledge, ready to modify its basic propositions—the former axioms—when the facts demand that sacrifice. Science recognizes eternal questions, but spurns eternal answers. Its challenge it not in the discovery of truth that is as dead as a known fact. The scientist is inspired by an idea of truth that is elusive, forever changing while warming his heart in affording him the conviction that his reach is coming closer and closer to his ideal, even though he knows he will never hold it fully in his grasp.

I am now ready to tell my version of the ending of the story of Sais. The youth, as he lifted the veil, saw engraved on the temple wall a message such as this:

> Only a fool looks for truth in a finite formula; only a knave would want to acquire it without toil and heartache. Final truth is tantamount to stagnant knowledge; there is no substitute for self-correcting, progressing, self-improving understanding. Dismiss your quest for truth in final formulation and embrace the greatest human virtue called Eternal Search for Truth.

Apparently the shock of this message destroyed a feeble soul that insisted on truth by revelation.

# PART IV

# ISSUES BEYOND THE BOUNDARIES OF PRESENT SCIENCE

# PART IV

The last part reviews philosophic issues widely current at the time the articles were written and ends with a sympathetic approach toward ethics and religion. The chapter entitled "Metaethics" is the precursor of a book entitled "Ethics and Science", Van Nostrand, 1964. Its contents are, I believe, especially relevant for our time.

CHAPTER 19

# PHENOMENOLOGY AND PHYSICS

### INTRODUCTION

The purpose of the present paper is twofold: in the first place, it is intended to acquaint the student of physics with the essential ideas professed and vigorously defended by one of the foremost schools of modern philosophic thought, phenomenology; in the second place, it is to state a reaction to phenomenological doctrine which may perhaps be regarded as typical of the working physicist with an interest in methodology.

It is essential at the outset to guard against a misunderstanding of the meaning of phenomenology which has occasionally arisen and is indeed suggested by the name. The physicist sometimes distinguishes between a phenomenalistic explanation and a dynamical explanation. A phenomenalistic theory is one which moves on the surface of physical phenomena and endeavors to describe them in terms of observed appearances rather than in the thoroughgoing manner provided by dynamic and causal theories. Thus the word phenomenology is likely to suggest to the physicist something bound to the appearance of things, rather than characteristic of their essential nature. The philosophic doctrine of phenomenology lays claim to thoroughgoing explanation. In fact, the term phenomenology is used in this movement as referring to the most fundamental back ground of all experience. In this sense, then, phenomenological explanation is the most basic type available.

Historically, the fundamental structure of phenomenology was established by Edmund Husserl, who was a pupil of Franz Brentano, and whose studies were strongly influenced by Kant. His two principal works are: *Logical Investigations* (1900), and *Ideas—Pure Phenomenology* (1913). It is in the latter book that the main thesis of phenomenology was developed. Most of the remarks in the present paper will be based on it.

### 1. GENERAL THESIS OF PHENOMENOLOGY

Husserl notes that our experience contains two types of facts, contingent and necessary ones. The proposition "this desk is brown" is contingent, as are most propositions expressing ordinary external or even scientifically accurate experience. But, it is claimed by phenomenology, there reside within all contingencies certain correlative necessities which it is the business of philosophy to uncover and study with care. Thus, while the proposition "the table is brown" is contingent, the statement "the table is not both blue and brown all over" is a necessary proposition which cannot be denied.

Examination of pure consciousness reveals a vast variety of necessary *a priori* truths of this general nature which science and philosophy in general have failed to consider adequately. To quote Husserl[1]: "It belongs to the meaning of everything *contingent* that it should have essential being and therewith an *Eidos*[2] to be apprehended in all its purity." "The essence (Eidos) is an object of a new type."

In further illustration of the fundamental difference between contingent and eidetic truth, let us consider two other examples taken from Husserl:[3] "All bodies are heavy" is a statement expressing contingency, whereas "All bodies have extension" implies an eidetic truth. From the point of view of modern physics these propositions appear perhaps unwisely chosen, and it may well be said that "bodies," such as electrons, appear in modern physical theories as point singularities and have no extension. One may well take the attitude that the distinction between the two types of proposition is not sharp and may be meaningless. This is the attitude of the empiricist. It gains evidence on observing that within the history of philosophical and scientific thought, propositions of the type here called eidetic have constantly been on the defensive against attacks upon their eidetic nature. They have been forced to give ground continually and their number seems to be diminishing. What once appeared as a necessary statement has been converted into a contingent one by the enlarged horizon of scientific alternatives. One wonders in this connection, therefore, whether all eidetic truth is destined to be ultimately converted into a matter of contingencies. The majority of physicists nowadays would probably take the attitude that this transformation is gradually taking place; however; few would be dogmatic in their attitude, and all would admit the transformation to be beyond logical proof.

How do we arrive at a philosophical verification of eidetic knowledge? Phenomenology gives the answer: by examination of pure consciousness and *Wesensschau*. What is seen confers immediate evidence, and seeing implies not merely sensory perception but also intuition of ideal objects, and of the essential structure of things and consciousness. To quote Husserl:[4] "Immediate 'seeing,' not merely the sensory seeing of experience, but seeing in general as primordial dator consciousness of any kind whatsoever, is the ultimate source of justification for all rational statements. Essential insight is a primordial dator act, and as such analogous to sensory perception, and not to imagination." The facts of logic, mathematics, and geometry are viewed by intuition and thereby attain sufficient evidence to be accepted without doubt. It is interesting in this connection that Husserl considers Euclidean geometry as being immediately evident and therefore indubitably correct. Here again, modern science has witnessed the transformation of a statement which was eidetically true in Husserl's day into one of merely contingent validity.

It is the conviction of the phenomenologist that the study of the eidos, which reveals itself in pre-analytic consciousness, will lead to a philosophy which is presuppositionless and therefore completely scientific. Some exponents of the phenomenological movement criticize such disciplines as traditional logic, which is reared upon preconceived structural bases. They regard logical axioms, like the law of contradiction, as constructions with which many experiences, to be sure, do conform. These axioms are imperfect, however, in that they are not derived through immediate *Wesensschau*, that they do not result from a study of pre-analytic immediate experience. This fault becomes apparent when the manifold relations of experience such as contrariety, becoming, change, which cannot be subsumed under the law of contradiction, are more closely inspected.

If phenomenology is to be a truly scientific discipline, it must exercise extreme care with regard to ontological problems. It must not, for example, accept elements which declare themselves within consciousness as really or objectively existing, even though this may be the status granted them in ordinary discourse. It must, in fact, refrain as far as possible from conferring real existence upon anything.

This concession has forced Husserl to adopt what he calls the "phenomenological epoché." The word epoché means abstinence. It implies the suspension of all existential judgments, the "bracketing" of the supposition that an external world is the cause, in a metaphysical sense, of our knowing about it. This bracketing, however, does not wipe out external things, and they are still present as objects of our thought.

To quote Husserl[5] again: "Consciousness in itself has a being of its own which in its absolute uniqueness of nature remains unaffected by the phenomenological disconnection (epoché). It therefore remains over as a 'phenomenological residuum,' as a region of Being which is in principle unique, and can become in fact the field of a new science—the science of Phenomenology."

Let us look more closely at the fate which befalls the external thing after it has been bracketed out of our philosophy in the manner here advocated. It is gone, to be sure, as a metaphysical entity, but it is still with us as an "intentional" object, that is, an object of awareness found within consciousness. The present terminology has been taken over by Husserl from Brentano, whose analysis of the meaning of the idea of "intentional" is well known to philosophers. Thoughts, desires, volitions, acts have objects (not necessarily external objects) and these, according to Brentano, are *"intended"*.[6] Intended objects can be studied through what Husserl calls "immanent perception." Thus it would be claimed that the property of a table's inability to be both blue and red all over belongs to the table as an immanent entity. It could be discerned by inspection of the phases of our consciousness even if a real table had never been seen.

Perhaps the physicist should be cautioned at this place against committing the genetic fallacy which would lead him to ask: How can a person who has never seen a table know anything about it? Would a new-born baby, if it were able to utter sense, have this particular knowledge? Even the phenomenologist would probably answer the questions in the negative, but he would hold that it does not matter how knowledge is genetically acquired. What counts in any analysis of articulate knowledge is not its history but careful discernment between its necessary and its accessory elements.

Immanent perception of intentional experience yields, in fact, indubitable truth, while "transcendent" perceptions, transcendent in the sense of leading beyond consciousness, may be beclouded by doubt. Husserl therefore speaks of the indubitability of immanent and the dubitability of transcendent perceptions. "Every immanent perception necessarily guarantees the existence of its object. If reflective apprehension is directed to my experience, I apprehend an absolute Self whose existence is, in principle, undeniable; that is, the insight that it does not exist is, in principle, impossible. It would be nonsense to maintain the possibility of an experience given in such a way *not truly existing*."[7]

After this brief characterization of the main thesis of phenomenology, let us turn to an examination of the essential method used in science. This will revolve about the central problem: the meaning of a physical thing. Having discussed its solution, we shall return to take a second look at phenomenology with a view to discovering its principal shortcomings in the face of the methodology of physics.

## 2. EPISTEMOLOGY OF PHYSICS

In presenting a theory of the fundamental method employed in physical explanation, it is well to adhere to the phenomenological epoché to this extent: our theory, while metaphysical in the sense that it includes epistemology, shall be non-ontological; it shall neither start with ontological premises nor aspire to ultimate ontological verdicts. The problems of the nature of being in their traditional setting will emerge, however, in one phase of the development; they will be shown to be solvable in a certain way by the prescriptions of the formalism to be presented. They will be shorn of their traditional importance and appear, we feel, as rather uninteresting in every scientific sense.

Let us start, then, in good epistemological fashion and in accord to a large extent with the precepts of Husserl, by looking about our experience in order to discover its essential elements. There will be discerned two classes of "things," the distinction between which, while not easily formulable in terms of definitions, is intuitively evident to all: sense data on the one hand; and on the other more abstract or formal things which somehow correspond to sense data but cannot be identified with them. E.g., a force belongs to the second class while the kinesthetic sensation of a push belongs to the former; heat belongs to the latter but the sensation of warmth to the former. Elements in the second class will be called "constructs" for reasons soon to be stated. They enter significantly into scientific theories.

It is not implied that consciousness can be exhaustively classified by this simple principle of division, for admittedly it leaves out of account many psychological elements of great importance (feelings, volitions, judgments, etc.). But it exposes clearly not only the essential basis of scientific knowledge, but also a contrast between its parts which is too frequently overlooked.

Little need here be said about sense data, for the term is well understood by all. Their character is exhaustively stated by a few simple propositions: we simply "have" them; they are beyond our essential control. It should also be pointed out that they are neutral to reason, that they emerge in our consciousness in a spontaneous manner in which we have but a spectator's limited power to interfere. They are perhaps most adequately described by Kant's poetic phrase, "the rhapsody of perceptions."

To the sum-total of all sense perceptions we shall apply the term "sensed nature," leaving open at present the question as to whether sensed nature shall include only presently actualized sense perceptions, (in which case the meaning of nature would be very limited indeed), or whether it shall include all present sense perceptions as well as past ones carried in memory, or finally the class of all possible sense perceptions, latent or actualized. It is important to recognize, however, that the physicist, in a process later to be called "verification of theories," rarely makes reference to anything more inclusive than sensed nature in the first and most limited sense.

To elucidate the meaning of the word "construct" as it is here used, I shall first give a few further examples. When Newton formulated his laws of motion, presumably on the occasion of seeing an apple fall, he did not perceive the mass of the apple. The mass, however, figures prominently in his statement of the second law of motion. It is true, to be sure, that Newton saw the color of the apple, its shape, and that he could have perceived its weight by handling it. Perhaps by striking it with his hand he could have introduced the element of inertia or "mass" of the apple into his sensed nature, but that is as far as he could go. The mass with which the second law of nature deals has none of these incidental, ephemeral, contingent aspects. It is, as it were, a symbol which is placed over against a certain complex of sense data, a symbol which is then subjected to rational rules of operation.

It is quite true that almost all people, including the physicist, make very little distinction between the construct mass, by which is meant the symbol just described, and the complex of sensory awareness, also called mass or matter in ordinary language, with which the symbol is somehow associated. But we hold that philosophically the cleavage between the sensory part of such a term as mass and its constructional aspect is very profound indeed, and that no adequate understanding of the meaning of physical theories can be attained without its recognition.

To use another example, let us consider the construct "electron." No one could properly regard it as a part of sensed nature because it cannot be sensed. In fact, it is known to the modern physicist that every process of seeing it, such as illumination by means of the famous gamma-ray microscope, will alter its state in a most puzzling manner. Modern physical theories, if they are correct, permit a proof of the essential unobservability of the electron. It is true, of course, that the electron can be seen indirectly in cloud-chambers, heard indirectly in counters, and so on. But it should be perfectly clear to anyone that this sense of seeing and hearing is not the same as that in which the color blue is seen or a certain noise is heard. The electron is a construct which has a certain *correspondence* to various sets of sense data but is not to be confused in an ontological manner with these data themselves.

It is easy to cite other examples of constructs from any exact science, in fact, exact sciences display far more interest in constructs than in sense data. All entities involved in physical theories are indeed constructs which relate somehow, by rules of correspondence about which more will be said later, to specific parts of sensed nature. Trouble is likely to arise only because of the bluntness of ordinary speech which frequently allows no distinction to be made between a complex of sense data and its correlate, the construct.

To clarify this, consider the entity "table." When I say, while looking at the object before me, "the table is brown," I may put upon this proposition two different interpretations. (a) I may mean by table the perceived complex of hardness with a certain sensory shape, size, and height; then I merely wish to state that I find, while looking, the sensation of brownness among the other sensory qualities. The statement is then entirely about sensed nature. (b) On the other hand, I may mean that there is a certain *external object* called "table" which has an underside, though I am not seeing it, which has an interior not open to my immediate inspection, has the property of being brown, a property which could most definitely be analyzed by certain physical procedures involving the spectroscope. In the latter sense I am viewing the table as a construct and not as a complex of sense data at all. The rules of correspondence (to be discussed later) by which I am permitted to pass from the sensory table to the table as construct are very obvious, very immediate indeed and hence easily overlooked, but logically their presence can hardly be denied.

Among other things, let it be pointed out that the person who fails to distinguish between the table as a set of sensory qualities on the one hand, and as a construct on the other, can see nothing but confusion in the statement of the modern physicist: "the table consists of invisible molecules." If the table is a complex of sense data, how can it consist of invisible entities? But if it is always a construct, how can it ever exhibit the quality of being brown?

The reader will undoubtedly exact an apology for the use of the term "construct" in the present situation. Let it be said in the first place that a term antithetic to "sensory object" is clearly needed. The word "concept" would not do, for it is a generic term reserved for classes of things rather than the individual entities which are here called constructs. Finally, it was intended to signify by the choice of term the fact that it represents a functional entity, one that is freely created and not forced upon the observer by sensed nature. Certainly, Newton constructed the idea of mass in much the same way as Aristotle constructed the notions of natural and violent motion, and the act of construction must be regarded as quite different from passive conception. For this reason, I regard the term "concept" in the present connection as entirely misleading.

It should not be inferred, however, that a physical construct is a fiction. The methodology of physics shows what is meant by valid constructs; it outlines unambiguous procedures whereby constructs may be shown to be valid or invalid. In other words, it permits the elimination, within the realm of constructs, of figments from valid constructs. This paper, however, it not to be devoted to an analysis of scientific methodology, which has been given elsewhere.[8]

Let us return to a more detailed consideration of the rules of correspondence which form the link between sensed nature and the universe of constructs. They are indeed confusing in their variety, their analysis still forms an interesting subject of study. It is not difficult, however, to single out a few special types. Since, paradoxically, the matter is less involved in the more advanced portions of physical science, we shall start there and later descend to the level of the interpretation of constructs denoting the simpler objects of daily life.

What the physicist means by "mass" is perfectly clear on the constructional side. Once he is given the mass of an object, he can perform mathematical processes and conclude results quite unrelated to what is involved in the idea of mass. But how is this mass related to sense perception? The answer here is simple—the rule of correspondence is an operational definition. The mass is ascertained through a set of procedures which link sensory experience with the realm of constructs.

From pointer readings on a balance, inspection of weights (all sense data!) an operational definition allows the experimenter to arrive at the construct *mass*. When he says he measures mass he means that, having had a complex of sense perceptions, he applies some operational definition leading to a certain *value* for the construct, and this procedure defines the construct itself. The measurement of every other "quantity" (the physicist's name for a certain type of construct) may be described in similar terms. This situation is thus quite general in physics.

But how about the construct table, tree, or any other external object? I hold that here, too, there is involved a rudimentary type of operational definition which allows the percipient to pass almost automatically, in reflex fashion, from the data of his perception to his construct "external table," "tree," etc. The fact that the transition is made automatically and almost always without the percipient's awareness of it is psychologically interesting, but provides no grounds for denying the essential difference between sensed nature and constructs.

When the logical dichotomy is overlooked or involuntarily inflated into extreme significance, the traditional problems of ontology appear. If we state the matter objectively and without too much metaphysical pretension, all that is involved here is a truly immanent situation. Within consciousness, we find, first of all, elements obviously belonging to the realm of sensed nature. Simultaneously with them there appears in our consciousness a construct, or a set of constructs, carrying within them germs of rationality, logical affinity, and other formal qualities which do not attach to sensed nature. On closer inspection we find rules of one-to-one correspondence between these entities. This is the state of affairs to which ontology in fact reduces. The subject might well be described as the systematics of the rules of correspondence.

This is not the place to describe in detail what the physicist means by "explanation." It is interesting, however, to observe, from the vantage point now gained, the general outlines of the methodological process of explanation. In all his theorizing, the physicist starts with a set of sense impressions, passes from them by rules of correspondence to the domain of constructs.[9] These are related by theories which are postulated relations between constructs. The theories allow constrained movement from one part of the domain of constructs to another. Having arrived at any suitable place within the domain, the physicist may return to sensed nature, using the same rules of correspondence. If, at the place thus indicated in sensed nature, the physicist verifies his "prediction," the theory or group of theories involved has been verified. This is the essential structure of every verification.

The methodology of physics, as I see it, is a body of axioms which regulate late the use of constructs and fix the rules of correspondence. These cannot, to be sure, be derived in an *a priori* manner from any more basic considerations. They must be gleaned by painstaking analysis from the successful scientific procedures of the past. Among these axioms may be found some of the categories of Kant; their function is indeed very much the same as that of his categories of pure reason; but exception must be made to his claim that these categories be deducible *a priori* and be immutable.

Foremost among the regulatory devices to be imposed on constructs is the principle of causality, about which we wish to insert here only a brief comment.

Causality as it is here understood and as it must indeed be understood in order to do justice to the exact sciences, has nothing to do with sensed nature, but with the constructs in terms of which sensed nature is to be analyzed. It requires that the relations between these constructs be of a special kind called "causal." Furthermore, when the analysis is carried through completely, it turns out that all physical theories, including modern quantum mechanics, are perfectly causal disciplines.[10]

Having outlined a general methodological background of the science of physics, let us now inquire into the status of its "absolutes." In view of the foregoing considerations, only one meaning can be given to the term "physical reality." Since our analysis has made no reference to any *Ding an sich*, the definition of physical reality may not include anything so esoteric. For scientific purposes, it is best defined as the totality of all valid constructs. Whether or not it shall include sensed nature, latent or actualized past or present, is a matter subject to anyone's arbitration. Certainly the safest attitude to take is that physical reality, or the physical universe, be defined to include all constructs known to be valid at present.

Note that this defines not a static but a dynamic universe, that the universe grows as valid constructs are being discovered. Physical entities do not exist in a stagnant and immutable sense but are constantly coming into being. It may not even be possible to say whether a certain "thing" exists. The quality of existence thus reflects imperfections of the physical theories of the day, but I do not see how it could possibly be otherwise.

Many profound thinkers of the present day would disagree with the attitude just formulated. They would maintain a faith in the convergence of the system of the entire set of physical explanations. They would feel that the unity of all science is a state capable of ultimate realization. They fix their eyes upon an *ideal*, unique, ultimate set of constructs for which they would reserve the name "reality." No one can deny the usefulness of this ideal, and I doubt if any scientist would willingly give up his faith in the ultimate unification of scientific constructs. But since the convergence in question is not capable of scientific proof, and since reality can be defined without reliance upon ultimate achievement of unification and since, finally, science is possible without it, I have preferred a less ambitious definition of physical reality.

As to the term "being," it would be best to eliminate if from scientific and philosophic discourse. The word is used in ordinary language with so many conflicting meanings that standardization of use could never escape the accusation of artificiality from some quarters.

The attitude just formulated with regard to reality must be admitted to be none too edifying. It is felt even by scientists to be emotionally somewhat unsatisfactory, but we shall grant that it is all which science can yield. With this vague feeling of discontent, the scientist is inclined to turn once more to phenomenology with the timid quest for a more satisfactory definition of what is real. Phenomenology promises to supplement that plane of immediate evidence which the scientist possesses in his sense awareness by a vastly larger and richer field of pure consciousness. Let us then turn to the problem which is thus posed and which may be put in the question, "Is sensed nature the only field of departure or arrival in the process of scientific verification, or will inspection of the eidetic structures of consciousness function in a similar way as dator of scientific fact?"

### 3. THE NOTION OF CERTAINTY IN PHENOMENOLOGY

If the question just posed is answered in the affirmative, then the realm of sensed nature, as this term was used in the foregoing section, has been enlarged in an extraordinary way, and it includes not only the realm of perceptions but also the experiences of pure consciousness. It can, indeed, hardly be denied that certain phases of introspection have degrees of spontaneity similar to that of sense data and are beyond our essential control. The etymological implication of the term "nature" would cover them as well. Closer analysis, however, will expose a point of difference between sense experience and introspection which may destroy the feasibility of the proposed enlargement in the meaning of sensed nature.

The difficulty concerns the selection of criteria for distinguishing trustworthy from untrustworthy data. It seems that if we grant that intuitive seeing is a dator act in the same sense as visual perception, then there must be discriminating features which permit distinction between true and hallucinatory seeing in the realm of intuition as well as that of visual perception. Let us, therefore, examine how science manages to reject from the totality of visual perceptions those which it brands spurious, such as hallucinations. It is important to analyze this process, for no science would arrive at a self-consistent body of explanatory constructs unless it had a way of eliminating spurious observations.

Here we find ourselves confronted with a situation the precariousness of which is not often realized by scientists themselves. We find on closer inspection that sensed nature itself contains no property upon which such rejection could be based. Hallucinations can be, and often are as impressive, as convincing, as vivid as are bonafide sense perceptions. They form, indeed, part of sensed nature. It would be incorrect to say that hallucinations are isolated phenomena which a single individual experiences and which cannot be checked by others, for there are numerous examples of mass hallucination. Nor can they be detected by substituting for sensory perception more reliable scientific processes; mirages can be photographed. How, then, are they weeded out? The answer is simple—spurious experiences are recognized as such not within sensed nature, but by reference to constructs. If a certain sense experience fails to correspond to constructs which enmesh themselves in an organic manner within an accepted system of constructs, then the observation is rejected by the scientist. This statement is indeed but a crude description of what is actually done, for the scientist, after all, is practiced in the use of constructs and can handle the criterion thus crudely stated with a precision which its statement perhaps does not convey. If the objection is made that the criterion is not unique, the scientist can only accept it, for there are numerous instances in which it is not clear whether or not recorded observations are to be rejected as untrustworthy.

It is also true that the criterion of authenticity of pure givenness in sensed nature suffers from a certain amount of circularity. Suppose a satisfactory set of constructs corresponding to the scientific experience in a given field has been set up and accepted. A new observation is then made and it fails to be explicable by the constructional devices available. Why should the scientist reject the observation rather than his theories? The answer to this query is of the same nature as the reply to the question, "How do we know that the inmates of an asylum are insane and the rest of the world sane?", for it could very well be the other way around. No clear-cut answer can be given except that logically the process of elimination is indeed fraught with uncertainty, an uncertainty, however, which rarely creates confusion in science itself save in periods of crises. We conclude, then, that visual perception, considered as dator act, does not carry with it any assurance of authenticity as datum, and that it can be certified as an acceptable datum only by reference to the general constructional scheme of things into which it has to fit. It is this process of fitting which makes science a self-corrective discipline.

It would appear that phenomenology possesses no similar criterion of validity, that it is forced to accept all eidetic phases of consciousness indiscriminately. It therefore lacks the advantage of being self-corrective. As a scientist, one feels that the development of principles whereby the results of pure *Wesensschau* need not be accepted indiscriminately is one of the most important problems of phenomenology, a problem in the solution of which exact science perhaps could point the way. It would also seem that phenomenological research is destined to be barren so long as it remains purely descriptive of inner experince in its infinite variety of forms. And if it ceases to do so, will it remain phenomenology in the sense of Husserl's specifications?

## NOTES

[1] Husserl, *Ideas*.
[2] Greek *eidos* = form.
[3] Husserl has borrowed these from Kant, who uses them to illustrate the difference between synthetic and analytic judgments.
[4] Husserl, *Ideas*.
[5] Husserl, *Ideas*.
[6] Note that the word is not used in its ordinary sense. For a complete exposition of the meaning of intentionality, and a clear, modern account of phenomenology as a whole, see M. Farber, *The Foundation of Phenomenology*, Harvard University Press, 1943.
[7] Husserl, *Ideas*.
[8] H. Margenau, *J. Philosophy of Science*, II, 1, 164, 1935; *Review of Modern Physics*, XIII, 176, 1941.
[9] Professor Northrop has given an interesting analysis of these rules, which he terms "epistemic correlations," in *Jour. Unified Science*, IX, 125–128, 1939.
[10] Cf. H. Margenau, *Monist*, XLII, 161, 1932.

CHAPTER 20

## PHYSICS AND ONTOLOGY

Philosophers whose judgment I respect and whose understanding I earnestly desire have criticized the use made of the term "construct" in a recent publication. In particular, Professor Werkmeister[1] skillfully advances an argument designed to refute the view[2] that verified constructs compose "physical reality." His argument is very beguiling but, I fear, somewhat verbal. Since, as it seems to me, his publication employs rather artificial means to establish a spurious difference between his view and mine, I present here a brief rejoinder.

Epistemology is a study of that part of our total experience which culminates in certain knowledge about what is called the external world. It must proceed without initial ontological commitments if it is to avoid the supposed paradoxes of modern physical theory, in particular relativity and quantum mechanics. Experience, however, must be taken in a larger sense than is admitted by the limited doctrine of empiricism; it includes sensory data as well as the constructive procedures often called rational, and everything intermediate between these. For the sake of precision I have called the incorrigible protocol experiences that form the psychological origin of knowledge, perceptions or data, their aggregate the P-plane or Nature. Other elements, like external objects and their well-defined properties, which enter into theories of nature and which always present aspects not wholly given in perception, were called constructs, the word being chosen to emphasize the inventive phase of experience to which they owe their being. In a sense, of course, these are "found" and not invented; but to use the word "found" would be to de-emphasize the corrigibility of constructs. As constructs are validated they progressively lose their fictive character and become verifacts. These verifacts compose physical reality.

Now Professor Werkmeister says that these constructs are merely our concepts of something else he wishes to call real. Language very strongly urges this transcendental leap, and most people (including myself) make it from habit and, I feel sure, from disinclination to argue about non-scientific matters. But let us see what happens if we refuse to make the leap.

According to Professor Werkmeister we are then caught in a dilemma. He asks: "If the *construct* flower *is* the *real* flower, does the construct wilt when frost destroys the flower?" Certainly, it is the construct that wilts and dies. Within our experience there is nothing else that exhibits these peculiarities in a manner significant for the process under description, unless we wish to ascribe the wilting and dying to our fading impressions of the flower. But clearly, the extinction of our perceptions is not uniquely correlated with the destruction of the flower; it can occur although the construct survives. Let us trace the epistemological situation more carefully.

Certain P-plane experiences suggest that we place a postulated entity—a construct in our experience, a flower—in correspondence with fleeting perceptions. Within the framework of these rules of correspondence, i.e., under the regime of these unaltered rules, later perceptions force us to modify the construct in a manner expressed by saying that it decays. Nowhere in this description have I been forced to go beyond experience to postulate a flower existing outside of the epistemological nexus. The price paid is a linguistic infelicity which the quotation from Professor Werkmeister so dramatically displays. The *logic* of the argument seems to me unimpeachable. And even the semantic difficulty can be removed if it is recognized that epistemological construction is different from manufacture. When we say: the table is constructed in the act of apprehension we do not mean to deny that the carpenter manufacturered it. Nor did I make the flower when I became aware of it. But the process whereby nature made the flower involves a sequence of constructs among which is the flower itself.

If the unfamiliar language here needed seems too high a price for consistency, then one must face the consequences of the transcendental leap. They are too well-known to require discussion here, having plagued the philosopher throughout history. The ontological flower which is not the construct I defined becomes an insular concept of the type unacceptable to science in its procedures thus far. Science can tolerate it buy only as an extra-scientific posit. It is not true, in my opinion, that science must postulate it in order to make its business meaningful, as Professor Werkmeister claims elsewhere in his writings.

The example of the wilting flower is conveniently drawn from everyday life where epistemological distinction often appears stilted. There are instances, however, where Werkmeister's realism is pushed against the linguistic wall. These abound in quantum theory but are occasionally encountered in common-sense situations. Suppose a star is known to have been destroyed by collision with another astronomical object which, sometime ago, passed near the earth and had its course charted. The star still shines brightly although it is no longer there. No one saw this star wilt and die; hence the phenomenally real star is present now. To say that the star was destroyed is to refer to the cessation of an entity postulated in a very round-about way in experience, *an object far more adroitly characterized by the word construct than by any term suggesting the magic of a transcendental essence.* Yet it is admitted that this example is no crucial argument against realism any more than Professor Werkmeister's flower is an argument against the view I have presented. A scientific case for or against insular constructs cannot be established.

*Physical* reality, as I conceive of it, must be capable of being known. I defined it as that which is known by valid procedures at a given finite stage in history. As an alternative, *The Nature of Physical Reality* gave attention to a conception of physical reality in terms of an ideal limit to which scientific knowledge tends. The latter is a wholly acceptable interpretation, but it is not in keeping with that minimum set of metaphysical principles which I recognized, the smallest set from which physics can in fact operate. For this "ideal" definition of reality involves a postulate of history rather than physics since it assumes the convergence of physical theory upon a unique terminus or a limit.

Now, if physical reality is that which is known at a given stage in history, then it is clear that reality changes with discovery. In other words, we never know, or have, a kind of reality that is permanent. Professor Werkmeister does not like this admission which, I am sure, comes as a shock to all who look upon science as the conveyor of eternal truth. He therefore complicates the situation by saying: reality is permanent but our knowledge of it changes, and habit of thought, linguistic prejudice, give a thundering applause. It seems to me to be a less presumptious view that holds physical reality to be *known* reality. On this view, physical reality changes its character in time; this is an unadorned way of saying, with Werkmeister: There is reality independent of our knowledge concerning it; our knowledge changes but the unknown object of that knowledge retains its essential attributes.

Reality undergoes many types of change, two of which are always recognized and deserve special mention. 1. There are variations within a constructional situation that are permitted by the logic of the constructs, changes requiring neither enlargement nor modification of accepted theory. These are usually called physical processes, motion, and the like. 2. The other type enlarges or modifies an accepted system of explanation; it might be termed methodological change. It is to emphasize the difference between these two types that the realist philosopher speaks of knowledge of reality in contradistinction to reality. This difference is meaningful from my point of view also, but it is not elevated to transcendental proportions.

Only on one point do I sharply disagree with Professor Werkmeister. He asserts that my definition of physical reality is self-contradictory, insofar as it attributes different, indeed mutually exclusive, properties, to the world in retrospect. Consider the methodological statement, taken from: *The Nature of Physical Reality*: "It is no longer paradoxical to say: after a construct is validated, it must be regarded as having been real before it was formed." This argument is taken to imply that the "real world was (at Newton's time) both relativistic and Newtonian." How this follows I do not see. Indeed it seems to me that the first quotation which portrays clearly what physicists are doing, expressly excludes Professor Werkmeister's allegation. For it says that, when relativity was discovered, relativity was used in retrospect to illuminate all earlier theories of physics, and Newtonian mechanics became what it is now conceived to be, an erroneous theory which is nevertheless a good approximation to the accepted one. Surely, relativity is not the last word either. Does that make experience ambiguous? Science has learned to deal with the contingency of change and the need of self-correction. What Professor Werkmeister calls a self-contradiction is uncertainty as to ultimates, which a philosophy of science must in all honesty countenance.

The account given in the *Nature of Physical Reality* presents a minimum basis for physical science. Hence the specific qualification *physical*, which seems to have escaped the notice of some critics. That account was as sparing as possible of metaphysical premises; it intended to show that some are needed and, furthermore, what these necessary presuppositions are. This is the job which seemed most important. Because of its minimal character, this kind of physical reality has a generality and an appeal which must be understood if further philosophizing is to be safe. More elaborate definitions of reality, such as Professor Werkmeister's. can be derived from it by adding to my "metaphysical requirements"[3] an ontological premise wholly compatible with the other items but not demanded by them. Perhaps it is because I left this elective final step as "an exercies for the reader" that Professor Werkmeister objects to my solution of the problem of *physical* reality.

## NOTES

[1] "Professor Margenau and the Problem of Physical Reality," *Philosophy of Science*, XVIII, 1951, p. 183.
[2] Margenau, Henry, *The Nature of Physical Reality*.
[3] Professor Feigl (paper presented at a Symposium of A.A.A.S., December 28, 1951) holds that these are methodological and logical rather than metaphysical.

## CHAPTER 21

## FAITH AND PHYSICS

A revolution is going on that far surpasses the material changes wrought by science in the development of atomic and hydrogen bombs. It is a revolution in fundamental relations of science and religion. The historic conflict between religion and science has driven scientists underground, rocked the foundations of churches, corrupted governments and sired interminable theological disputes. It has always come about because religion speculates about the universe, its beginning and its end, and this overlaps the domain of natural science. In many realms of thought, contradictions within religious expression are tolerated as honest differences of opinion. Such consideration is rarely applied to the conflict between science and religion.

There are tow basic aspects of religion: the so-called cosmological in which religion theorizes about the universe, and the moral phase in which religion develops a code of human behavior and commits man to it by an act of faith. The conflict arises because men join with fervor those dual areas of religion. If science can show that the cosmological claims of religion are wrong, religion's case in the moral field is weakened. This is an ideological guilt-by-association which causes men to turn away from religion even in the moral domain where science is at present powerless.

This seems to be precisely the dilemma facing us at present. There is a widespread feeling that science has overpowered religion in the natural realm, and people look to science for guidance in the sphere of human action and in the spiritual values. If perchance their expectations were wrong, our time might easily be the eve of doom.

I do not believe that the contest between religion and science has been decided in the cosmological field, nor that it ever will be decided. This belief is based, first of all, upon the simple fact of history. Science is not an unchanging, static set of propositions, not a permanent body of proved facts. Quite obviously it changes, and the changes are not merely additions of knowledge. Revision of basic tenets, overthrow of assumptions that prove to be erroneous are the marching orders of science. The recent exposure of error in the "mirror theory"[1] of physics is ample reference to prove this point. Religion, too, is in a state of progress. The evident fact is that both science and religion are involved in a process of growth, and if one were pitted against the other and were said to be the winner, who could guarantee the finality of that decision?

333

There is a need for continual reappraisal of the relation between religion and science, and never was this need greater than it is today, for science has quite recently undergone a revolution of its fundamental concepts that is unique in history. These changes are more striking and of far greater philosophical moment than the scientific devices that are altering our external lives; they are more noteworthy than the atomic and hydrogen bombs. In a brief space the new philosophy of science may perhaps be illuminated in terms of the following three propositions.

1. Modern physics has completely refuted old-style mechanistic materialism over which the 20th century conflict of science and religion has raged. Modern physics teaches that the ultimate constituents of the universe, the so-called elementary particles, do not have the simple properties which we are accustomed to ascribe to moving billiard balls. Their attributes are difficult to picture and depend for their description on abstract considerations. Reliance upon the mechanical models suggested by common sense has been found fruitless and misleading. Formal principles of mathematics and logic must often replace direct intuition.

2. We are witnessing at present, and shall doubtless continue to see, an enormous widening of the horizon of knowledge and of scientific tolerance. That men will learn more facts is trite and hardly worth recording. But there are two ways of expanding knowledge. One is the acquisition of data in the manner of the physicist at the end of the last century, the busybody who understood his subject completely and looked for the next decimal place in the numerical constants of nature. The other is the open-minded reception of novel truths with full realization of their contradiction of past convictions. It is the latter attitude which characterizes the progress of recent science. The smug and complacent attitude of the 19th century physicist who, to choose a trivial example, denied the existence of the human soul because he could not weigh it or locate its whereabouts, is gone forever. Whitehead's reference to the fallacy of simple location, the reminder that existence need not be tied to localizability, was strange in its day but is almost a commonplace to modern physics.

There is another way to put the story: Science has lost its dogmatism. Present developments show it to be continually engaged in modifying and improving its theory, a Heraclitic flow of facts and ideas. Science now disavows static and final truths, giving in every age a different answer to man's eternal questions. It knows that its principles as well as its facts are changing, and it has renounced the error of believing in the possibility of explaining all human experiences, past and future, in terms of those principles and laws which are now called science.

3. The mechanistic philosophy of the 19th century presented human freedom as a paradox. Man felt within himself that he was free, yet the laws of nature seemed to demand strict determinism in his actions. This contradiction, rather artificially resolved in the past, can now yield to tested scientific theory. The contradiction gone, the paradox presented by earlier science has vanished, and while the issue of freedom is not settled, it has been transferred from the wastebasket of paradoxes to the shelf of fascinating and challenging problems to be solved. And the solution will certainly require a collaboration between scientists and men versed in the ways of the human spirit.

Quite apart from these recent developments, which cast a new light upon the relation between religion and science, that relation continues to be distorted by a misunderstanding of the nature of science, which is widely regarded as dealing exclusively with facts. Let me say with all the emphasis at my command that science is no more interested in bare facts than is any other discipline. It is interested in *relevant* facts, that is to say, in facts with *meaning* through their relation to concepts and laws. It is not enough for the scientist to know. What he wants is *understanding*, and understanding requires both *facts* and *ideas*, observation as well as conjecture.

Discovery has been, and is, the essential business of science. But discovery does not mean turning up rocks to see what is under them. It involves the creation of new insights, linked to observations having crucial significance with respect to these insights. Observation alone is not enough, isolated facts imply nothing and command no interest whatever.

What science achieves is a correlation of facts with ideas. It needs facts as our body needs food, but within the organism of science, facts are processed, combined, organized and connected by a texture of reason. It is the whole of the organism, including that texture of reason, of ideas and conjectures, which is science. In a very deep sense, science has its origin in the circumstance that the observations of our senses, the facts are not sufficiently orderly to satisfy our desire for simplicity and consistency. Science is an elaborate answer to the paradox of the bruteness of our experience. To summarize: incoherent facts are unified by science into a consistent whole with the use of reason.

To achieve this goal, science transcends the facts. It employs what is known as the deductive method, which goes from assumption or premise to conclusion. The premises of science are called postulates, and they cannot be demonstrated to be true by facts. Let me illustrate. When I explain the observation of a falling object by saying that the earth attracts it, I am making an abstract statement which cannot be directly verified. Nobody has seen this force of physical attraction, and no one ever will. What the statement means is this. If I assume, following Newton, an inverse square law of force, then consequences of this assumption agree with the facts observed. The same facts may possibly also be consequences of some other assumption or theory not yet considered, and so it is with most fundamental theses of science, with all great generalizations. They are never proved directly. Their ultimate validity is never assured. Being postulates, which can only be verified through their consequences, their acceptance requires more than a correct knowledge of facts. It requires a *commitment*, an acceptance of unproved postulates, often called axioms. This intellectual acceptance is the same thing as what we call *faith* in religion.

Failure to realize this element of faith in science leads to scientific dogmatism. There are those who regard the present stage of scientific knowledge as ultimate and refuse to consider phenomena or experiences outside its momentary competence. They make a distinction between what science is now able to explain and what escapes its grasp; the former they call natural, the latter supernatural, and they believe this partition to remain meaningful. According to this unreasonable notion, radio and television were once supernatural phenomena.

And dogmatism in religion, equally indefensible and equally mistaken, rears itself upon the arrogant conviction that religious truth is laid down once and for all in a static pattern, rigid, lifeless and inexorable, incapable of progress and improvement. These bone-dry dogmatisms always clash and clatter, and the noise they make through the centuries is taken as the sign of conflict between science and religion. It is a misleading sign.

The purpose of science is to organize the unilluminated facts of our experience into a texture of reason, simplicity, orderliness and multiple perspectives. Science selects for its chief domain the facts leading to concrete knowledge of the external world. The scientist, because of his closeness to the problem and the narrowness of his attention, sometimes loses sight of the strange circumstances that such orderliness can be achieved, and takes what he calls the uniformity of nature for granted. He is often immune to the sense of wonder which prompted Schleiermacher to speak of the one miracle before which all others lose their meaning, that miracle being the absence of breaks in the lawfulness of nature, the absence of miracles.

It seems to me that Schleiermacher's point in portraying uniformity of nature as a divine gift to man is confirmed by the Biblical story of the flood. First, you recall, there was chaos, terminated by a divine act of creation. Then followed a period lawlessness and confusion that ended in the great flood. One interpretation of the turbulent days prior to Noah's Ark, which is elaborated in the Jewish Talmud, emphasizes that during this period nature, and nature's God, did not act in accordance with consistent principles; that there were no natural laws and, hence, no possibility for natural science. Lawfulness, behavior in conformity with reasonable principles, was God's gift to Noah, made in the beautiful covenant of the rainbow.

Jehovah smelled the sweet savor; and Jehovah said in his heart, "I will not again curse the ground any more for man's sake, for that the imagination of man's heart is evil from his youth; neither will I again smite any more everything living, as I have done. While the earth remaineth, seedtime and harvest, and cold and heat, and summer and winter, and day and night shall not cease." And God said, "This is the token of the covenant which I make between me and you ... for perpetual generations: I do set my bow in the cloud, and it shall be for a token of a covenant between me and the earth. And it shall come to pass, when I bring a cloud over the earth, that the bow shall be seen in the cloud: And I will remember my covenant, which is between me and you and every living creature of all flesh."

If I understand this passage correctly it means to say that the order of the universe is a divine gift. Religion here acknowledges the legitimacy of science. If science has the purpose of organizing the facts of our cognitive experience into an orderly pattern, religion, it seems to me, endeavors to supply the rational bonds for another kind of facts, or immediate experiences, that are of equal concern to living man as the facts of science. As facts they are crude, meaningless and unrelated. I see them residing in those experiences most of us acknowledge to be peculiarly religious, in the spontaneous feeling of gratitude that wells up in man's heart at sight of a kind deed, the feeling of awe in the face of overwhelming beauty, the guiltful contrition that follows a sinful experience, the sentiments of misery and abandon at the insufficiency of human power before fate, the longing for grace and redemption. Just like the facts of science, they require a texture of rational organization. And this, I take it, is what formalized religion or theology aims to provide. That its theory is replete with intangible ideas, that in the terminology of its detractors it bristles with the "technicalities of salvation" is small wonder to one who is familiar with the intangibles of science. Their presence in itself is no objection. The success of religion is measured by the degree of rational coherence which it bestows upon these singular religious experiences that assail the sensitive mortal.

In Biblical language, the uncoordinated experiences which religion attempts to conquer and transform into a pattern of reasonability are well expressed in the message to Adam, Genesis 3:17-19: "Cursed is the ground for thy sake; in sorrow shalt thou eat of it all the days of thy life; thorns also and thistles shall it bring forth to thee; and thou shalt eat the herb of the field, in the sweat of thy face shalt thou eat bread ... "

On the other hand, a vague glimpse of the coordinating pattern, its culmination perhaps, but still void of detail and in need of implementation, is found in the words of Jesus in Matthew 11:28: "Come unto me, all ye that labor and are heavy laden, and I will give you rest." Here is a religious theme of supreme satisfaction, an organizing idea of power and simplicity in terms of which many crude experiences make beautiful sense.

To bridge the gap between Genesis and Matthew by a texture of rational connections is one of the important tasks of professional religion. And in essence, that task parallels the one of science.

### NOTE

[1] By this is meant the view that physics simply mirrors facts.

# CHAPTER 22

# METAETHICS

## 1. SCIENCE AND ETHICS

Scientists have often claimed that scientific knowledge itself, when fully grasped, will generate the rules of proper human conduct. Beginning with Socrates' doctrine that knowledge is virtue they have endeavored through the ages to squeeze the *is*, hoping that it would yield an *ought*. The futility of such an undertaking needs to be unmasked and clearly exposed to view; for it involves the naturalistic fallacy which has been rightly criticised by many writers. To put it simply: even if we knew everything about the physical universe, about human physiology, about man's natural dispositions, his drives, his instincts and his normal reaction to all stimuli; even if we could predict how average men will in scientific fact behave under all specified circumstances (at a given time of the evolutionary process), we should still have no basis for judging the moral quality of his actions. Even if the drive for survival or for individual happiness were absolutely universal we could still not prove, by using the laws of science, that man *ought* not to die or *ought* to be unhappy in certain situations. This absence of affinity between the substance of science and the substance of ethics must be recognized at the outset. Although I shall suggest in the sequel that we are in possession of principles which may, if properly applied, engender facilities for judging objectively the moral behavior of men, and to judge them independently of local standards of value, I do not refer to *any generalizations of scientific fact*.

I am, however, suggesting that the abstract methodology of science, which enables us to rise from particular fact to universal law, contains important hints relating to the possible conversion of the factual moral *is* into the regulative *ought*, to the transformation of values conceived culturally and relatively, into values embodying transcultural norms. Before presenting the details of this suggestion it seems desirable to analyze the semantics of the word value.

## 2. THE MEANINGS OF VALUE

Traditionally, the center of attention among moral philosophers is the concept of *value*. Human actions are said to strive toward the acquisition, or attainment, or realization of "values", and the hope is held out that if values can only be classified according to some reliable principles or measured on some universal scale, then the moral quality of an action can at once be determined by the kind or the loftiness of the value which the action realizes or intends to realize. Such a procedure is suggested, albeit without clear promise of success, by the simpler natural sciences to which the advocates of this theory of value often refer.

However, in actual fact science offers neither support nor refutation of the thesis that human actions have reference to a scale of values and that values are measurably correlated with some property engendered by or accompanying behavior. The thesis must therefore be subjected to logical scrutiny and to the test of factual success. It could fail because the concept, value, might *not* be capable of sufficiently precise definition, might *not* be referrable to any unique scale; or values may not be the purposes (or the causes) of human action; or the chosen scale itself might prove unreliable. It is my belief that all these misadventures have occurred and are secretly troubling those who travel into ethics through the theory of values; that the concept of value itself is at present the object of a global rescue operation involving most of the resources of what is traditionally called moral philosophy. However, I shall not press that belief on this occasion.

Another kind of question can be asked concerning values. Quite apart from any success or failure of the attempt to assign value to actions and then measure value by reference to some other perhaps more easily discernible quality, one needs also to make sure that this correspondence when established is sufficient to complete the task of ethics, which is to set *normative*, not merely de facto standards of conduct. It is conceivable, for example, that it could satisfy the scientific query as to what men do, how they in fact behave, but leave unanswered the question whether it is right or wrong that they should thus behave. In that case the theory of values might serve as a basis for cultural anthropology and for sociology, but not for a kind of ethics which requires a standard of judgment higher than observed behavior.

Let us consider briefly the current meanings of the word value. Since the word "value" is at the centre of much traditional philosophy of ethics it may be well to review its linguistic origin. Somehow, our comprehension of things, situations and people tends to be deepened, our interest in them quickened by a knowledge of their ancestral histories, and this should certainly be true for words like value that are employed as pack animals. In this instance, we discover an interesting variety of meanings among its ancestors, only one of which—and not the primary meaning—has come to be reflected in the present usage of the word.

The Latin *valere*, from which value is derived, meant primarily to enjoy physical strength. The adjective *validus* is frequently applied to a bull (taurus validus). But the word took on related meanings: to be capable in a more general sense, to be powerful, influential. The state of health was always included in its Latin connotation, as is evident in the Roman farewell: *vale*. The metaphoric extension goes on from "powerful" to "fitting" and "appropriate" and finally to the notion of "equivalence" of coins ("dum pro argenteis decem aureus unus valeret"). Here the original comes close to our present usage, as when we speak of the equi*valence* of words. For in Latin, synonyms are words of "equal value" ("verbum, quod idem valeat"). Rarely, however, would the Roman have used valere to refer to value in our present primary sense; this would have been more accurately rendered by *pretium, honos,* or *virtus.*

Interestingly, if value has been upgraded from physical strength to general worth, the word "price" has suffered the opposite fate. It started approximately where value ended, with the Latin *pretio*, and has now come down to its mundane usage in economics. The French *prix* and the Spanish *precio* still retain a little of the more "precious" flavor.

We now inspect the meanings of value in today's context. Simple words implying value judgments are good, pleasant, beautiful, genuine, honorable, virtuous, and their opposites. They can be applied to several different types of noun, those denoting things, persons and indeed processes or courses of action. For the present we restrict our discussion to things and persons. One peculiar aspect about the assignment of a value to an object, often noted in the literature and greatly emphasized by G. E. Moore,[1] is its indirectness of reference. When we say an object is good we do not mean it is blue, large, heavy or that it possesses any other particular sensory qualities; yet in another sense we mean all of these. Goodness is not a natural property of objects but still it expresses itself *through* their natural properties. For this reason, goodness, and more generally value, are often termed non-natural qualities of things. Here then arises one of the much discussed problems of value theory: are non-natural qualities reducible to natural ones; are they a collective function of all the natural qualities of an object; do they refer to something aside from the natural qualities such as the reaction of a human mind or the use to which the object can be put; or are non-natural qualities irreducible and *sui generis*? The last proposition is affirmed by Moore.

The present approach to values and ethics lies in a different direction, for we are less interested in the logic of usages than in the dynamics of ethical procedures. Let us therefore abandon the study of values as non-natural qualities of things.

Values are properties or attributes of *things*. They attach, first of all, to concrete objects. Here, value is generated by the use to which they can be put, and this use is either a limited one in the status of loan or lease, or it is unlimited in outright ownership.

Values are also attributed to *intangibles* like health, happiness, friendship, security and leisure. Different principles begin to operate here, primarily because one of the components which determine the value of commodities ceases to be significant; these intangibles can be created at will, they are not subject to laws of material supply and thus indefinitely available. Hence these intangible goods defy the laws of arithmetic, which are applicable only to discrete things. We have no calculus suitable for an assignment of value in these instances. They are, at least at present, unmeasurable and are for that reason often called *qualities* in contradistinction to *quantities*.

Hence other, less precise and more intuitive means are forced upon us in the endeavor to appraise such things as health, love and happiness. One possibility is to select arbitrarily one of them, say happiness, as primary, and then judge the competence of others, such as health and love, to create this primary good. I used the word "arbitrarily" with deliberation for I know of no rigorous criterion that would accent one "good" more heavily than another. Attempts go forth in many places to show that there is indeed a primary good, definable in more basic terms, such as human self-fulfilment, or the goal of human evolution or the survival of the race. All of these "more basic" principles, it seems to me, are forced to beg the final question as to why they should be regarded as good at all.

It is utterly astonishing to see how many moral philosophers, though rigorously trained in logic, and often exclusively in logic, persistently balk at the need for *choice* which confronts us here. One of the deepest insights conveyed to us by this branch of modern philosophy concerns the impossibility of establishing a formal system that is assuredly self-consistent and complete, and certainly none that can do without chosen axioms and primitives. It should be evident, therefore, that at some place one must meet a legitimate occasion for arbitrary choice, and apparently we have met one here.

We thus conclude that the value of intangibles is estimated intuitively by the degree to which they contribute to the creation or maintenance of some primitively defined and wisely chosen "absolute" or primary good.

The preceding comments dealt first with the value of common objects, limited and unlimited in their availability, and secondly with the value of intangibles. At the next stage we encounter living creatures and we ask about *their* value. Life itself, of course, can be said to have value, but this alone cannot be used to establish a scale of values, for then all living creatures would have equal value. Nor will the addition of other intangibles beside life yield a measure of value: when one speaks of a good person his goodness is clearly not meant to be merely a function of the desirable qualities with which he is endowed. A happy mood does not make him good, neither will health, nor love of his parents or his fellowmen. These do not constitute his goodness even though they may be a necessary condition for it. No, the goodness of a person is clearly dependent on what he *does*. An important change is thus seen to occur as we pass from the study of value in inanimate objects or in abstract qualities to the domain of the living, where 'free' decision is possible together with voluntary action. It is through the accident of freedom that the meaning of good, and indeed of the entire value concept, is transferred from its lodgings in external qualities — where it resides for things — to the internal dynamics of willing and acting. A person is good because *what he does is in accord with certain rules*.

In the shift to the living, then, we have made the transition to ethics. A good egg and a good boy are incomparably different in the connotations of good. Having encountered and recognized what constitutes value in man, we have likewise uncovered the main springs of ethics: will and action in accordance with certain rules.

We have dealt with the nature of values and the ways by which they are established and made manifest. Each value, however, when established, still prompts us to ask whether indeed it *ought* to be a value and, if it is, whether our assessment of it is 'right'. A fact is what it is and must be accepted as such; its essence is fulfilled in its being, not in its being right or wrong. Values, on the other hand, have both a factual and a normative component, the latter declaring itself in the judgment of an ought.

We encounter the second component along with the first on all levels of valuation. A thing may be valued or priced in accordance with custom, subjective want or the laws of supply and demand. Its factual value is thus fixed. Nonetheless we wonder whether the value or the price is *right*.

As one moves to higher planes of value, the problem of the ought presents itself with greater urgency and at the same time, as if in compensation, the area of dispute concerning it is lessened. Life, health, honesty are all acclaimed as desirable in the overriding judgment of most men. Little need therefore be said here about the normative aspects of these intangible values.

It is in the vast arena of human actions that the ought becomes imperative. And precisely there the distinction between the factual and the normative is widely disregarded. Value in the social sciences is too frequently identified with the actual behavior, with observed preferences of people within a group, and this is then often tacitly elevated to a norm. The reason for this oversimplified treatment of social situations is easy to see: preferences are observable, statistically measurable while norms are not, and the view prevails that what is measurable becomes *ipso facto* scientific, and everybody wants sociology to be a science. Nevertheless, when common practice is accepted as normative, the ideas of obligation, honor, guilt, remorse and retribution undergo an erosion which transforms them into shallow psychological phenomena, and leaves their human victims at the mercy of psychiatrists.

There is what one might call a scale of oughts. This phrase is intended to convey, first that the normative note rings audibly in every value judgment, and second, that its intensity rises in a crescendo from mere detectability in the assignment of material values to imperative urgency in evaluating human actions. In the sequel, when the word value carries its accent on the factual, as in comparative anthropology, I shall call it an "est value"; when its meaning is normative, superfactual, I shall use the term "esto-value". The important question then is: how can esto values be objectively established? The easy way, which involves reliance upon revealed doctrinal truth, is evidently not open to people and societies which reject such truth. We therefore look for an alternate, perhaps a complementary method of establishing esto values by both formal and empirical means.

## 3. THE PARALLELISM BETWEEN SCIENCE AND ETHICS VIEWED AS EMPIRICAL ENTERPRISES

This section is intended to show that, while ethics and science are completely different in their substances (what ought to be vs. what is) and their languages (imperative vs. indicative), their abstract methodologies are similar. This similarity can be used to clarify many problems of traditional ethics.[2]

For the sake of brevity, let me explain the relevant aspects of the scientific method, conceived in its widest sense, by means of a simplifying diagram. An extended discussion of it is given in *The Nature of Physical Reality*. In figure 1 are depicted symbolically various stages of the process by which scientific knowledge is acquired and verified. At the top we encounter the level of protocol experiences, the sense data, the observations which, being incoherent and devoid of order in themselves, require "explanations" and rationalization by supplementary concepts which are not directly given in the protocol domain. Explanation involves conceptual procedures, schematically outlined in the figure by the structure of levels below the protocol experiences. They function in the following way.

At the base every science exhibits very general propositions called axioms and denoted by $A$. These axioms differ in the different sciences; only

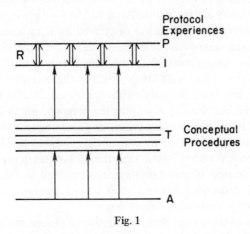

Fig. 1

occasionally, when branches of inquiry attain a very high degree of development, do their bases coalesce. This happened recently, for example, in certain parts of physics and chemistry when quantum mechanics was discovered. Whatever the axioms of a given science (or of a part of science) at a given time may be, they give rise by deductive formal explication to higher level, i.e. less general propositions which are ordinarily called laws or theorems and are designated by $T$ in the diagram. From the various $T$ one derives still more particular inferences, called $I$ in figure 1. For instance in geometry $A$ might represent the axioms of Euclid, $T$ the various theorems about plane figures, $I$ statements about the properties of right triangles such as Pythagoras' theorem. These final inferences are in general still devoid of empirical content, since they refer to formal elements which have no necessary counterpart in the world of sense, that is, among protocol experiences. Indeed the entire range of conceptual procedures from $A$ to $I$ represents a formal system, of interest to the logician and the mathematician. To convert it into an applied science one must place it into correspondence with the $P$-plane of the diagram, and this requires the introduction of a set of relations, labelled $R$, which permit $I$ to be compared with $P$.

In some sciences, (e.g. geometry) the propositions composing $I$ sure fairly simple and obvious. They involve such concepts as sides of triangles, angles, lengths of lines which immediately suggest comparison with actual observable objects. In others the connection is very remote. For instance, a great deal of analysis and insight was required before Max Born recognized that the simple construct $\psi$, which appears as the solution of the Schrödinger equation, has reference to a probability distribution of observations in the protocol domain. A special set of *rules* stating that relation had to be discovered, a set called "rules of correspondence"; these are designated by $R$ in figure 1. Closer inspection reveals that every science requires rules of correspondence, although their presence has long gone undetected. This is true even in geometry; for the lines and angles of an ideal triangle, which is the object of the conceptual procedures, are not truly identical with the elements of concrete figures drawn on material blackboards. Nor is the temperature, which functions as an abstract symbol in the propositions concerning heat phenomena, propositions which flow ultimately from the axioms of thermodynamics, recognizable as in any sense identical with the mere sensation of hotness in my skin. The two are related by an operational definition, as explained in *Open Vistas*, and that operational definition, like the other connecting links just mentioned, is what I have called a rule of correspondence.

It is by virtue of these rules that the conceptual procedures can be brought into contact with $P$-experiences; through them, scientific verification of theories becomes possible.

Thus far, our attention has been confined to the description of the connections between various levels of our diagram. Let us now raise the question of *entailment*. Clearly, when $A$ is given we can rise without injection of further postulates to the level $I$. The conceptual procedures are more or less self-contained. But to go from $I$ to $P$ we need the rules of correspondence, and these are not entailed by what is below or what is above them. They are *chosen*, much as one chooses axioms, with an eye upon how the entire scheme of explanation is most likely to work successfully. The gap between $I$ and $P$ is a logical hiatus which a special postulational fiat must bridge.

The axioms, of course, are likewise unentailed; they are subject to human choice. Again, this was not clearly understood in earlier periods of science, when axioms were regarded as ultimate, unchangeable truths. We now know that they do change as science develops, and that their flexibility imparts to science the dynamism, the self-corrective qualities which are so generally admired and which a static basis can not provide.

We conclude our survey of the method of science by emphasizing once more the postulational character of two elements or levels in figure 1: $A$ and $R$. If scientists were not free to choose these important elements of their method, or if they entertained major disagreements concerning them, their enterprise could not be successful: it would probably be in the same state as current theories of value.

Having studied the methodology or, in a large sense, the fundamental language of science, we now examine the language of values with special reference to human actions. I propose to develop that language by means of a diagram very similar to figure 1. Our basic concern shifts from the goal of *explanation* to the goal of *suasive control of human actions*, from a descriptive analysis of what happens to a hortative language of how men ought to act. In view of this shift the grammatical form of all relevant statements must be altered. Whereas the basis of figure 1, *A*, contained declarative sentences, the basis of figure 2 depicting the methodology of ethics speaks in imperatives or commandments, *C*. To be sure, other starting points of ethics have at

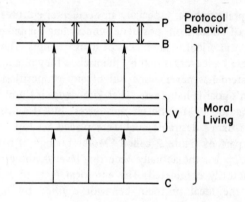

Fig. 2

times been proposed by philosophers, but the testimony of history demonstrates impressively, I think, that all effective moral philosophies have begun with imperatives. The commands, *C, imply* specific rules of conduct, propositions relating to particular human situations. Thus, the commandment "thou shalt not kill" entails that one should preserve life; hence human *life* becomes a value. If "thou shalt not steal" is included in *C, property* becomes a value. For this reason I have labelled such deductive consequences *V*, for value, in figure 2. But I would rather not regard *V* as a *logical* explication of *C* since I do not look upon logic as the primary determinant of ethical systems: *V* follows from *C* through the process of *living*; a social group dedicated to *C* evolves values, not by speculation but by concrete actions. In further detail this living process engenders particular patterns of behavior, *B*. The ethical enterprise from the basis *C* to its vital implications *B* is therefore continuous, but this continuity ends at *B*.

Science became an empirical, verifiable system of reasoning because the rules of correspondence permitted a confrontation of the level *I* with protocol experiences. Their absence would have left science without application, like some parts of pure mathematics. In a similar way, ethics remains univerifiable, devoid of universal oughts if analysis is forced to stop at $B$. The behavior $B$ itself satisfies no criterion of external validation, for it follows from $C$ and can therefore not be used to validate $C$. The passage from $C$ to $B$ is "vitally tautological", that is to say, a given group of people, a culture, living in accordance with $C$ will automatically exhibit a behavior pattern labelled $B$. How, then, do we get from $B$ to the level above, called protocol behavior? What is this ethical protocol, and what are the rules of correspondence joining it with $B$?

Many moral philosophers, especially in occidental cultures, have tried to develop systems of ethics from postulates concerning human goals. Eudemonism, hedonism, utilitarianism are names of such attempts. The telling characteristic of all these endeavors is that by themselves they remain ineffective. A living ethical system has never come out of any proposition which merely records that man's goal is happiness. That knowledge is simply not sufficient to guide man's action in specific circumstances. But if a principle claiming that happiness is man's desirable goal were placed on top of the elements composing that part of figure 2 called "Moral Living", if it were used as a *criterion of validity* for the actually occurring level $B$, the ethical enterprise could be satisfactorily completed. The empirical facts of $B$ could then be compared with the ideal protocol behavior *defined* by the principle of happiness.

For this reason I am loth to accept eudemonism and all the other human goals which have been offered as *the* basic principles of ethics. They are principles in terms of which chosen sets of imperatives can be validated. I therefore prefer to call them principles of validation, or postulated *primary* values (in contradistinction to the values V which automatically result from the commands). If this understanding is accepted, figure 2 takes on a remarkable similarity to figure 1. The primary values function in a role comparable with the rules of correspondence, enabling comparison between $B$ and $P$. There are differences, to be sure, but their discussion will be omitted in this concentrated account.

The scientific process is successful when $I$ agrees with $P$ in figure 1. The ethical process is successful when $B$ agrees with $P$ in figure 2. In the former case, scientific theory is verified; in the latter, ethical norms are validated. Before verification scientific constructs form hypotheses, afterwards they become true, confirmed theories or laws which state universal (though not ultimate) truth. Before validation, ethical imperatives are tentative, reflect local patterns of behavior and make no universal normative pretensions; after validation they transcend the est and take on the esto character of an ought.

Scholars versed primarily in the humanities tend to regard the theory I have proposed as unsatisfactory because it leaves the origin of the commands as well as the origin of the principles of validation obscure and the principles themselves subject to human choice. To me this is encouraging, since it reflects precisely the epistemological status of the axioms and the rules of correspondence in the scientific scheme. Indeed it increases my hope that the ethical enterprise can be successful inasmuch as it involves all those ingredients which have made science successful within its realm. And I am further delighted that it gives man choices instead of a rigid and immutable structure of norms.

Let us also note the strong affinity between ethics and religion. If ethics has the methodological structure here outlined, then it is open ended like every other human enterprise. Postulates (the Commandments) form its base, and other postulates (principles of validation or primary values) enter near the summit. It is but natural for man to regard their injection into the stream of history by blessed individuals as acts of revelation. It is here that ethics and religion come together.

## 4. ON THE POSSIBILITY OF TRANSCULTURAL ETHICAL NORMS

In the preceding section we surveyed the infelicitous and hopeless state of a language which talks about values per se. Values are not useful starting points of any discussion that aims to deal with human behavior in terms which claim relevance for the total multicultural panorama. As our diagram shows, they are situated in the middle between the imperatives and the validated behavior (figure 2); they draw their meaning from $C$, their authority and their ought from the agreement between $B$ and $P$.

Perhaps we can now see the reason why international communication about values has been blocked. Values as ordinarily conceived make no reference to the methodology of science, indeed they spurn it by a pretense of aloofness which arrogantly disclaims all connections between value and fact. Hence those who place their entire trust in the language of science are repelled by such talk about values; in particular they see no opportunities for meaningful comparisons of different cultural standards. In the context here developed, values are the counterparts of the tested laws of science and their claim for consideration can become objective in a quasi-scientific sense by virtue of the possibility of validation, as outlined. Hence there is a chance, at least, that this new view of ethics, which converts that ancient branch of philosophy into an empirical and a challenging undertaking, will restore communication about values between scholars in alien ideologies.³

NOTES

¹ G.E. Moore, *Principia Ethica*, The Home University Library, Oxford University Press (1912).
² A confrontation between science and ethics is offered in a significant, more linguistically oriented paper by Mario Bunge ('Ethics as a Science', *Phil, and Phenom. Research*, **22** (1961) 139). It has points of contact with the present treatment, but also displays important differences from our point of view. To facilitate comparison, let it be noted that what I call imperatives Bunge calls norms. My primary values or goals are his desiderata. Among the differences is this: Bunge holds that the relation between the two is completely rational, one entailing the other, whereas I believe that, in general, compatibility is established by living the commandments to see whether, in fact, they lead to the norms. We note, too, that compatibility through living adherence to imperatives does not rule out logical entailment of goals.
³ This view is further developed in the author's *Ethics and Science*, Van Nostrand, 1964.

CHAPTER 23

# THE PURSUIT OF SIGNIFICANCE

## INTRODUCTION

The problem of significance is not limited to our era nor to our station in the world; it springs from man's essential being and troubles him in all ages and wherever he exists. It is timeless, yet it has peculiar relevance to our times, whose confusion is characterized by loss of perspective and by misconceptions about the role of science in society and its threat to human values.

Correct perspective can be gained only if we realize that events which we have ourselves experienced never seem as glorious or as dramatic as those conveyed to us in the songs of poets and through the eloquence of historians, who can afford to ignore the commonplace. The common impression that our age is undistinguished and decadent has therefore considerable psychological likelihood of being false. Consonant with this observation is the claim made by historians, the finding that every age has had its lamented maladies, and that the widely advertised gloom hovering over the present is hardly unique.

On these grounds it is justifiable to dismiss, for example, the complaint that we have lost our spiritual values. In truth these values are as evident as ever—perhaps even more evident in this day of public soul-searching over radio and television—but they are different; because we know more about the world older values have lost their validity and their appeal, and our critical appraisal of previous standards is not only necessary but good.

In a similar vein it is said that we are preoccupied with material things. If this means that we have more goods available for enjoyment the statement is true, but it has no point because earlier civilizations were without them by necessity and not by choice; furthermore, let it be noted that we also have more churches and more poetry.

The most timely worry seems to arise from the rapid growth of science, which is said to threaten man's soul, to reduce him to a cog in a machine. Since this fearful prospect is often identified with a major loss of significance it seems important here to set the record straight. Whether science will ultimately bring about a state of regimentation in which initiative has fallen prey to control by automatic computers is by no means certain and depends on our own reaction to recent discoveries: what *is* certain, however, is that human society started in a machine-like state and was progressively released from it by the blessings of science. Contrary to general belief, the *human machine*, an institution partly preserved in today's so-called political machine, is historically older than all but the simplest inanimate artifacts. An excellent account of its workings and its gradual disappearance in parts of the world is found in a recent article by Lewis Mumford (*Daedalus,* **94**, 271 (1965)), and I take the liberty to quote from it as follows.

"In the period when the institution of kingship arose, no ordinary machine, except the bow and arrow, yet existed: even the wagon wheel had not yet been invented. With the small desultory labor force a village could command, and with the simple tools available for digging and cutting, none of the great utilities that were constructed in the Fertile Crescent could have been built. Power machinery was needed to move the vast masses of earth, to cut the huge blocks of stone, to transport heavy materials long distances, to set whole cities on an artificial mound forty feet high. These operations were performed at an incredible speed: without a superb machine at his command, no king could have built a pyramid or a ziggurat, still less a whole city, in his own lifetime.

"By royal command, the necessary machine was created: a machine that concentrated energy in great assemblages of men, each unit shaped, graded, trained, regimented, articulated, to perform its particular function in a unified working whole.... The assemblage and the direction of these labor machines were the prerogative of kings and an evidence of their supreme power; for it was only by exacting unflagging effort and mechanical obedience from each of the operative parts of the machine that the whole mechanism could so efficiently function. The division of tasks and the specialization of labor to which Adam Smith imputes so much of the success of the the so-called industrial revolution actually were already in evidence in the Pyramid Age, with a graded bureaucracy to supervise the whole process....

"This new kind of complex power mechanism achieved its maximum efficiency in the era when it was first invented: in the case of the hundred thousand workers who built the Great Pyramid at Giza, that machine could develop ten thousand horse-power; and every part of its colossal job was performed with machine-like precision....

"Though the lock step discipline of the labor machine was happily alleviated by the art and ritual of the city, this power system was kept in operation by threats and penalties, rather than by rewards. Not for nothing was the king's authority represented by a scepter, for this was only a polite substitute for the mace, that fearful weapon by which the king would kill, with a single blow on the head, anyone who opposed his will.... The price of utopia, if I read the record correctly, was total submission to a central authority, forced labor, lifetime specialization, inflexible regimentation, one-way communication, and readiness for war. In short, a community of frightened men, galvanized into corpselike obedience with the constant aid of the mace, the whip, and the truncheon. An ideal commonwealth indeed!"

Science has freed mankind from this condition of servitude. Contrary to the feeling of many humanists, it is not simply a one-way ticket to perdition; its impact upon our society is much more subtle and more complex.

Science is morally neutral, as will be indicated in later parts of this essay. It increases man's capacity for good and evil. Science provides the means for destroying and for saving mankind, yet it cannot say whether it is good to destroy or to preserve. This, in short, is modern man's predicament: enabled by science to accomplish ever greater and more massive tasks, saddled with an increasingly awesome responsibility, he discovers that his very source of power leaves him without moral guidance. This is why the scientist today, more than ever, needs the council and the guidance of the humanist.

The problem of living significantly is not new in our day; it is ageless. But the increased potencies for living significantly and insignificantly give it a timely cast. There are different spheres or, since the principles controlling them can be placed in ascending order, let me say levels of significance, and these will be treated under four headings. The first, or lowest level is largely concerned with meaning. It deals with words and other kinds of symbols; traces *semantic* significance all the way from noise to art. The second deals with scientific or *epistemic* significance, the sort of meaning that leads to understanding and makes prediction possible. A brief review of scientific methodology, which serves to establish this kind of relevance, is indispensable there. Under the third heading, *moral* significance, I shall review the rules by which ethical significance is conferred on human action and indicate, by the way, how the pursuit of moral significance runs parallel to the quest for scientific relevance. Lessons can be learned from this parallelism. The fourth and last section is devoted to the highest form of significance, which involves personal dedication beyond the guidance of scientific and even moral norms.

## 1. MEANING

One kind of significance, perhaps the most commonplace and universal, arises in the field of communication, when the contents of one mind or its intentions are to be transferred or made effective in another. What is communicated is a *symbol*, usually a picture, a sound or a word, and it is the nature of a symbol that it has a meaning other than itself. What the word, book, ordinarily signifies is not the four letters that compose it, nor the sound made when it is uttered, but an object which is called its *referent*. Words can, of course, be used in self-referential fashion. External reference and self-reference occur, for instance, in the sentence taken from a discourse on the classical heritage of science: "A scientist who uses expressions and ideas like expressions and ideas thereby acknowledges his indebtedness to the classical past."[1] Strictly speaking, external significance, sometimes called semantic significance, involves a three-term relation between the idea the speaker has in mind, the symbol (e.g., book), and the external object which it designates, and it requires that this relation be in accord with linguistic rules.

This is hardly the place to expound fully the elaborate and to my mind somewhat dull subject of semantics, which deals with the manifold correspondences between symbols and their referents. Broadly speaking, symbols have cognitive, emotive and hortative significance when they are meant respectively to convey knowledge, feeling or directions for behavior. To focus attention let me first limit discussion to cognitive symbols in the form of words occurring in a natural language like English. We can then distinguish several levels of cognitive significance.

A world has, first of all, its dictionary meaning, sometimes called lexical meaning. This attaches to isolated words and distinguishes them from arbitrary sequences of letters or sounds. To be sure, this distinction, like all the others, is not absolute, for there are expressions like "blah blah" which are on the verge of having cognitive dictionary meaning. Living speech, however, rarely employs isolated words, its unit being the sentence. It is, as it were, a linguistic molecule in which the atoms—words with their isolated meanings—can be combined in a variety of ways. There are certain highly inflected languages like Latin and Greek in which the same words arranged in any order carry the same sense. English is not one of these, for the sentence: the cat bites the dog, means something quite different from: the dog bites the cat. In German, for example, case endings avoid this change in meaning. Every language, however, is able to confer an extra sense, or at least additional emphasis by using different arrangements of the same words in a sentence and this supplementary significance, this molecular meaning, is called *syntactical*.

Very few artificially constructed linguistic molecules have syntactical meaning. A few words arbitrarily selected and strung together as a phrase rarely make sense, for they defy the rules of syntax (even if they obey those of grammar) which are perhaps the most important agencies for establishing linguistic significance. But there are others. The sentence, "my book is lost," is constructed properly and has syntactical meaning. Yet its significance is minimal because it is without context. It would convey much more if I knew that a student had been called upon to recite in class but did not know his lesson and used the sentence as an alibi, or if the story made clear what book was lost. This final level of cognitive meaning, which supersedes the lexical and the syntactical, may be termed *contextual*.

A proper study of the pursuit of significance must include, I suppose, the foregoing rather trivial account of semantics. Certainly, the attainment of semantic clarity, the avoidance of ambiguity in communication is a first step to more desirable goals. But must ambiguity always be avoided? Is it perhaps possible for a deliberate miscarriage of ordinary semantics to heighten the significance of sentences? This is indeed accomplished by a device called *metaphor,* and it is through metaphor that language and other kinds of symbolism approach the domain of art.

The word means transfer, carrying a meaning from one situation, where syntax tolerates it, to another, where it functions as the suggestive symbol of a symbol. Often, the metaphor amounts to an extension of the primary meaning of a word, as in the phrases: foot of a mountain, bed of a river. Many scientific terms like force, momentum, field, inertia owe their technical sense to simple transfer. Emphasis, emotional charge and even poetic beauty are often the result of a single word metaphorically employed; in fact the most impressive phrases exemplify this device: a rising or falling market, dashing manners, a brilliant style, or the horns of a dilemma.

Metaphor is not limited, however, to simple transfer; it can rise to a crescendo of artistic significance in passages that sustain appropriate stimulation through several words, or fall flat as a mixed metaphor in unfortunate cases. The poetry of every nation is full of successful uses, "wind writing its saga in the dunes of sand," "Can death be sleep when life is but a dream?" "Thou still unravished bride of quietness, thou foster child of silence and slow time." But often these are only a hairbreadth away from inept and ludicrous malformations, one of which I remember being taught as a horrid example in German schools. A consolation to the sorrowful ran like this: "The tooth of time, which dries so many tears, will soon cause grass to grow over this wound."

Metaphor can range from high linguistic significance to barren lack of meaning. There is a tendency among scientific semanticists to write metaphors off as aberrations, mainly, I suppose, because they are difficult to handle and to classify. To do so deprives language of its major charm and it lowers a curtain between prose used merely for communication and poetry. And it may well be that metaphoric symbolism is the essence of all great art. Let me speak briefly to this point.

There are those who regard a work of art as providing a unique experience without relevance to anything else. Goethe's Faust senses danger in man's tendency to arrest and linger in the ecstasy of beauty because, meaning nothing, it saps his strength. He tells the devil: "Werd ich zum Augenblicke sagen, 'Verweile doch, du bist so schön,' dann magst du mich in Fesseln schlagen, dann will ich gern zugrunde gehn." [Were I ever to say to a moment in time: stay, you are so beautiful—then you may imprison me and I shall gladly die."] Somerset Maugham spoke of beauty as a mountain peak; all roads stop there and do not lead beyond.

If this is the correct interpretation of beauty, and if art is the creation of beauty, art is strictly without significance. This does not mean it has no value for man, or its pursuit is inferior to the intellectual enterprise which does create cognitive significance. It simply makes art and beauty ends in themselves. But I sometimes doubt the adequacy of this theory, and it is a reflection upon the role of metaphor in language which prompts me to question it.

For it seems that a beautiful poem or a painting, aside from causing pleasure, usually *has* a metaphorical significance, is a symbol for something else. Certainly we admire Greek sculpture because of the beauty of its form and its anatomical precision, which are indeed ends in themselves. But its appeal is heightened because it recalls in our minds the glamorous figures of the Olympos and the stories we learned in our youth. The paintings of the great Renaissance masters, Titian, Michelangelo, Leonardo, are beautiful from every point of view, but their appeal is enhanced and they attain significance because they depict sacred scenes. My home is graced by a picture of Fujiyama, a Japanese painting on silk, which I greatly admire because of its simplicity of line and its boldness of style. However, not until I visited Japan and learned of the lore surrounding Mt. Fuji did I experience its full appeal.

Great art, it seems, does more than capture a fleeting experience of beauty. It arrests it, yes, so that our mind can revel at leisure in its richness, but it also gives it symbolic significance. In short, it plays the role of a powerful metaphor. Semantic meaning, to which this section of the essay was devoted, may therefore be said to culminate in art.

## 2. RELEVANCE

Scientific significance, or relevance, is a quality of knowledge which sets understanding apart from a mere awareness of facts. All science begins with observations, with collecting facts. To be sure the word fact is in need of careful definition, for in common language it designates both that which can be regarded as certain and also the result of an observation, and these are not the same. In the former sense, it opposes the uncertain, the dubious; in the latter sense its counterpart is a theory or an idea. I wish to use the word fact here in its second sense.

Some people believe that the business of science is the discovery, the careful description and classification of observed facts, nothing less nor more. They feel that this activity will expose the laws of nature, which they take to be generalizations of facts. In their view a scientist solves an enormous picture puzzle; he discovers the pieces and trusts that nature or providence has shaped and adjusted them in such a way that they will neatly fit together. Finally, when enough piecemeal facts are collected and put in their proper places, a recognizable pattern is thought to result, and the problem at hand has been solved. The problem or pattern is then finished once and for all and a new one in some adjacent field is started. Coupled with this favorite interpretation of science is often the expectation that some day, when all facts are known and the universal pattern is finally assembled, man will be able to regulate all his affairs in ideal fashion and the golden millennium will have arrived.

This view just sketched commits grave errors. For it takes facts as given and significant in themselves. On the other hand, however, it is evident to every scientist that a bare and isolated fact, an unrelated experience, a single observation commands no one's sustained attention; the philosopher Steinbeck, in his "Travels with Charley" notes in passing: "Everything in the world must have design or the human mind rejects it."

An isolated fact is condemned to be *un*satisfying, to clamor for context and fulfilment. But a set of related and suggestive facts, a significant experience, a set of observations often called an experiment which has been designed for a certain purpose—these are the building blocks of science. And what is it that causes a fact to be related to other facts, that makes an experience significant, a perception suggestive, a set of observations an experiment? Clearly these require a certain background of interpretation to take on meaning, a medium of expectations which they confirm or confute, a texture of theory which they illuminate. A forest of facts unordered by concepts and constructive relations may be cherished for its existential appeal, its vividness, its pleasure or its nausea; yet it is meaningless, insignificant, and usually uninteresting unless it is organized by reason. We therefore arrive at this important conclusion: Facts are not interesting or important ingredients of science unless they point to relations, unless they suggest ideas combined into what is called a theory.

A thesis which often accompanies the view here criticized, i.e., the view ascribing scientific importance to mere factualness, involves the assertion that theories are built *out of* facts. This claim is equally mistaken. At issue here is one of the oldest problems of philosophy, the apparent contrast between the uncertainty of facts and the certainties of reason. Observation is always fraught with error; it is never infinitely precise and its results will fluctuate on repetition, as every freshman learns when he first tries to perform laboratory measurements. Yet the laws of nature speak in precise, categorical and universal terms. Measurement can never completely establish the inverse square law of universal gravitation; in spite of this infelicity, we believe in the literal validity of laws, clearly on grounds which transcend the fallibility of measuring techniques.

The number $\pi$ was recently computed by two mathematicians to 100,000 decimal places. If the value of $\pi$ were established only on the basis of observations made in the real world of observations it could have no meaning, no *significance*. The observational definition of $\pi$ represents it as the ratio of the circumference to the diameter of a circle. But no circle has even been drawn and no diameter has ever been constructed which allows the specification of this ratio to an accuracy greater than is given by six decimal places. What, then, is the meaning of 100,000 places?

One of the first and perhaps the most fascinating solutions of this fact-law dilemma, the apparent incompatibility between the claims of reason and the limited competence of observation, was proposed 2,000 years ago by Plato, who ascribed to the ideas of reason a life of their own and assigned to them an august realm of independent existence. Facts, he held, were imperfect representations of pure ideas; man's knowledge of ideas, while aided by his exposure to facts, was made possible by direct contacts he experienced with ideas in earlier, purer existences. This beautiful Platonic theory of reminiscences clearly recognized the problem, but its proposed solution is no longer acceptable today.

The picture puzzle view of science has other faults. By making science an arithmetic of facts it leads us to expect a finite sum of knowledge, hinting at a future state when all facts are known and cosmic bliss prevails. This expectation reveals itself as false in every scientific discovery. No discovery ever terminates an inquiry; it may answer a question of fact, but in doing so it raises further problems because it progresses into a continuous field of logical relations studded with kernels of fact, and when a fact is conquered the relational web beyond it suggests further facts as challenges to further progress. Science has no boundaries. It has an horizon which widens as science advances, placing in view more and more unknown terrain, but never a wall which ends inquiry.

I have sketched the age-old problem of how man comes to recognize theoretical structure in a welter of unrelated facts because it is the problem of scientific significance. Facts by themselves lack significance; they gain it through their relation with ideas. Hence it is proper that I sketch this relation, which is at the heart of philosophy of science, in some detail. It has been treated very extensively in my book *The Nature of Physical Reality*. If in this part of my essay I wax a little technical, I hope you will forgive me.

From the vantage point of method one may distinguish two kinds of science, descriptive and theoretical. The distinction is rough; every science has descriptive and theoretical elements, and it would be better to speak of sciences that are primarily descriptive and others which are primarily theoretical. Geography, botony, zoology, large parts of sociology and political science belong to the former class, physics, chemistry, astronomy, modern economics to the latter.

In the descriptive sciences observable facts are sought, discovered and correlated, but their method of correlation is *inductive*. One set of facts is directly connected with another set by their invariability or frequency of occurrence. A species of plant is always found to grow in a certain climate, or in a certain soil; crime is often correlated with unemployment, etc. These are modes of correlational exposition, short of full explanation, practiced in the first-named sciences and they confer a scant measure of significance on the correlated facts. However, as every logician knows, an inductive inference which, like the regularities just noted, links one set of observations with another on the basis of frequencies without the intervention of theoretical constructs, or causal models, is always subject to probabilities; it is never certain. The conclusion that all swans are white, based on observations made in the United States, is contradicted by the existence of black swans in Australia. To summarize these findings let me speak of the kind of significance which is found in the descriptive sciences as *statistical relevance*. It is the type sought by empiricist philosophers like Bacon, not by Plato and Einstein, and it represents the lowest form of what was collectively called scientific relevance.

The method of the theoretical sciences is different. Here facts are first placed in correspondence with concepts, constructs or ideas, and then *logical* or even mathematical relations are set up among these concepts. For instance, the chemist observes facts when he handles chemicals in test tubes and notes colors of substances in reactions, observes temperatures, volumes, weights, acidities, etc. But instead of correlating all these and making tables of concomitant variations, he invents the ideas of atom and molecule, ascribes to them certain, often very abstract, properties in terms of which he can reason. Notice that atoms, molecules, electrons are not facts in the ordinary sense; they are not objects of direct observation. One reason for this assertion is that they are much too small to be seen, even in microscopes. Some of these constructs of explanation, like electric or magnetic fields, the $\psi$-functions of quantum mechanics, are so conceived that they can never submit to observation no matter how indirect. They are somehow *related* to observable effects, but their principal role is to aid reason while remaining in correspondence with the facts. They reside in the Platonic realm of ideas, where $\pi$ to 100,000 decimal places exists.

What, then, is it that makes these constructs, originally invented by the human mind as counterparts of facts, useful and acceptable? What entitles us to say that electrons, which nobody has ever seen and which are not facts in the consistent sense of the word here employed, are nevertheless real? The answers to these questions are not as simple as might first appear, for the constructs of explanation are not "given" by nature, not even suggested with any cogency, and clearly their acceptability changes in time. The desk before me was not always thought to be made up of atoms and molecules as they are now understood, nor will it be conceived to be so structured in the indefinite future, for surely our knowledge of atomic reality will change. The light impinging on my retina was once regarded as composed of visual beams, then of corpuscles, then of elastic waves; these were superseded by electromagnetic waves and now we think of light as photons. What prompts these changes in conception, these *changes in the scientific significance* of the facts of vision?

Briefly, the relevance of a scientific experience, its understanding, depends on two sets of criteria. One is called empirical, and it requires confirmation in observational practice. Let me explain. The facts observed by the chemists led to the construct, atom. It is endowed with a specific mass, atomic number, valence, a detailed electric composition. All these are postulated on the basis of a given group of facts. By their very nature, however, these constructs (atom, valence, electrical composition) logically entail consequences, not by inductive but by *deductive* implication, and these consequences can be tested. Thus the postulated structure of an atom might entail certain optical consequences not inherent in the original facts, and these consequences can be proved true or false by spectroscopic observations. The empirical requirement under discussion demands that they shall be confirmed.

Empirical confirmation, however, is not a sufficient requirement for the validity of a scientific hypothesis (which is a logically connected group of constructs). Indeed the history of science teaches that occasionally two different hypotheses seem equally confirmed. Hence another set of criteria must be called into play, and these are non-empirical. They involve the imposition of such purely ideal demands as simplicity, economy of concepts, and last but not least, elegance in their formulation. Aesthetic considerations here fashion the very substance of science. Reliance upon these etherial rules is increasing in modern physics, where speculation, indeed fruitful speculation at the forefront of knowledge, is largely devoid of effective guidance by earthy observations. Truth in science rests upon a useful combination of facts with fruitful concepts, i.e., with constructs subjected to both the empirical and the metaphysical requirements here outlined. And relevance, i.e., the most satisfying form of scientific significance, lies in the understanding which this combination permits.

To close this section, let me return to Plato's problem and offer the solution implied by modern science. Plato was right in supposing that ideas have a life of their own, that they are not mere distillations from observable facts.

But they do not inhabit a permanent heaven, nor does man remember them from earlier lives. He constructs them in creative scientific acts, then reasons back from them in a deductive manner and attempts to verify their consequences in observational experience. Elegant construction plus empirical success confer significance.

Thus $\pi$, as a mathematical construct created by man, can be computed with unlimited precision. Whether it bears a fruitful relation to actual observation in our world, i.e., to the facts, is not guaranteed by its genesis. The circumstance that it does, that observation within its limited accuracy agrees with the calculated value, makes the observational material significant. Such significance was not present to the earliest geometers for whom the ratio of most circles to their diameters had the contingent value of approximately 22/7, unrelated to other mathematical situations in which $\pi$ occurs.

## 3. MORAL SIGNIFICANCE

It is now time to face the third level of significance, that which places the accent on *value*, in contrast to meaning, upon experiences, actions, statements and things. Here, too, we must first remove a popular oversimplification which suggests that science deals exclusively with facts, whereas the arts, liberal and fine, are concerned with values. The vital inclusion of constructive ideas (beside facts) into the domain of science has already been emphasized, and we note further that the concern of science is also with values. Insofar as scientists choose the problems they investigate, decide upon courses of action, they surely make value judgments. Such observations are almost beside the point, for it is generally conceded that nobody can perform a deliberate action—deliberate in contrast to thoughtless or reflexive—without some sort of motivation, and motivation weighs values.

But while science is clearly pervaded by a texture of values, it is strangely true that it cannot generate values. There is much confusion here. It is claimed that anthropology shows that happiness (if it can be defined), the desire to live, to procreate and so forth are universal traits of man and *therefore* values. This is faulty reasoning, however, as can be illustrated very simply by reference to a linguistic infelicity tolerated by the English language, a logical slip which often escapes detection. We say perfectly properly: What can be noticed is noticeable. This tempts us to say: What can be desired is desirable. But in English we do not use the word desirable as defining what *can* be desired; it means what *ought* to be desired and this is something wholly different. There is an inadvertent slip from "can" to "should." It is not possible to commit this error in German, where noticeable means bemerk*bar*, but desirable wünschens*wert*. And *Wert* is the word for value.

This bit of semantics forces us, on reflection, to recognize that the word value has a double meaning. The distinction between what can be desired and what is worth desiring, so often overlooked, requires that we make allowance for two kinds of values: first, those which manifest themselves in the actual behavior of people, in their conduct as it is observed by anthropologists and sociologists, which I shall term *de facto* values; second, those higher values, called normative, in terms of which *de facto* values can be judged. Contrary to common belief, which likes to derive comfort for a bad conscience from the wide spread of a sinful practice, there is no moral significance in descriptive accounts of actual behavior, whether they be anthropological records about primitive tribes or Kinsey reports. Behavior—de facto values—needs justification in order to be morally significant, and at this point we face the crucial problem of all ethics, which culminates in the questions: What is the origin of those principles by means of which behavior may be judged right or wrong? How do men attain their *normative* values?

We met a situation not unlike this in our discussion of science, when we faced the transition from the contingent, meaningless facts of incidental observation to the significance-charged universals of theoretical understanding. There the transition was made possible by a peculiar feature of the methodology of science, which couples postulated basic concepts with the pragmatic yields of sensory experience. In ethics, men often turn to religion to find a solution of the corresponding problem; they assume the normative principles to be revealed by God and disclaim for themselves all responsibility for their creation. I regard this as a perfectly rational and satisfying approach, indeed one to which in the end all others may have to be reduced. Unfortunately, however, this easy and radical disposal of the normative value problem is under suspicion among great numbers of thoughtful people in today's world; their number is considerable in the West and much larger in Eastern cultures, where certain forms of atheism, or non-belief in a personal God, prevail. The religious solution is therefore unlikely to produce *trans-cultural* values, normative principles which can claim universal allegiance and respect. For this reason, I shall sketch here the way in which all cultures, regardless of their religions, actually arrive at moral norms. The process to be described is often accompanied by metaphysical and religious conceptualizations which enhance its dignity and its appeal, but it could, I think, go forward without them as it does, for instance, in communist and Confucian lands.

Permit me to voice at this juncture a personal plea for suspension of judgment regarding my own religious beliefs. When a person attempts, for the sake of objectivity, to design a methodology of ethics without benefit of theological ideas he runs the risk of being called an atheist. I know this from responses to a recent book, *Ethics and Science*, which undertook this task. The objection to a neutral development of ethics is often so vehement that people will not read on to see what the author later has to say about religion. In my own case, the conviction that ethical significance and normative value can be established in terms of human experience without reference to higher authority does not ultimately exclude religious faith. But I trace the source of faith to other situations where it asserts itself imperiously and with unqualified necessity, not to the problems of ethics where historical evidence shows that one can get along without it.

To put the matter in another way, ethics and religion have complementary functions, each rendering aid to the other. They are like man and wife. In large parts of the globe ethics has become a widower, and we are about to inquire how he manages to survive. And in the center of our inquiry stands the question of the ethical "ought," of the way in which normative values emerge within de facto behavior.

Historically successful moral systems should provide the answers. All of these, from India through Judea to the Western world, whether religiously oriented or not, seem to begin with an acceptance of ethical precepts or imperatives. In some instances, as already noted, these commands are self-authenticating because they rely upon the authority of a God or a ruler. In others, such as Confucianism, they are acknowledged because of their reasonableness or because ancestors have found them satisfying. Mere commands, however, do not form an ethical system; their directives need justification to be acceptable.

One therefore finds in every working value system some conception of ultimate goals. To name a few: some men want mere survival; others desire happiness, either personal or collective; others seek self-fulfillment; others, brotherly love among all people, or the peace that passes understanding. Let me call these *primary values. It is our human lot that we must choose among them*, for there is no way of deriving them from anything anterior by logical means.

To state the case more accurately, I should say that the choice is made for us when we are born. Our culture is charged with a tradition which is dedicated to one or more of such primary values, and they enter our bloodstream as we grow up. But how do we know that our culture has made a better choice than any other? What lifts our ideas out of the parochialism of mere custom and preference? Since these goals by themselves clearly contain no *a priori* evidence of universal validity, they, too, are without normative force. They cannot be shown in their own terms to be right or wrong, good or bad.

A normative factor appears only when, through the living of men, primary values are successfully coupled with original commands. A given set of imperatives, governing the behavior of a group, may cause it to survive or to die, may cause happiness or unhappiness, love or hatred. If it results in survival, happiness and love, and if these are indeed among the primary values chosen by the group, the behavior is somehow distinguished as being in conformity with the beginning and the end of the ethical pursuit. Commands and goals are then empirically validated through the living of men.

For the sake of brevity and precision let me denote a set of commandments like the decalogue by $C$; the behavior implied by $C$ shall be $B$, and the goal, such as collective happiness, $G$. We face the two alternatives that $B$ in accordance with $C$ leads, or does not lead, to $G$. If it does, the ethical system defined by $C$ and $G$ is validated; if not, it is confuted. Mere postulation of $C$ and $G$ represents the posing of an ethical hypothesis; when a civilization lives $C$ and attains $G$ the hypothesis becomes a valid ethical system; when it lives $C$ and fails to attain $G$ the ethical hypothesis is to be rejected as invalid.

All this seems extremely formal and remote from orthodox ethical theory. Still, I cannot help feeling that this is in fact the way in which history validates or eliminates ethical systems. The whole of human history is, in a certain sense, an ethical laboratory in which our race has performed numerous experiments, some of which failed while others succeeded. Success, according to this view, means consistency between $C$ and $G$.

In a moment I shall argue that it is this consistency which confers normative quality upon the behavior implied by $C$. First, however, it seems useful to observe a certain similarity, a parallelism between the methodologies of ethics and and of science. In science, as we have seen, cognitive significance was achieved in terms of a relation between facts and concepts. In ethics, significance attends upon a relation between commands and goals, but a relation of a different sort. While in science the relation (or correspondence) is an intellectual, i.e., a logical or mathematical one, in ethics it is a *living* link between imperatives and purposes of life.

This difference has often been misrepresented; the parallelism has been drawn too closely. For it has been suggested that the ethical imperatives *imply* or suggest the primary values, i.e., the purposes they are to serve, just as the constructs of a scientific theory imply, that is to say, allow a logical inference of, the observable facts. Or conversely, the purposes are said to define the commands. If this were true there would only be need for postulating ethical commands *or* primary values, not both. The difficulty with this position moves into view when one realizes that commands do not always achieve their ends, that their relation to goals is not a strictly logical one. No one can be sure *in abstracto*, for instance, that honesty will always lead to happiness. Or consider the claim made by Jesus in the beatitudes that to be meek is to inherit the earth; logically this is a non-sequitur. Yet people have lived the command and succeeded in attaining the stated goal. Commands and primary values are logically independent to so high a degree that the process of living must be interposed if consistency is to be established. If ethics were viewed as a science, it would have to be conceived as a highly empirical one.

The philosophical theory here outlined does not satisfy all the demands which one would like to make upon ethics. While it is able to discriminate between successful and unsuccessful systems it cannot establish a single ethical system as unique. Its competence is limited to a judgment of consistency between an arbitrarily given $C$ and an arbitrarily chosen $G$. Thus it is entirely conceivable that several pairs of $C$ and $G$ could be validated by history. In that case, our theory provides no criterion for preferring one pair to the others. In principle this is the state of affairs in science, too, where the uniqueness of a given explanation cannot be certified. Indeed at every epoch in science there are competing theories which vie for acceptance near the forefront of research, and there exists no clear-cut method of rejection. Yet somehow, under the stress of further observation and the operation of the metaphysical principles I have mentioned, one mode of explanation ultimately wins out. Will this be true in ethics, too?

The answer lies in the future of the human race. But there are already hopeful indications that it may be positive. Amazing agreement exists among the basic ethical imperatives of all peoples: The commandments of Moses and the Golden Rule are matched in essence everywhere and they enjoy, in principle at least, global acceptance. On the other hand, the goals, the primary values, seem more varied and more debatable, but the debates take place among philosophers rather than among moral people. Technical terms abound. Those who argue that the purpose of ethics is to promote pleasure are called hedonists, those who seek collective happiness are eudaimonists, those dedicated to the fulfillment of man's highest abilities (whatever they may be!) are humanists, and so on. In truth, however, their differences are academic. They originate for the most part from lack of clarity in the meaning of such words as pleasure, happiness, brotherly love, human ability. Far more important than the differences are the points of agreement. None of the goals here mentioned excludes any of the others; indeed it often seems to me that when one philosopher argues for brotherly love, another for Stoic contentment and a third for universal happiness all may be talking about the same desirable state of affairs, the first two being a bit more specific than the last. The upshot of such reflections is not despair over conflicts in this basic realm but a feeling of encouragement at the practical unanimity which already exists in accepting the foundations of ethics.

Let me now restate how moral significance, the normative element, enters the human scene. Behavior, even when in conformity with untested precepts or commandments, is without it. The precepts become obligatory, take on normative quality, when in the course of living they have proved effective in the realization of ethical goals. Action then is ethically significant.

Earlier we observed a certain parallelism in the formal structures of science and of ethics. Both start with unproved and unprovable assumptions. In science they are the postulates, in ethics they are imperatives for action. Both can be verified or validated, but in ethics the process of validation is of long duration. Neither discipline offers rules for creating the assumptions from which it proceeds. There is a sense in which both are open-ended. Whenever a discipline displays this character of open-endedness man's mind turns elsewhere in search of substance to fill the gap. In science, this tendency is less pronounced; a rigid mentality is often content to accept basic principles as postulates and be done with it. Some, however, do speak of the genius who formulates sweeping scientific principles as being inspired. In ethics, the injection of precepts into the course of history is more often regarded as a divine event. One should see in this historic coupling of ethics and religion a natural affinity between the two. But the need to ask a religious question is not confined to ethics; it occurs in science as well. The sudden burst of creative scientific insight in a human mind is just as mysterious and remarkable as the birth of a moral vision.

### 4. PERSONAL DEDICATION

The types of significance achieved by the rules, processes or strategies I have thus far surveyed share two noteworthy features:

1) Their pursuit is guided by rational or pragmatic principles whose validity or usefulness can in some manner be demonstrated. One can discuss them with a measure of cogency in objective terms; in seeking them an individual relies on reason and experience more than on intuition and inspired insights, on the commonplace more than on some unique startling illumination.

2) The quest for significance in all three instances is a group endeavor; the controlling principles enjoy intersubjective, and in that sense objective, acceptance by groups of men. Their collective function excludes to a large extent the exercise of intensely personal motives, rules out acts which break the bonds of tradition. If every form of significance were to be defined in

accordance with these collective norms, sacrificial acts, individual heroism, the selflessness of saints—and even unprecedented crime, which evidently cannot be judged in such terms—would therefore fall outside the realm of significance.

Man is indeed a rational and a social animal, and these aspects of his nature find expression in the cravings for significance through symbolism, through science and through ethics. The adjectives rational and social do not, however, describe man completely. They ignore that vast, obscure reservoir of human potentialities which, like the submerged part of an iceberg, lies below the small visible area that is governed by known rules. Existential philosophers keep reminding us in a rising chorus (to my taste somewhat ad nauseam), that this phase of our being is supremely important. Psychiatrists following Freud insist that the subconscious, the irrational, is the decisive factor in all human actions. I mean to give these advocates their due, but I hope to do so without losing my balance. Although it is contrary to custom in existential literature, I am not content with unresolved paradoxes, with vague allusions to some undefinable ground of being, with matters that are beyond comprehension. Even if we admit kinds of significance which transcend reason, that is to say, which are not caught within the net of established categories of thinking and collective norms, their admission must not contradict or flaunt the principles of reason. There is a difference between transcending reason and talking nonsense.

To anticipate what is involved in this final, this highest and most personal form of significance, let me name its features before showing how they arise in our human situation; I speak here of singular acts of dedication or personal sacrifice, of transmoral altruism which reason can neither justify nor condemn, of the choice of ideals far beyond human reach.

Let us first ask the question: Why are the concerns about lawfulness, which led to the previous kinds of significance, unable to satisfy man's deepest longing for perfection? What is the intimately personal, existential side of his nature that forces him beyond the collective sphere of ordinary moral behavior? The answers, though onesided and extreme and sometimes incoherent, may be found in the writings of philosophers like Kierkegaard, Marcel, and Tillich, and in mystical literature of East and West. They are epitomized with singular impressiveness in one of Albrecht Durer's beautiful allegorical engravings "Ritter, Tod und Teufel" (The Knight, Death and the Devil).

The knight who rides through a gloom-filled gorge, his dog by his side, appears confident, resolute, fearless. Clad in full armor, the visible personification of what is strong, good and safe in the world, he moves toward his known and consciously chosen goal. But he is accompanied by two sinister figures, death with his hourglass and the horned and hooved devil. Somehow, these two represent dangers, forms of anxiety—to use the existentialist cliché—additional to and different from those which knightly reason and prowess can conquer. Yet the countenance of the horse-man shows that he has come to terms with them; he is self-assured, and therefore without fear. Here is the total symbolism of man's predicament and its conquest.

The knight is the embodiment of all those principles which confer significance in the first three realms—breeding, knowledge and virtue. He is the product of his culture, sensitive to the demands of reason and of collective social norms. And he has overcome the spectres which flank him—death symbolizing uncontrollable fate on one side, and on the other the devil, embodying the deep and wounding knowledge of man's inadequacy before the highest morality. These two represent man's inescapable afflictions, the anxiety induced by the thought of non-being and the anxiety expressed by the tragic concept of original sin, the inevitability of evil. How can individual man, symbolized by the knight, having attained the three lower levels of significance, rise to the fourth, the summit, overcoming the twin spectres?

First we ask, by what means do men conquer death and fate? How do we manage to remain self-assured and confident, to display the "courage to be" in view of the finiteness of our existence, confronting the "abyss of non-being"? Although I have used here the language of the existentialists, I must admit that I fail completely to understand the meaning of such dramatic phrases as "courage to be" and "abyss of non-being." The latter, it seems to me, is best translated as "fear of insignificance"; it transcends the fear of death or non-being. And courage to be—a figure of speech which can hardly be reconciled with the patent fact that being requires no courage—I should much prefer to render as "assertion of personal significance." (Somehow I cannot help feeling that Seneca had more courage when he committed suicide to secure release from an intolerable moral dilemma than if he had remained alive.) This latter quality, affirmation of personal significance in the face of impending annihilation, is the object of the following discussion.

The risk to be overcome is a double one: it may be, to quote Tillich's memorable passage, "the risk of losing oneself and becoming a thing in the whole of things or of losing one's world in an empty self-relatedness." One can sink to the level of a machine or withdraw into the cocoon of a solipsist. These two extremes are perhaps not unrelated; the English language even puts them rather close together—doubtless by accident—when it permits us to speak of a person who is *a part* of the world and of another who is *apart* from the world, suggesting that neither has really found himself.

In my thinking and reading about this problem I have come upon a variety of approaches which seem to accomplish the release from the symbol of death, that is, from existential insignificance. Each involves a hazard of personal dedication, a cast of the die against non-being. For we are now put to casting dice, since we are beyond the help of proof by coercive reason and societal constraint.

One approach is by way of the ordinary Christian and Islamic belief, or rather faith, in immortality. There is no solid scientific ground for such belief, no guarantee of truth. We risk the refutation of brutal extinction in accepting it, but the very affirmation of immortality in the face of our experience of change and death has redemptive qualities that bestow significance. To what extent can one make this faith seem demonstrably sound? If I were asked this question as a scientist my answer would, strangely and perhaps paradoxically, be grounded on a premise of ignorance, and my argument would run as follows.

Permanence, indestructibility, eternal existence, I would have to say, are invoked by the scientist whenever he comes to an ultimate, the nature of which he does not understand. Here are three examples: Matter was once held to be eternal. The whole cosmos had no beginning and no end from the point of view of the science of the last century, which knew nothing of its expansion. Elementary particles like electrons were regarded as uncreated and indestructible so long as antimatter had not been discovered. Without much question, the scientist treats his unexplained ultimates as if they were eternal. It seems that every mystery which defies explanation is endowed by human reason with infinite duration.

Now the greatest mystery in all of science to this day is consciousness or, if you please, individual awareness of existence. Other concepts, like elementary particles, matter, have been reduced to more fundamental ones or are on the verge of being so reduced, and they are thereby deprived of permanence. Consciousness—in spite of the operationalist and the occasional behaviorist who laugh it off as a trivial accompaniment of organized matter—has never been reduced to anything more fundamental. The supposition that it endures, though by no means proved, is therefore compatible with the principles of reason.

Still, the risk is there, and it is this risk which makes faith in immortality more than a fond hope, more than acquiescence in a plausible belief; it converts that faith into a self-assertive gesture of personal significance. That significance is heightened by the consequences of the faith: Even if the belief is false, it bestows a greater fullness and richness of life.

Another historically successful bid for significance beyond death is recorded by Western mystics and Oriental sages in the experience described as an encounter with the divine. Here again, as in the belief in immortality, deeply religious and personal elements come into view. The encounter appears in theistic forms in the West; in the East it is said to be a union with a universal soul or transpersonal reality. Both experiences, however, have as their common effect the bestowal of confidence, the conquest of nothingness and the capture of vital meaning.

Buddhism, with its worldly wisdom and its highly practical orientation, achieves the conquest of non-being by an almost paradoxical affirmation of it, by its striving for nirvana. The word, nirvana, means literally "blown out." Some hold that its reference is to the extinction of individual personality, akin to the Vedantic union with the all-essence, Brahman, but the Buddha, when asked, did not commit himself. Certainly it means the extinction of the fire of passion, and through this a release from the anxiety of death and fate.

In the West, a similar quest for quiet significance was led by the Stoic philosophers of Greece and Rome. Their highest aim was equanimity, the serenity of mind which comes from a deliberate acquiescence to fate, best expressed by the beautiful line of Cleanthes: "Trahunt fata nolentem, ducunt volentem."

While these quiet encounters with the divine display tacit resignation and attach supreme value to a negative attitude toward life, some of the Christian mystics describe the event as a momentous, unforgettable and spectacular experience which sets man on a positive course of action. Meister Eckhart, a theologian at the University of Cologne who lived around 1300, speaks dramatically of man's encounter with God. In one instance he likens it to a lightning stroke which polarizes man toward the divine light. This simile is based on the then popular belief that lightning somehow attracts, turns the leaves of trees and the faces of animals or men toward its flash before it strikes. His other example is seeing the sun with fully open eyes and this, he says, will cause one to see the sun wherever one looks thereafter. These are beautiful and undoubtedly faithful portrayals of inner experiences, even though they are scientifically incorrect. (For lightning does not attract, and the after image is black, not bright.)

Some men, spurning faith in immortality and the spiritual encounter, seek significance in high adventure. They cast their lot in supremely reckless fashion, they wrest significance from the sky to implant it in a finite life. They shine like meteors and die in ashes. What matters with them is not success, but personal commitment, the kind of dedication which in the last analysis is the most sacrificial and the greatest human act.

Such dedication may be to a scientific enterprise, like space exploration, but its motives are not scientific. It may be to the ideal of human betterment, but it is not inspired by thought of gain, nor by practicalities, perhaps not even by love. It may be the decision to die for an idea, for a fellow man, or in the Promethean defiance of the gods. All of these can be described as selfish acts, but only in the sense that rescue from personal insignificance is selfish. Dedication takes us outside the realm of good and evil, beyond the control of principles. Hence is it equally possible for man to commit himself to a significant act of crime. This, I take it, is the tragedy of the human situation: man has choice; he can overcome anonymity by crime, he can fashion for himself a sinister sort of significance by espousing evil. In terms of Durer's allegory, he has defeated Death, but only with the aid of the Devil.

I have surveyed three major ways in which the chasm of non-being has been overpassed by bridges to significance. Let me finish by noting that these ways are not mutually exclusive or contradictory. In fact, the wisest men in our culture employ all three. Buoyed by a faith in immortality, sensitive and alert to the encounter with transcendence in the form of the spiritual or the good or the beautiful, each makes his life a sacrifice to what he has chosen as his highest aim. Some do it in the spirit of Christ, some in the spirit of Seneca, others in the spirit of Nietzsche.

So much, then, for the significance that conquers Death; I now turn to the Devil.

Earlier we saw how laws of reason and respect for human dignity engender within our collective experience a knowable code of moral precepts, of binding ethical commands. But alas, while the mind is knowing and willing, the flesh is weak; there is incompatibility between our *knowledge* of the good and our *ability* to satisfy it. This creates the problem of sin, the sense of guilt which defies our claim to moral significance.

The problem is inescapable and it troubles the minds of all normal people, in spite of the efforts of Freudian psychoanalysts and would-be social scientists who try to explain a bad conscience as a pathological remnant of an unenlightened past. The image of the devil, the human need to cleanse one's soul, the liberating effect of the "pater peccavi," the redemptive quality of the confession, are as old as recorded history and as new as the latest fad. They are agonized indications of man's cosmic frustration, his fear of the moral abyss of unmeaning, of moral insignificance.

To surmount it, he has found aid in the ideas of personal atonement, divine grace and redemption. The major chord here has distinctly religious overtones, but I do not see how they can be avoided if we seek ultimate significance.

Expiation of guilt by personal atonement is the most primitive release from the consciousness of sin, and is practiced by most older civilizations and religious cultures. Even in wholly secular societies it functions as punitive justice. The Roman church has institutionalized this method of release in its confessional and its penances, which provide wholesome and massive, though sometimes thoughtless, liberation from the significance-denying oppression of guilt. In time, however, these quasi-automatic measures of restoring self-assertion and self-esteem came to be challenged by sensitive minds who strove to raise the processes of redemption to a more metaphysical plane, clearly in the hope of making them more significant. They were the protestant reformers.

Luther's contribution to our theme centers about his favored word *trotz*, "in spite of," as Tillich has observed. The guilt complex is not mechanically dispelled by works of atonement, it persists *in spite of* the most elaborate penances, as Luther's own experiences showed. Conversely, redemption from sin is possible according to the apostle Paul and Luther *in spite of* our inability to make sufficient amends. In spite of the despair brought by self-condemnation there is an avenue of salvation through acceptance of the mercy of God. This is the doctrine of justification by faith.

To quote Tillich again, it is "the courage to accept oneself as accepted in spite of being unacceptable." The Lutheran formula that "he who is unjust is just" expresses the victory over the anxiety of guilt and self-condemnation, the idea of grace beyond human deserts. It points to the existence of an infinite reservoir of mercy and spiritual regeneration. "And if the world were full of devils that threaten to devour us," Luther wrote, "we are not afraid. One small word can destroy the prince of evil." That word is grace.

Thus, the highest form of significance comes to us through personal dedication to the goals we have chosen in response to the challenges of ultimate annihilation and of moral despair. It is acquired by travelling an ascending road which has, as way stations, altars erected to faith in immortality, encounter with the divine, high adventure, self-dedication, defeat of personal ignorance and sin, grace. At one or several of these it is our human lot to pause and worship.

And so I want to leave with you, as the figure of supreme significance, Durer's knight after he has escaped his two sombre companions, the knight confidant, alone, immune to the threat of death and devil.

*The article that appears above was given as a lecture during the Centennial Program of the College of Wooster, February 18, 1966.*

## NOTES

[1] Taken from a lecture by Professor E. Panofski.

CHAPTER 24

# NOTE ON QUANTUM MECHANICS AND CONSCIOUSNESS

To say that quantum mechanics is not complete is to affirm its tolerance, indeed its need, for progressive deepening and refinement. It seems unlikely that its radical probability character can be altered by the installation of physical hidden variables. The introduction of consciousness as a variable complementing a quantum state to render it classically deterministic, perhaps even as a submanifold of Hilbert space or an ingredient of the Hamiltonian, seems fantastic though not contradictory to anything we know. Wigner, himself, when suggesting the need for implementation of quantum mechanics to render it applicable to physiological and psychological processes, has hinted sympathetically at such contingencies. There is a philosophic argument which, though highly metaphysical, gives it a measure of credibility, but in a somewhat surprising, quasireligious sense. At the risk of inviting condemnation by many colleagues in physics I make bold to voice it here as a most unconventional but defensible suggestion transcending, to be sure, the currently accepted principles of quantum theory.

The argument is this. Human freedom of action declares itself as a genuine, undeniable experience in every deliberate, conscious decision. Its possibility involves two analytic elements: the presence of an actual set of physical alternatives admitted by the physiological state of the human brain on the one hand, and the action of some principle of choice supervening upon the dynamically indeterminate alternatives offered by the state of the brain, on the other. In a previous publication (Wimmer lecture, Journal of Philosophy) I have labelled these simply, chance and choice. The chance seems guaranteed by quantum mechanics, the principle of choice is at present an enigma. We call it will. But will is a supreme, a peak manifestation of human consciousness. Its inclusion as a "hidden variable" may occasion a return to a form of causality which bears the old fashioned name of determinism. If this be the resolution of the problem of freedom, and I know of no other that takes the problem seriously, then the application of the argument to the external world takes a somewhat unexpected form.

Human consciousness, i.e. human will, clearly does not select among the alternatives presented by the ket of the universe, since this would cntradict the stochastic character of quantum mechanical description. If, then, a consciousness analogous to the human will which enacts a choice in the process of deciding, supervenes upon the possibilities left open by a quantum mechanical measurement and selects a particular outcome, that consciousness can not be a human one. We thus confront two possibilities: one, there resides within every physical system a sort of sovereign will, i.e. a degree of consciousness. This view of pan-consciousness is one of the oldest philosophic doctrines, both East and West. The other possibility amounts to postulating a superhuman will and leads straight into religion.

This vastly and vaguely extended metatheory, if it could be more fully developed, might restore validity to Einstein's statement: God does not play dice. Our present state of knowledge, tied to the currently accepted interpretation of quantum mechanics, however, invites but does not require that affirmation.

CHAPTER 25

# RELIGIOUS DOCTRINE AND NATURAL SCIENCE

## 1. METHOD OF SCIENCE

The appropriate introduction to this paper would be a full discussion of the method of science. My version of this was presented in the *Nature of Physical Reality*. A summary of it may be found in the first chapter of *Open Vistas*. Its essence, crudely illustrated in Fig. 1 of this paper, will here be very briefly stated.

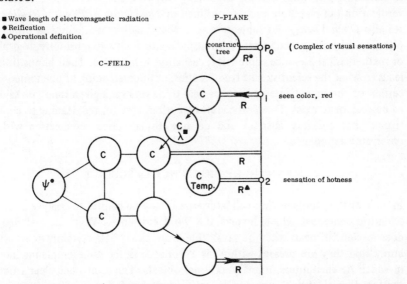

Fig. 1 $\psi^*$, the quantum mechanical state function, is an abstract construct arrived at by a sequence of operational definitions and logical relations. Single lines connecting constructs denote logical relations, some of which function as (constitutive) definitions.
Distance of a construct from the P-plane symbolizes degree of abstractness.

Every science connects a certain kind of experience (immediate, sensory, directly given), here called protocol experience because it controls in large part the development of science, with other elements which are somehow constructed (less immediate) vis à vis the protocols. For reasons stated elsewhere, the protocols are schematically represented as a plane, extending at

375

right angles to the figure, called the $P$-plane, for they form a sort of limiting boundary toward which our valid concepts tend. The "constructs," so called because they are not "given by Nature", from a large domain called the $C$-field. The connections between them, symbolized as double lines in fig. 1, are provided by "rules of correspondence", of which the simple act of reifying a set of sense perceptions and the more elaborate use of operational definitions are examples. They are called rules of correspondence, $(R)$. Our thesis is that $P$-experiences are subjective, contingent, qualitative and incoherent; they form in part what Kant called the rhapsody of perceptions. Because of the lack of relations among them, $P$-facts require to be organized, made stable, objective, rationally tractable, by translation into constructs via rules of correspondence. The constructs themselves, forming the $C$-field, are subject to validating principles that are essentially of two kinds: successful empirical verification (an elaborate process outlined in *The Nature of Physical Reality;* see also *Open Vistas*) and subjection to certain guiding, metaphysical principles (logical fertility, consistency, multiple connectivity, extensibility, logical or mathematical elegance, simplicity, casuality, invariance). Their imposition leads to what the scientist calls true theories, or understanding of phenomena; certain of the ingredients of theories held to be true (at a given time) are said to be real, or to exist. These procedures are often very formal, leading to constructs that are very abstract and do not display direct connection with observable experience.

## 2. MEANING OF EXISTENCE IN SCIENCE

In view of the problems that will later arise it may be helpful to analyze a few scientific sentences which involve the word existence. What, for instance, does a scientist mean when he says: "Electrons exist?" He certainly does not imply that they are present within his experience in the same simple manner in which he encounters ordinary external objects. This is at once clear from the fact that electrons can never be seen or apprehended in the direct manner in which we assure ourselves of the existence of large visible things. Furthermore, according to modern physics, electrons do not even have positions at all times nor some other visual attributes ordinarily assigned to objects, for they partake of the renowned dualism which is said to make them sometimes appear as particles and sometimes as waves. These are qualities unheard of in ordinary things. Evidently the electron is a physical construct which does not lie very close to the $P$-plane but is something rather abstract, not to be conceived in a simple intuitively imageful way.

The claim that electrons are real means this. Suppose we postulate such entities, endowing them with a certain charge, a certain mass, and certain other qualities such as spin and Hamiltonian. These constructs now have the unique quality of being correlated with a variety of *P*-facts in a simple manner. Among these facts are observations on the flow of electricity in wires, observations on the flow of electricity in liquids or in gaseous conductors, the production of heat and light by currents, the peculiar appearance of certain cloud chamber tracts, the behavior of photocells and a host of other physical facts. In terms of the diagram of Figure 2 the electron,

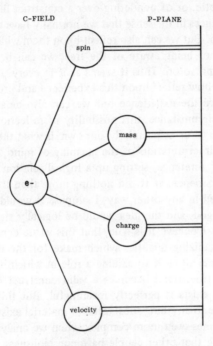

Fig 2. The electron (e⁻) is not directly related to the *P*-plane. Its existence is defined through its attributes (observables) such as its spin, mass, charge, velocity, etc. Only these latter constructs are connected with *P* through rules of correspondence.

together with its qualities and properties, corresponds to a small set of constructs some distance from the *P*-plane, and from this set there emerge heavy lines connecting the set with several parts of the *P*-plane. Because this compact theoretical structure is logically fertile, allows so many successful circuits of empirical confirmation, we have no hesitation in declaring the set of constructs, i.e., the electron together with its mass, its charge, etc., part of physical reality. That is the sense of the statement; the electron exists. In particular, no claim is made that the electron is directly perceptible, nor that it has the qualities of ordinary things, e.g., that of occupying a definite point of space at every moment of time.

Let us now turn to another proposition which may be called scientific: namely, the statement that man has a mind, or that human minds exist. Here the meaning of existence is even less obvious than in the foregoing example. For what it asserts, if anything, is this. Man is capable of simple sensations like seeing a tree, or hearing a sound, or being engaged in a reasoning process like doing arithmetic, or of pondering over a construct like a number. These are *simple* experiences in the sense that we normally have them without being conscious of them. But we can also reflect upon them while engaged in them, we can be aware of being aware of the tree, we can be aware of thinking about numbers, and so on. Thus it seems as if in every phase of our experience, we can somehow reflect upon this experience and pronounce it ours; we may not only have the experience but we can also be aware of having the experience. This circumstance, this possibility of reflection which represents a given experience to us as peculiarly our own, is what the reference to mind exposes. On this interpretation of the meaning of mind, the construct establishes a universal reference, setting up a logical relation between itself and every possible experience. If it did nothing more than this, if its usefulness could not be tested in any other way, I suppose it would occupy the status of the Berkeleian god, and the idea would be logically sterile. It is conceivable, however, and I believe it is true, that this vague construct may also be endowed with organizing qualities which make for the unity of our entire experience and thus allow it to assume a role in which it becomes logically fertile. If this is true, mind becomes a valid construct of science and the statement that it exists is perfectly meaningful. But this existence has no necessary relevance to anything substantial or material existing in space.

The story becomes even more complex when we analyze such statements as the one alleging that other people have consciousness. As far as my own experience is concerned, it is perfectly meaningful for me to say that I am conscious, or at least that I am conscious of something. There is a direct introspective way in whcih this construct of consciousness can be empirically tested. It is part of everyone's $P$-plane. But when consciousness is assigned to others, every avenue of *direct* testing is closed. In a sense the definition of consciousness then changes from its previous, first-person meaning, it becomes a construct, takes on an indirectness which deprives it of immediate significance and transfers what residue of meaning the concept retains to the manner in which it relates itself to other constructs. The consciousness of others does not enter into immediate correspondence with our $P$-plane under any circumstances, but the idea makes plausible why other people behave as I do under similar circumstances. The construct, while immune to direct verification, renders coherent a host of other constructs and provides logical fertility for a set that would otherwise be barren. It is only in this more circuitous sense that the consciousness of others can be maintained, and that this construct can be said to be valid or to exist.

Perhaps it is clear from these examples that existence, even in science, is not a simple thing to comprehend or to demonstrate. The meaning of statements alleging existence, validity, and truth may be very far from anything that common sense conveys.

## 3. THE ROLE OF INTUITION IN SCIENCE

Since the claim is sometimes made that non-scientific knowledge is based upon intuition while science uses the inductive or the rational approach it may be helpful to discuss briefly the role played by intuition in science. The word intuition requires clarification in order that misunderstanding be avoided. It seems to me that it has two meanings which I shall set forth as meaning $a$ and meaning $b$.

As we have seen, the process of scientific explanation aims as setting up correspondences between valid constructs in the $C$-field and data on the $P$-plane. To attain this goal, a scientist can move in either of two directions. He can start on the $P$-plane and drive his inquiry into the $C$-field. This is called the *inductive* approach. Alternatively, a scientist may begin with the postulation of some conjecture involving simple or otherwise appealing constructs, derive from them their implications and see whether these implications, with the use of certain rules of correspondence, are verified in $P$-experiences. This is called the *deductive* approach. Historically, significant discoveries have attended either process. Which approach works best in a given problematic situation is a psychological question; it cannot be answered in general and depends largely upon the personal disposition or endowment of the investigator. Some minds are given to induction. They are able to survey with minute care a set of data or facts, abstract from them with the use of Baconian or other inductive rules their residue of significance, and connect them successfully with certain constructs which the data in a sense "suggest." Another type of mind, more rationalistic in its aptitude, will begin with speculations about certain abstract constructs, perhaps certain particularly simple and beautiful differential equations. It will solve these equations, will develop the implication of initial abstract guesses, and see whether or not they have any significance with respect to $P$-experience. Men of genius like Einstein and Dirac belong to this group. When the connection between the $C$-field and the $P$-plane has been made it can be traversed in either direction and the question as to deduction or induction and their relative merits has ceased to be important.

Now there are scientists who have an uncanny ability. By training or natural endowment they have come to possess the faculty of contemplating a set of $P$-plane facts, pondering over them, and then arriving at once at a valid set of constructs which, they would say, are "suggested" by the phenomena. The ordinary scientific mind would have to go painfully and slowly through all stages of the process of linking constructs with $P$-experiences. The person having cultivated this remarkable faculty, however, is capable of performing what may be called the "inductive leap," a passage which seems to soar over the intervening gap between abstractions and facts with an ease not given to the ordinary mortal. This "inductive leap" is often spoken of as *intuition* in science. It is what I referred to as meaning $a$ of that word. There is something striking, incomprehensible, psychologically miraculous about this leap, something akin to revelation in religion. In the terminology of Pitirim Sorokin, it involves a supraconscious act.

While this kind of intuition is strange and perhaps remarkable to the psychologist, it creates no difficulty for the method of science. For although it generates scientific correlations in a marvelous and seemingly unorthodox way, it does not set itself above the ordinary criteria of validity in science. When the leap has been performed, the creative investigator, or at any rate his contemporaries or successors, will set to work filling in the details of the intuitive process, and if the details are contradictory, or if the results of the completed process of explanation fail in other respects to conform with the maxims of scientific validity (i.e., with the metaphysical requirements and the requirement of empirical verification), the inductive leap is declared a failure.

Less spectacular instances of intuition in this sense $a$) also occur in ordinary cognition, where they are called instinctive guesses, or successful conjectures.

There is also meaning $b$ of intuition: it designates a certain kind of introspection for which the claim is made that it delivers insights and truths. The advocates of this mode of obtaining knowledge often hold that true knowledge, scientific or otherwise, is somehow lodged within our mental makeup, is hidden in nuclear from within our own mental faculties and lies there ready to be grasped. Thus it has been maintained, for example, that the properties of space and time need not be discovered by painstaking investigations concerning the behavior of actual bodies in space and time; they may be revealed by "intuition", by a careful analysis of the ideas of space or time as they present themselves to the unindoctrinated mind. This was in part the view of Kant, who held time and space to be pure forms of intuition and used that term approximately in the sense *b* here under scrutiny. The character of space thus determined was Euclidean, as is well known; any doubt concerning the validity of Euclidean space would, in view of the unerring powers of intuition that revealed this truth, be a challenge to the very essence of human understanding.

In a similar way Husserl and the phenomenologists have advocated the use of intuitive knowledge obtained by what they call *Wesensschau*. This is supposed to be some sort of principled introspection leading to indubitable truth, or to use their favored term, *eidetic* truth. Such intuition is said to reveal the laws of logic, the principles of the human mind, and even some of the basic knowledge of the physical world. It is what I have called intuition in its meaning *b*.

When claiming validity independently of the requirements of valid constructs, intuition *b* brushes aside the main controls of the method of science. Its results become promiscuous and untrustworthy, as can be seen from the elementary fact that on the basis of this kind of intuition a bona fide sense perception is indistinguishable from an hallucination. Both proclaim themselves to be eidetically true, and unless one seeks for validity in terms of pervasive logical coherence and empirical testability, both deliverances, the true and the illusory, must be accepted without discrimination. Even on a higher scientific plane the faults and errors of intuition in its meaning *b* are quite apparent. So far as the concepts of space and time are concerned, this intuition proved entirely deceptive, for they are not Euclidean according to present evidence, and it is now well known that doubt as to the Euclidean nature of space and time does not constitute a basic challenge to all human understanding. The mind of the modern scientist, trained in non-Euclidean geometries, can very well have intuitions quite different from those of Kant and Husserl. We thus see that intuition in the mere sense of clear discernment of introspective facts does not provide the safeguard which scientific methodology requires.

What has been said is that intuition alone does not guarantee scientific truth. There is nothing in the methodology of science which ostracizes intuition, even in sense *b*. Psychology should and largely does regard it as a kind of a *P*-experience to be reckoned with, as providing data which must somehow form the termini of associations with constructs, ruled by the methodology of science, in terms of which the facts of intuition themselves become understandable. It may also be that intuitions of type *b* form a *P*-plane for introspective psychology. But if they do, and if psychology is to have the structure of a science, it must provide in addition to these intuitions, rules and criteria in terms of which they may be consistently judged to be valid or false.

## 4. THE NEW ASPECT OF SCIENCE AND ITS LESSENED STRAIN UPON RELIGION

Part 1 has concerned itself with some very general features of scientific method, features so general as to be applicable to almost any kind of science, ancient, classical and modern. The treatment, though induced by contemporary developments of physical science, did not take them into explicit consideration. It behooves us, therefore, to comment briefly upon the *novel* aspects of physical science, especially since they facilitate the passage from the strict field of science to the less formalized domains that lie around science. These include religion.

Fifty years ago a physicist would have been amazed, indeed dismayed and shocked, at a statement I made earlier in this lecture. I said that the ultimate constituents of the physical world, like electrons, may not have definite positions at all instants of time. How can we conceive of particles without loci in space and time? Does not this claim contradict the most fundamental tenets of common sense? It does indeed, and in so far as it conflicts with common sense, common sense is in error. We have learned with some pain, perhaps, that the ultimates of nature need not have picturable attributes. This lesson is now obvious, especially when it is learned in easy stages. Common sense once thought that every object had a definite size which is independent of its state of motion. Then the theory of relativity taught us that size is not an inalienable quality of physical objects. Common sense once thought that everything existing had to have a color. Meanwhile we have become accustomed to the idea that entities smaller than a wave length of light can naturally not be the carriers of color. This means that atoms cannot be said to be blue, or green, or red, they simply have no color. Common sense once thought that every object, no matter how small, must occupy a definite region of space. This insistence was a facile generalization based upon observations in the molar world of ordinary human experience. But clearly when an object like an electron is far too small ever to be seen, far too small ever to be grasped or to be experienced in kinesthetic or tactile fashion, the attribute of localizability may very well disappear. There is no *logical* difficulty in supposing that something which is too small to be seen may not have a position at all. At this point we simply have to ignore the bidding of common sense, free ourselves of its beguilement, and proceed on the basis of logical and mathematical rules alone. When this is done we arrive at the science of quantum mechanics, which provides a very successful set of constructs in terms of which atomic experience can be uniquely understood. Quantum mechanics says that there are circumstances under which electrons cannot be credited with determinate positions in space. All this reminds us strongly of what Whitehead said about the "fallacy of simple location." He called attention to the fact that there may well be existences which cannot be localized. Quantum mechanics has proved him right with respect to the denizens of the atomic world.

From this and similar changes in the concepts of modern atomic physics there has resulted a freer and more tolerant view of the requirements of scientific explanation. It was only natural that the materialistic mechanist of the late nineteenth century should tie his definition of existence and reality to the hard little particles, the substances, the pellets of stuff with which his science filled the world. For him it would have been absurd to concede the reality of atoms which cannot be definitely localized in space and time; he would have been a fool to admit the reality of mind or of consciousness in defiance of the requirement of simple location. For him the questions, where is the mind, where is the place of consciousness, were entirely proper. But times have changed, and science now acknowledges as real a host of entities that cannot be described completely in mechanistic or materialistic terms. For these reasons the demand which science makes upon religion when it examines its claims to truth have become distinctly more modest; the conflict between science and religion has become less sharp, and the strain of science upon religion has been relieved. In fact a situation seems to prevail in which the theologian can seriously listen to a scientist expounding his methodology with some expectation that it may ring a sympathetic chord. It is not altogether out of the question that the rules of scientific methodology are now sufficiently wide and flexible to embrace some forms of religion within the scientific domain. At any rate, science has become a widely open field and, as I see it, there are several ways in which it can adjust itself to the concerns of religion. In the following I shall point to three such ways.

## 5. RELIGION AS A METASCIENCE

The openminded and perceptive scientist, even if he has no desire to ask religious questions, cannot help but marvel at the success of his own method. As he ponders over the infinite and unruly mass of his factual experiences, as he contrasts it with the striking simplicity and elegance of the constructional scheme whereby he is able to explain the formidable contingencies of $P$-facts, he succumbs to a feeling of surprise, as though he confronted a miracle. His amazement concerns the circumstances that it should be possible at all for man to comprehend so vast a domain of unorganized happenings, to comprehend them in a manner that makes rational sense. Excepting a few died-in-the-wool empiricists who make it their business to proclaim the *nil mirari*, scientists feel wonder and awe at the realization that our experiences are not a chaotic welter, but display that measure of order and consistency which expresses itself in the use of simple constructs. It is true, of course, that most scientific laws are mere approximations, but even if this is granted, it is eminently still worth notice that phenomena in the world behave approximately with the regularity they are observed to possess. Paradoxically, such amazement does not spring from the occurrence of breaches in natural order which are often called miracles; on the contrary, it attaches to what seems to be the greatest miracle of all, namely, the *lack* of interruption of the natural order which expresses itself in the continued and perhaps expanding simplicity of human explanations. The theologian Scheiermacher phrased this sentiment concerning the one supreme miracle, namely natural order, with unforgettable beauty in his speeches to the German nation. If this sentiment be religious, science does indeed engender it.

Yet I doubt if this form of religion, cosmic religion if you please, will satisfy the desires of the theologian. He may wish to take it as a basis and go on from there, postulating a *cause* for order and a deity to maintain it. In doing so he goes, of course, beyond the confines of science; his religion becomes what I should like to call a metascience; but I see nothing in the methodology of science which forbids this expansion, this obvious extrapolation upon the method of science. Most scientists readily admit that their methods have limits, and that beyond these limits procedures controlled by other principles may well take hold.

Let me explain once more why I have called this a metascience. It seems proper to assume that everything that can be depicted by means of items in the diagram of figure 1 is comprised within the domain of science. The view in question contemplates this diagram as a whole and then seeks a cause for its structure. In doing so it necessarily transcends the diagram itself, it rises so to speak to a dimension from which science can be viewed in its outlines and its totality, much as theoretical science views the plane of facts. To represent this situation graphically in conformity with figures 1 and 2, the diagram would have to be three-dimensional. It is this transcendence out of the domain of science into a region from which science itself can be appraised that I meant to expose and emphasize by employing the word metascience.

## 6. RELIGION AS AN ENLARGEMENT OF EXPERIENCE IN THE "EXISTENTIAL" DOMAIN

Further examination of figure 1 leads to another interesting conjecture. That diagram is open to the left, closed by the $P$-plane to the right. A given $P$-phenomenon can be explained by a series of steps to the left that has apparently no end. Thus, for example, in answering the question why an object falls near the surface of the earth, one may refer to the Galilean theory of free fall which says merely that all bodies fall with equal and constant acceleration. This takes us to a set of constructs not far removed from the $P$-plane. But this law of free fall is nothing more than a special instance of a more general set of constructs known as the theory of Universal gravitation, which in our symbolic sense of distance lies further away from the $P$-plane. Again we need not stop there. It is possible to view the theory of universal gravitation as a special case of Einstein's law of general relativity. We have thus taken a third step back to the left, away from the $P$-plane. To be sure, at the present stage of science it is necessary for us to stop at this stage. But there is nothing to block further progress into the more abstract. Indeed if the past development of science allows a prediction, it is that we shall some day find an even more general law in terms of which the law of Einstein and others can be jointly comprehended. No limit seems to be set to man's progress to the left in the $C$-field. But as we reverse this procedure, going from the general to the more particular, we end in the $P$-plane, we conclude by saying that the stone simply falls. This is a brute fact, grotesque, final and meaningless.

Thus arise two questions. The first has often been asked in the history of philosophy as follows: If the $P$-plane limits experience, is there anything beyond experience? If so, and if science is limited to experience, then the affirmative answer transcends science. But I doubt if it necessarily involves religion. What lies beyond may be the Kantian thing in itself, that essence which, being no part of experience, is never knowable. Or there may be some mystical kind of nonscientific reality which, lying beyond experience, can never be fathomed. If it is thought that we may encounter the divine in this passage beyond the $P$-plane, that divine, since it excludes the possibility of experience, is not likely to interest the theologian.

But the closure of the field of experience raises still another question. Perhaps it arises from the circumstance that in our entire epistemology we have limited ourselves to vehicles which are rational procedures. We have used induction and deduction in traveling back and forth through the demain of figure 1. Could it not be that, in order to fathom and probe the fullness of what is actually present on or near the $P$-plane, we are required to abandon reason and give ourselves wholeheartedly and without restraint to basking in the sensation of the immediately given? Here we encounter a new emphasis, different from what we called intuition. The fundamental essence of the ebb and flow of sensations, the richness of the immediacy of our direct experience, the metaphysical substance of what assails our being in the act of sensation and affection, may after all escape the net of rational analysis. This is the view of the existentialist who feels that our representation of the $P$-plane as merely the limiting surface of the scientific domain cannot do it justice, and that greater emphasis upon the purely existential, upon the contingent and spontaneous features of our total experience are necessary.

Some philosophers of this school, notably Heidegger, go so far as to claim that the scientific process in mapping the $P$-plane upon constructs, actually falsifies experience. Only conscious and detached attention to immediacy will undo the damage of science. Heidegger's "Sein" or being, which is largely identical with our $P$-plane experience, is said to be the center of interest of enlightened man. To capture it by scientific means is to violate it, to set traps for it, and in the process of capture the scientist injures or kills "being." Thus, what be finally comes to hold is not truth or being, but the corpse of truth. Existentialist analysis, on the other hand, claims to stalk being like rare game and to watch it without disturbance. For only in its natural setting can one comprehend the essence of being without degrading it.

I think it is true that science, particularly modern science, by its coldly logical and analytic attitude, and what is more by its frequent disregard of philosophic questions, has stimulated the reactionary view just mentioned. Despite its many faults, despite its anti-intellectual flavor which I deplore, this view has nevertheless the virtue of calling attention to a few limitations which the method of science should clearly acknowledge. For there are questions which science with its present methodology will probably never answer; the full drama of existence cannot be enacted on the stage of science with its contemporary setting. Questions like those raised by Kierkegaard and Heidegger, questions like: why am I, why is there anything at all, why the phenomenon of experience which science analyzes, what is the basis of art, what is the quintessence of beauty? Questions such as these appear as idle vaporings when viewed as problems of science. Yet they bespeak an intense human concern and contain a powerful appeal that defies the positivistic insistence that they are meaningless or insignificant. If science does not answer them is it not reasonable that at this point we resign ourselves to other hands? This is indeed affirmed by those who see religion as an extension of experience into the existential domain. They feel that the $P$-plane must somehow be opend up by a new kind of analysis, an analysis not scientific, an analysis for which science offers no help. What happens when this extension is permitted can hardly be predicted in detail. One can go the way of Sartre and dwell in non-religious fashion upon the nausea of existence. Or one can go the way of Kierkegaard and Gabriel Marcel and couple the existential affirmation with an excursion into the domain of religion. At any rate the $P$-plane quite obviously is an area of contact between science and religion, as the widespread acceptance of existentialist philosophy today clearly shows.

## 7. RELIGION AS PART OF AN ENLARGED SCIENCE

My fondest hope lies in the direction of amalgamating religion with science. For it is by no means out of the question that a theory of religion, *i.e.*, theology, when fully developed, may exhibit the same formal structure as science itself. A suggestion affirming such conjectures is already present in the writings of William James, who regards a body of religious beliefs as a doctrine capable of pragmatic verification. He is vague, to be sure, when discussing the precise manner in which such beliefs are to be tested, but the general idea is certainly there.

If such an approach is to be started, the first question to be answered is: what is the $P$-plane of religious experience? A possible and probably correct answer appears to be: the kind of immediate experience which is often regarded as distinctly religious. I mean such things as the feeling of gratitude for our human existence that is clearly not expressible to any man; mystical communion with the infinite, the sense of the eternal, despair at the prospect of irrevocable annihilation, the inexplicable feeling of the relevance of an occasion which the saints describe as an encounter with the holy, our frightened exposure to the tremendum—to say that these are peculiarly religious experiences is not argue that they are *exclusively* religious. For they are also $P$-facts for several of the so-called social sciences, and I hesitate to suggest that psychology, psychiatry, sociology, and anthropology should not be concerned with them and endeavor to show how they can be organized in the constructional schemes of these sciences. This however, does not cast out the possibility of an analysis in religious terms, nor does it show it to be illegitimate. For a given simple sensation may very well be the starting point of several inquiries, one into a physical the others into a biological or a pychological domain of constructs. The fact that a given experience can be a $P$-datum for a variety of sciences must always be recognized and is no argument against the validity of the various explanatory schemes. And in this context religion, too, can claim its due.

What follows next in the development of a "science" of religion is a little difficult to predict, though probably not more difficult than it would have been to forecast the structure of modern science in Aristotle's day. Sciences grow when people become convinced of their importance and their necessity, and they develop their methodology as they mature. There are those who believe that theology already provides a $C$-field in terms of which a concatenation and a logical nexus between the experiences I have named can be achieved. If this is to be accepted, the idea of theology must be subjected to the same metaphysical requirements which we impose on scientific theories. That is to say, they must partake of logical fertility, multiple connection, extensibility, simplicity, etc. Nor is this often denied by workers in the field.

Moreover, if religion is to have the structure of science, it must also expose itself to tests in the manner of our circuits of empirical verification. This forces us to reject at once certain peculiar kinds of theology, such as the deism of the enlightenment and probably also predetermination of the Calvinistic type. For these theories could never be tested. Any tests man could devise would be foreordained, would have been included in the creator's foresight at the very beginning. It would, therefore, be futile to regard the outcome of the tests as significant. But such criticisms do not affect most major theological systems.[1]

You see, natural science is not wholly without suggestions as to the structure of a religion based on the grounds of its own methodology. But it offers no detailed material aid. Least of all does it require the slavish adherence of theological doctrine to the constructs of physics, chemistry, or biology. Not even the social sciences, notably psychology, deem it necessary any longer to ape the physicist. This does not imply contrasts or contradictions—for surely, if a concept applicable in one field has no application in another it does not contradict it; the notion of temperature is entirely in harmony with that of an atom, although it has no relevance for a single atom. It is the methodological structure of science that might be transferable; I do not advocate "physicalism" in religion.

## 8. POINTS OF CONTACT BETWEEN MODERN SCIENCE AND TRADITIONAL RELIGION

A few instances demonstrating approaches between some fairly universal religious tenets and modern science will be cited in conclusion of this article. An act of creation is a typical ingredient of many religious beliefs. Old-style science denied this by an appeal to certain conservation principles (matter, energy). Thomas' *creatio ex nihilo* was regarded as absurd.

Today we have the big-bang theory of the origin of the universe, and we see a certain background radiation as testimony to a violent act of creation. Furthermore, the creation of matter no longer violates the principle of conservation of energy, which has become equivalent to matter. A uniform dense sphere of matter, having mass M and radius R, possesses energy mainly of two kinds: relativistic mass energy of amount $Mc^2$, and gravitational potential energy given by $-kGM^2/R$. Here G is the gravitational constant and k a numerical constant close to 1. The second term is negative because gravitation is an attractive force. The sum $Mc^2 - kGM^2/R$ can very well be zero, the condition being $R \approx GM/c^2$. If it is obeyed, St. Thomas can be satisfied.

Two other remarkable features of the relation $R \approx GM/c^2$ are worth recording. If we insert for M a good estimate of the mass of the total universe (as given by Shapley) and for R the radius of the expanding universe we find the relation approximately satisfied. This might be interpreted to mean that no contradiction with the law of energy conservation would ensue if our universe were created or destroyed today.

The other noteworthy fact, mentioned here parenthetically, is that relation (1) defines the Schwarzschild radius, the condition for the existence of a black hole. But I shall forego any religious speculation that might arise from that.

Science and religion touch each other on a very basic plane in the biblical account of the great flood, for in its aftermath Jehova grants as it were a charter to science, with an implication that the two shall live in peace. Reference is here, not to the first act of creation in Genesis 1, which resulted in the existence of a physical universe, but to a second, equally important one, which established the lawfulness of the universe. First, we are told, there was chaos, tohu vabohu; then followed a period of lawlessness and confusion terminated by the flood.

And when the waters receded God pledged His subsequent adherence to lawfulness in the beautiful covenant of the Rainbow.

Finally, I venture to include a somewhat irrelevant comment on a prevalent popular view which seems to hold that most scientists are somehow antireligious. Statistics (of which I am not aware!) might well bear this out. But it is my distinct impression that truly creative scientists, especially those of our era whom I had the privilege to know, belie this popular belief.

The present paper has dealt with possible connections between religion and science, as its title indicates. In the historical development of most religions, they have drawn greater support from areas of human concern other than science, notably ethics. The connection there is easily seen, at least in its fundamental aspects. Every living culture develops ethical imperatives which are expected to permit it to survive and to achieve other desirable goals.

In many cultures, though not in all, this ethical code is thought to be divinely inspired. The residue of the code within the individual is man's conscience. Now it is the inevitable tragedy of the human race that under many circumstances man is unable to resist the so-called demonic forces which impel him to violate the code. Thus arises the concept of original sin, symbolized in the Judaeo-Christian religions by the final scene in the garden of Eden. Man finds himself frustrated before the moral law and, having offended his conscience, seeks redemption. This universal need culminates in the postulation of a deity who can inflict punishment or grant redemption.

The possible religious implications of ethics form a secondary accompaniment to this major theme in most Western religions.

## NOTE

[1] In this context see H. Schilling, *Science and Religion*, Scribner and Sons, N.Y., 1962.

## PUBLICATIONS BY HENRY MARGENAU

'The Zeeman Effect in, the Cerium Spectrum between 3000 and 5000A units, *Phys. Rev.* **30**, 458–465 (1927).
'The Problem of Physical Explanation', *The Monist,* **39**, 321–349 (1929).
'Dependence of Ultra-violet Reflection of Silver on Plastic Deformation', *Phy. Rev.* **33**, 1035–1045 (1929).
'Die Abweichungen vom Ohmschen Gesetz bei hohen Stromdichten in Lichte der Sommerfeldschen Elektronentheorie', *Zeits. f. Phys.* **56**, 230–231 (1929).
'Zur Abweichung vom Ohmschen Gesetz bei hohen Feldstarken, *Zeits. f. Phys.* **60**, 234–236 (1930).
'Über die Veranderlichkeit der Anzahl freier Metallelektronen mit der Temperatur', *Phys. Zeit.* **31**, 540–546 (1930).
'Zur Theorie der Molekularkrafte bei Dipolagsen', *Zeits. f. Phys.* **64**, 584–597 (1930).
'Second Virial Coefficient for Gases: A Critical Comparison between Theoretical and Experimental Results', *Phys. Rev.* **36**, 1782–1790 (1930).
'Causality and Modern Physics', *The Monist* **41**, 1–36 (1931).
'Note on the Calculation of Van der Waals Forces', *Phys. Rev.* **57**, 1425–1430 (1931).
'Surface Energy of Liquids', *Phys. Rev.* **38**, 365–371 (1931).
'Role of Quadrupole Forces in Van der Waals Attractions', *Phys. Rev.* **38**, 747–756 (1931).
'The Equation of State of Real Gases', *Phys. Rev.* **38**, 1785–1796 (1931).
'The Uncertainty Principle and Free Will', *Science,* **74**, 596 (1931).
'Quantum Dynamical Correction of the Equation of State of Real Gases', *Proc. Nat. Acad. Sci.* **18**, 56–62; 230 (1932).
'Probability and Causality in Quantum Physics', *The Monist,* **42**, 161–168 (1932).
'Pressure Shift and Broadening of Spectral Lines', *Phys. Rev.* **40**, 387–408 (1932).
'Shift of the Transmission Band of Silver by Cold Working', *Phys. Rev.* **40**, 800–01 (1932).
'Pressure Broadening of Spectral Lines II', *Phys. Rev.* **43**, 129–134 (1933).
'Pressure Effects of Foreign Gases on the Sodium D-Lines', *Phys. Rev.* **44**, 92–98 (1933) (with W.W. Watson).
'Pressure Effects of Nitrogen on Potassium Absorption Lines', *Phys. Rev.* **44**, 748–752 (1933) (with W.W. Watson).
'Zur Theorie der Verbreiterung von Spektrallinien', *Zeits. f. Phys.* **86**, 523–529 (1933).
'Asymmetries of Pressure Broadened Lines', *Phys. Rev.* **44**, 931–934 (1933).
'Application of Many-Valued Systems of Logic to Physics', *Jour. Phil. of Science,* **1**, 118–121 (1934).
'Meaning and Scientific Status of Causality', *Jour. Phil. of Science,* **1**, 133–148 (1934).
'The Complex Neutron', *Phys. Rev.* **46**, 107–110 (1934).
'Progression of Nuclear Resonance Levels with Atomic Number', *Phys. Rev.* **46**, 228 (1934) (with Ernest Pollard).
'Nuclear Energy Levels and the Model of a Potential Hole', *Phys. Rev.* **46**, 613–615 (1934).
'Flexibility of Scientific Truth', *Phil. of Science,* **1**, No. 4 (1934).
'Methodology of Modern Physics I', *Phil. of Science,* **2**, 48–72 (1935).

'Methodology of Modern Physics II', *Phil. of Science*, **2**, 164–187 (1935).
'Evidence Regarding the Field of the Deuteron', *Nature*, **135**, 393 (1935) (with Ernest Pollard).
'Resonance Interaction between Deuterons and Alpha-Particles', *Phys. Rev.* **47**, 571 (1935) (with E. Pollard).
'Collisions of Alpha-Particles in Deuterium', *Phys. Rev.* **47**, 833 (1935) (with E. Pollard).
'Natural Width of K-2 Lines', *Phys. Rev.* **47**, 89 (1935).
'Collisions of Alpha-Particles in Hydrogen', *Phys. Rev.* **48**, 402 (1935).
'Theory of Pressure Effects of Foreign Gases on Spectral Lines', *Phys. Rev.* **48**, 755 (1935).
'Pressure Effects on Spectral Lines', *Rev. Mod. Phys.* **8**, 22 (1936) (with W.W. Watson).
'Quantum Mechanical Description', *Phys. Rev.* **49**, 240 (1936).
'Note on Pressure Effects in Band Spectra', *Phys. Rev.* **49**, 596 (1936).
*Foundations of Physics*. John Wiley and Sons, Inc. pp. xiii + 537 (1936) (with R.B. Lindsay).
'Discussion on Nagel's Paper', *J. Am. Stat. Assoc.* **31**, 27 (1936).
'Relativity and Nuclear Forces', *Phys. Rev.* **50**, 342 (1936).
'The Teaching of Intermediate Physics'. (Invited paper read at the Brunswick Meeting of the New England Section of Am. Phys. Soc.) *Phys. Rev.* **50**, 872 (1936).
'Critical Points in Modern Physical Theory', *Phil. of Science* **4**, 337 (1937).
'Long Range Interactions Between Dipole Molecules', *Phys. Rev.* **51**, 748 (1937) (with D.T. Warren).
'Pressure Broadening in Bands of Dipole Molecules', *Phys. Rev.* **51**, 48 (1937) (with W.W. Watson).
'Pressure Shifts of Krypton Lines', *Phys. Rev.* **52**, 384 (1937) (with W.W. Watson).
'Normal States of Nuclear Three and Four Body Systems', *Phys. Rev.* **52**, 790 (1937) (with D.T. Warren).
'Note on the Validity of Methods in Nuclear Calculations', *Phys. Rev.* **52**, 1027 (1937) (with D.T. Warren).
'Excited States of the Alpha-Particle', *Phys. Rev.* **53**, 198 (1938).
'Pressure Effects on Spectral Lines', *Proc. of 5th Conference on Spectroscopy*, M.I.T., 123 (1938) (with W.W. Watson).
'The Teaching of Intermediate Physics', *Am. Phys. Teacher* **6**, 295 (1938).
'Variational Theory of the Alpha-Particle', *Phys. Rev.* **54**, 422 (1938) (with W.A. Tyrrell, Jr.).
'Binding Energy of $Li^6$', *Phys. Rev.* **54**, 705 (1938) (with K.G. Carroll).
'Quadrupole Contributions to London's Dispersion Forces', *J. Chem. Phys.* **9**, 896 (1938).
'Van der Waals Forces', *Rev. Mod. Phys.* **11**, 1 (1939).
'Binding Energies of Light Nuclei', *Phys. Rev.* **55**, 790 (1939). (with W.A. Tyrrell, Jr. and K.G. Carroll).
'Binding Energy of $He^6$ and Nuclear Forces', *Phys. Rev.* **55**, 1173 (1939).
'Probability of Modern Physics', (Invited paper; read before 5th International Congress for Unity of Sciences) Harvard (1939).
'Probability, Many-Values Logics and Physics', *Phil. of Science* **6**, 65 (1939).
'Van der Waals Potential in Helium', *Phys. Rev.* **56**, 1000 (1939).
'Relativistic Magnetic Moment of a Charged Particle', *Phys. Rev.* **57**, 383 (1940).
'Magnetic Moments of Odd Nuclei', *Phys. Rev.* **58**, 103 (1940) (with E. Wigner).

'Interaction of Alpha-Particles', *Phys. Rev.* **59**, 37 (1941).
'Statistics of Excited Energy States of Nuclei', *Phys. Rev.* **59**, 627 (1941).
*Development of the Sciences* (Chapter on Physics), pp. 91–120; Yale University Press (1941).
'Foundations of the Unity of Science', *Philos. Rev.* **1**, 431 (1941).
'Forces Between Neutral Molecules and Metallic Surface', *Phys. Rev.* **60**, 128 (1941) (with W.G. Pollard).
'Metaphysical Elements in Physics', *Rev. Mod. Phys.* **13**, 176 (1941).
'On the Forces Between Positive Ions and Neutral Molecules', *Phil. Sci. Tech.* Section **4**, 603 (1941).
'The Role of Definitions in Physical Science, with Remarks on the Frequency Definition of Probability', *Am J. of Phys.* **10**, 224 (1942).
*The Development of the Sciences.* (Physics, Chapter III,) Yale University Press (1942).
*The Mathematics of Physics and Chemistry.* (with G. Murphy) D. Van Nostrand Co, Inc., pp. 581 (1943).
'Theory and Scientific Development', *Scien. Mo.* **LVII**, 63 (1943).
'The Forces Between Hydrogen Molecules', *Phys. Rev.* **63**, 131 (1943). *Phys. Rev.* **63**, 385 (1943).
'The Forces Between Water Molecules and the Second Virial Coefficient for Water', *Phys. Rev.* **66**, 307 (1944) (with V.W. Myers).
'The Forces Between a Hydrogen Molecule and a Hydrogen Atom', *Phys. Rev.* **66**, 303 (1944).
'Phenomenology and Physics' (Invited paper, Am. Phys. Soc. Pittsburgh, 1943) *Phil. and Phenomenol. Research* **5**, 286 (1944).
'The Exclusion Principle and its Philosophical Importance', *J. Philo. Sci.* **11**, 187 (1944).
'Atomic and Molecular Theory Since Bohr. Article I: Historical Survey', *Am. J. Phys.* **12**, 119 (1944) (with A. Wightman).
    'Article II: Logic and Mathematical Survey', *Am. J. Phys.* **12**, 247 (1944) (with A. Wightman).
    'Article III: Résumé of Specific Results', *Am. J. Phys.* **13**, 73 (1945) (with R.B. Setlow).
'Film Formation of Water Flowing Through Thin Cracks', *Am. J. Sci.* **243**, 192 (1945) (with R.E. Meyerott).
'On the Frequency Theory of Probability', *Philo. and Phenomeno. Research* **6**, 11 (1945).
'Conduction and Dispersion of Ionized Gases at High Frequencies', *Phys. Rev.* **69**, 508 (1946).
'Physical Processes in the Recovery of TR Tubes', *Phys. Rev.* **70**, 349 (1946) (with F. McMilbank, Jr., I.H. Dearnley, C.S. Pearsall, and C.G. Montgomery).
'Fysikkens Oprinnelse Og Utvikling', Hefte I–II, *Saertrykk Av Fra Fysikkens Verden*, 49–97 (1947).
'Particle and Field Concepts in Biology', *Scientific Monthly* **44**, (3) March 1947.
'Western Culture, Scientific Method and the Problem of Ethics', *Am. J. Phys.* **15**, 218 (1947).
'Theory of High Frequency Gas Discharges. I. Methods for Calculating Electron Distribution Function', *Phys. Rev.* **73**, 297 (1948).
'Theory of High Frequency Gas Discharges. II. Harmonic Components of the Distribution Function', *Phys. Rev.* **73**, 309 (1948) (with L. Hartmann).

'Theory of High Frequency Gas Dicharges. IV. Note on the Similarity Principle', *Phys. Rev.* **73**, 326 (1948).
'Pressure Broadening in the Inversion Spectrum of Ammonia', *Phys. Rev.* **76**, 131 (1949).
'Collision Theories of Pressure Broadening of Spectral Lines', *Phys. Rev.* **76**, 1211 (1949).
'Inversion Frequency of Ammonia and Molecular Interaction', *Phys. Rev.* **76**, 1432 (1949).
'Reality in Quantum Mechanics', *Phil. Sci.* **16**, 287 (1949).
'Einstein's Conception of Reality', Einstein Volume, *Library of Living Philosophers* (1949), Paul A. Schilpp (Ed).
'Ethical Science', *Scientific Monthly* **LXIX**, 290 (1949).
'Theory of Magnetic Resonance in Nitric Oxide', *Phys. Rev.* **78**, 587 (1950) (with A.F. Henry).
'Electric Conductivity and Mean Free Paths', *Phys. Rev.* **79**, 970 (1950).
*Nature of Physical Reality.* McGraw-Hill Book Co., Inc. pp. xiii + 479 (1950).
*The Nature of Concepts.* (Two Chapters) University of Oklahoma Press, pp. 139 (1950) (edited with F.S.C. Northrop).
'The Meaning of "Elementary Particle" ', *Am. Scientist* **39**, 422 (1951).
'Statistical Theory of Pressure Broadening', *Phys. Rev.* **82**, 156 (1951).
'Conceptual Foundations of the Quantum Theory', *Science* **113**, 95 (1951).
'Report on Recent Developments in Philosophy of Quantum Mechanics' (with J. Compton).
'Reply to Professor Beck on Kantianism', *Philos. and Phenom. Research* **XI**, no. 4 (1951).
Book Reviews and Minor Articles Replying to Comments Elicited by *The Nature of Physical Reality.*
'Physics and Ontology', *Phil. of Sci.* **19**, 342 (1952).
'Physical Versus Historical Reality', *Phi. of Sci.* **19**, 193 (1952).
'Intermolecular Forces in Helium', Note 21. Conf. on Quantum Mechanical Methods in Valence Theory, 8–10, Sept. 1951. (Published 1952).
'On a Possible Method for Estimating the Repulsive Potential Between Closed Shells', Note 22. Conf. on Quantum Mechanical Methods in Valence Theory, 8–10 Sept. 1951. (Published 1952).
'Note on the Calculation of Exchange Forces', Note 23. Conf. on Quantum Mechanical Method in Valence Theory, 8–10 Sept. 1951. (Published 1952).
'Ion Clustering', *Phys. Rev.* **85**, 670 (1952) (with S. Bloom).
'Scientific Bases of Ethics', (Lecture) *Main Currents* (1952).
'Quantum Theory of Spectral Line Broadening', *Phys. Rev.* **90**, 791 (1953) (with S. Bloom).
'On the Interaction of Closed Shells', *J. Chem. Phys.* **21**, 394 (1953) (with P. Rosen).
'The Deductive Method in the Physical Sciences', *Electrical Engineering,* April, 1953.
'The Forces Between Hydrogen Molecules', (with A.A. Evett). *Phys. Rev.* **90**, 1021 (1953). *J. Chem. Phys.* **21**. 958 (1953).
'Integrative Education in the Sciences', *Am. Assoc. Colleges for Teacher Education,* 132 (1953).
'The New Faith of Science', *Carleton College*, pp. 1–20 (1953).
'The Deductive Methods in the Physical Sciences', *Main Currents,* pp. 131, March 1953.
*Physics.* McGraw-Hill Book Co. pp. xii–814, 2nd Ed. (1953) (with W.W. Watson and C.C. Montgomery).

'Can Time Flow Backwards?', *Phil. Sci.* **21**, 79 (1954).
'Advantages and Disadvantages, of Various Interpretations of the Quantum Theory, *J. Washington Academy of Sciences* **44**, 265 (1954). *Physics Today* **7**, 6 (1954).
'On Interpretations and Misinterpretations of Operationalism', *Scientific Monthly* **79**, 209 (1954).
'Causality in Quantum Electro-dynamics', *Diogenes* **6**, (1954).
'Quantum Theory of Line Broadening by an Ionic Plasma', *Astrophys. J.* **121**, 194 (1955) (with R. Meyerott).
'Electron Impact Broadening of Spectral Lines', *Phys. Rev.* **98**, 495 (1955) (with B. Kivel and S. Bloom).
'Knowledge, Faith and Physics', *Main Currents* **11**, 108 (1955).
'The Competence and Limitations of Scientific Method', *J. ORSA* **3**, No. 2, 135 (1955).
'Effects of Electron Collisions on the Width of Spectral Lines', *Phys. Rev.* **98**, 1822 (1955) (with B. Kivel).
'Line Broadening by Electrons: Validity of Simple Theories', *Phys. Rev.* **99**, 1851 (1955) (with R. Meyerott).
'Facts and Values', *Brown University Papers* **XXXI**, pp. 1–21, (1955) (Commencement Address).
'Present Status and Needs of the Philosophy of Science', *Proc. Am. Phil. Soc.* **99**, 334 (1955).
*The Mathematics of Physics and Chemistry*, (2nd Edition). D. Van Nostrand Co., pp. 604 (1956) (with G. Murphy).
'Estimate of Pressure Effects on $N_2$-Band Lines', *U.S.A.F. Rand Research Memo.* 1669, 1–S (1956).
'The Validity of the Statistical Theory of Pressure Broadening, *Rand Research Memo* RM-1670-AEC (1956).
'Stark Effects in Line Broadening', *Rand Research Memo* RM-1779-AEC (1956).
'Unified Method of Science: Can it be Applied to Religion?' *Christian Register* (1956).
'Line Broadening of an Impurity Spectrum in Silicon', *Phys. Rev.* **103**, 879 (1956) (with D. Sampson).
'Why Teach Philosophy of Science?', *Age of Science* **2**, (1956).
'Moderna Primer', *Congreso de la Sociedad Interamericana de Filosofia*, Santiago, Chile (1956).
'The New View of Man in His Physical Environment', *Centennial Review of Arts and Science*, **1**, No. 1 (1957).
'The Meaning and The Faith of Science', *Main Currents*, **13**, No. 3 (1957).
'Stark Effects of Line Broadening', *Phys. Rev.* **106**, 244 (1957) (with M. Lewis).
'Philosophy of Science in the Twentieth Century', *Junior World History*, Summer (1957).
'Relativity: An Epistemological Appraisal', *Phil. of Sci.* **24**, 297 (1957).
'Van der Waals Forces'. Article in *Encyclopedia of Chemistry*, G. Clark editor Reinhold Publishing Co., New York (1957).
'The Modern Predicament', *Main Currents*, **13**, (1957).
'El Nuevo Concepto de Hombre en su Ambiente Fisico', *Episteme, Anuario de Filosofia*, Universidad Central de Venezuela, Caracas, 312 (1957).
'Cyclotron Resonance: Method for Determining Collision Cross Sections for Low Energy Electrons', *Phys. Rev.* **108**, 1368 (1957) (with D. Kelly).
'Philosophical Problems Concerning the Meaning of Measurement in Physics', *Phil. Sci.* **25**, 23 (1958).

'Conductivity of Plasmas to Microwaves', *Phys. Rev.* **109**, 6 (1958).
'Statistical Broadening of Spectral Lines Emitted by Ions in a Plasma', *Phys. Rev.* **109**, 842 (1958).
'Conductivity of Plasmas to Microwaves', *Phys. Rev.* **112**, 1437 (1958).
'Structure of Spectral Lines from Plasmas', *Rev. Mod. Phys.* **31**, 569 (1959) (with M. Lewis).
'Perspectives of Science', *The KEY Reporter,* Phil Beta Kappa, July (1959).
'Philosophy of Physical Science in the Twentieth Century', *J. World History* **IV**, No. 3 (1958).
'Fields in Physics and Biology', *Main Currents* **15**, (1959).
'Theory of Pressure Effects on Alkali Doublet Lines', *J. Chem. Phys.* **30**, 1556 (1959) (with L. Klein).
'Frequency Shifts in Hyperfine Splitting of Alkalis Caused by Foreign Gases', *Phys. Rev.* **115**, 87 (1959) (with Fontana and Klein).
'Microwave Conductivity of Slightly Ionized Air', *J. App. Phys.* **30**, 1385 (1959) (with D. Stillinger).
*Naturphilosophie: Die Philosophie in XX. Jahrhundert* Ernst Klett Verlag, Stuttgart (1959).
'Scientific Basis of Value Theory', Chapter in *New Knowledge in Human Values,* edited by Maslow. Harper's (1959).
'The Structure of Spectral Lines from Plasmas', *Proc. of IV International Conference on Ionization Phenomena in Gases.* Uppsala 17–21 August 1959.
'Formulas for Estimating Widths of Spectral Lines, Emitted from Plasmas and Their Limits of Validity', *Proc. of IV International Conference on Ionization Phenomena in Gases,* Uppsala 17–21 August 1959.
'Elargissement des Raies Spectrales Produites par des Ions dans un Plasma', Colloquest Internationaux du Centre National de la Recherche Scientifique. No. LXXVII, Bellevue, 1–6 Juillet 1957. Published 1959 (with M. Lewis).
'Essais D'Interpretation Theorique des Bandes Satellites Induites par la Pression', Colloques Internationaux du Centre National de la Recherche Scientifique, Bellevue, 1–6 Juillet 1957. Publishes 1959. (with L. Klein).
'High Frequency Breakdown of Air', *J. App. Phys.* **31**, 1617 (1960) (with D. Kelly).
'Forbidden Helium Line in a Plasma Spectrum', *J.Q.S.R.T.* **1**, 46 (1960) (with Sadjian and Wimmel).
'Meaning and Scientific Status of Causality' Chapter in book *Philosophy of Science,* edited by A. Danto; Meridan Books, New York (1960).
'Science, Philosophy and Religion', *Dalhousie Rev.* **39**, 447 (1960).
'Does Physical "Knowledge" Require *a priori* or Undemonstrable Presuppositions', Chapter in *The Nature of Physical Knowledge,* edited by L.W. Friedrich, *S.J.;* Marquette University Press, Milwaukee, Wisc. (1960).
'Causality', *Main Currents* **16**, (1960).
'Frequency Shifts, in Hyperfine Splitting of Alkalis: A Correction', *Phys. Rev.* **122**, 1204 (1961) (with R. Herman).
'Bacon and Modern Physics: A Confrontation', *Proc. Am. Phil. Soc.* **105**, 5 (1961).
'Correlations Between Measurements in Quantum Theory', *Prog. Theo. Phys.* **26**, 722 (1961) (with R. Hill).
'Preliminaries', Chapter I in *Quantum Theory,* Volume I, edited by D.R. Bates; Academic Press, New York (1961).

'Fundamental Principles of Quantum Mechanics', Chapter II in *Quantum Theory*, Volume I, edited by D.R. Bates, Academic Press, New York (1961).
'Causality in Quantum Electrodynamics', Chapter in *A Science Reader* edited by Ryan. Holt, Rinehart and Winston (1961).
*Open Vistas.* Yale University Press (1961).
'Is the Mathematical Explanation of Physical Data Unique?', *Proc. of 1960 International Congress.* Published 1962.
'Discussion: Comments on Professor Putnam's Comments', *Philo. of Sci.* **29**, 292 (1962) (with E. Wigner).
'The New Style of Science', *Main Currents* **19**, (1963). *Yale Alumni Magazine* (1963).
'Theory of Pressure Shifts of HCl Lines Caused by Noble Gases', *J. Chem. Phys.* **38**, 1 (1963) (with H. Jacobson). *J.Q.S.R.T.* **3**, 35 (1963) (with H. Jacobson).
'Measurement in Quantum States', Part I. *Philo. of Sci.* **30**, 1 (1963). Part II. *Philo of Sci.* **30**, 138 (1963).
'Intercultural Communication of Ethical Judgments', XIII Congreso Interancional de Filosofia, Universidad Nacional Autonoma de Mexico, 65 (1963).
'Absorption Coefficient and Microwave Conductivity of Plasma', *Astrophys. J.* **137**, 851 (1963) (with P. Mallozzi).
'Measurement in Quantum Mechanics', *Ann. of Phys.* **23**, 469 (1963).
'Theory of Pressure Shifts of HCl Lines Caused by Noble Gases', *J.Q.S.R.T.* **3**, 445 (1963).
'Philosophy of Physical Science in the Twentieth Century', Chapter in *Evolution of Science,* ed. G. Metraux, Mentor Books, New York (1963) (with J. Smith).
'Discussion: Reply to Professor Putnam', *Philo. of Sci.* **31**, 7 (1964) (with E. Wigner).
*Mathematics of Physics and Chemistry,* Volume II. D. Van Nostrand Co., Inc., Princeton, N.J. (1964) (with G. Murphy).
*Ethics and Science.* D. Van Nostrand Co., Inc., Princeton N.J. (1964).
Book Review of *Physics in the Soviet Union* by A.S. Kompaneyets. *Am. J. Sci.* **263** (1965).
'On Interpretations and Misinterpretations of Operationalism', Chapter in *Humanistic Viewpoints in Psychology,* ed. F.T. Severin, McGraw-Hill Book Co., New York (1965).
*The Scientist.* Life Science Series, Time, Inc., New York (1965) (with D. Bergamini).
'Role of Scholarship in the Age of Science' (Speech on inauguration of new President of University of Wyoming) published by University of Wyoming, Laramie (1965).
'Man in Nature', *Hartwick Review,* Oneonta, New York, Vol. I. No. 1 (1965).
'Modern Science in Philosophical Perspective' (Speech given at Symposium at University of Western Ontario) published in *Science in Industry and the Universities,* (1965).
'What is a Theory', Essay in *The Structure of Economic Science,* ed. Sherman Krupp, Prentice-Hall, Englewood Cliffs, New Jersey (1966).
'The Philosophical Legacy of Contemporary Quantum Theory', Essay in *Mind and Cosmos,* Vol. III, ed. Robert Colodny, University of Pittsburgh Press, (1966).
'Theory of Plasma Radiation: Part I: Mathematical Development', Proceedings of the 7th International Conference on Phenomena in Ionized Gases. Beograd, Yugoslavia (1966) (with P. Mallozzi).
'Theory of Plasma Radiation: Part II: Physical Applications', Proceedings of the 7th International Conference on Phenomena in Ionized Gases, Beograd, Yugoslavia (1966) (with P. Mallozzi).

'Statistical Theory of Radiative Processes in Plasmas', *Ann. of Phys.* **38**, 177 (1966) (with P. Mallozzi).
'Simultaneous Action of Resonance and van der Waals Forces', Technical Report No. AFWL-TR-66-35, Vol. II (1966).
'Exclusion Principle and Measurement Theory', Article in *Quantum Theory of Atoms, Molecules and the Solid State*, ed. P.O. Lowdin, Academic Press, New York (1966).
'ESP in the Framework of Modern Science', *Psychical Research*, **6**, 214 (1966).
'Science and Philosophical Perspectives', Proceedings of the 1966 National Junior Science and Humanities Symposium, Ft. Monmouth, New Jersey (1966).
'Nonadditivity of Intermolecular Forces', Article in *Quantum Theory of Chemistry*, Vol. 3, ed. P.O. Lowdin, Academic Press, New York (1967) (with James Stamper).
'Pursuit of Significance', *Main Currents*, **23**, No. 3 (1967).
'Objectivity in Quantum Mechanics', Chapter 10 in *Foundations of Physics*, ed. Mario Bunge, Springer-Verlag, Berlin (1967) (with James Park).
'Seneca's Ethics Viewed from a Modern Standpoint', Congreso Internacional de Filosofia, Libreria Editorial Augustinus, Madrid, Spain (1967).
'Probabilities in Quantum Mechanics', Chapter 4 in *Studies in the Foundations, Methodology and Philosophy of Sciences*, Vol. 2, Springer-Verlag, Berlin (1967) (with Leon Cohen).
'Quantum Mechanics, Free Will, and Determinism', Symposium 64th Annual Meeting, American Philosophy Association, Published in *J. Phil.* **LXIV**, 714, (1967).
'Integrative Education in the Sciences', *Main Currents*, **24**, No. 2, 36 (1967).
'Science in Perspective', (Gerstein Lecture given at York University, Toronto) Published in *Science and The University*, Macmillan of Canada, Toronto (1967).
Book Review of 'Symmetries and Reflections, Scientific Essays' by Eugene Wigner, *Am. J. Phys.* **35**, 1169 (1967).
Book Review of The Cosmos of Arthur Holly Compton by Marjorie Johnston (Ed.). *Science*, **159**, 865 (1968).
'The Search for Meaning in a Senseless World', *Hartwick Review*, **4**, No. 1, 56 (1968).
'Causality', Article in *Contemporary Philosophy*, Ed. R. Klibansky, La Nuova Italia Editrice, Firenze, (1968). (with B. van Fraassen).
'Philosophy of Science', Article in *Contemporary Philosophy*, Ed. R. Klibansky, La Nuova Italia Editrice, Firenze (1968) (with B. van Fraassen).
'Simultaneous Measurability in Quantum Theory', *Inter. J. Theor. Phys.* **1**, No. 3, (1968) (with J. Park).
Book Review of Philosophy of Space and Time by Michael Whiteman. *Inter. J. Parapsychology*, **X**, No. 4, 410, (1968).
'Quantum Mechanics, Free Will and Determinism',, *J. Phil.* **LXIV**, 714 (1967).
'Integrative Education in the Sciences', *Main Currents* **24**, (1967).
'The Search for Meaning in a Senseless World', *Hartwick Review* **4**, 56 (1968).
'Scientific Indeterminism and Human Freedom', *Wimmer Lecture*. Archabby Press, Lafrobe, Pa. (1968).
'Causality', *Contemporary Philosophy*, ed. R. Klibanski La Nova Italia Editrice (1968) Firenze (with B. van Fraassen).
'Simultaneous Measurability in Quantum Theory' *Intern. J. Theor. Phys.* **1**, 211, 1968 (with J. Park).
*Theory of Intermolecular Forces* Pergamon, Oxford, 1969 (with N. Kestner) 2nd ed. 1971.

*Integrative Principles of Modern Thought* Gordon and Breach, N.Y. 1971 (Ed).
'Emergence of Integrative Concepts in Contemporary Science' *J. Phil. Sci.* **39**, 252 (1972) (with E. Laszlo).
'Physics and Semantics of Quantum Measurement' *Foundations of Physics* **3**, 19, 1973 (with J. L. Park).
'On Popper's Philosophy of Science', *The Philosophy of Karl Popper;* Library of Living Philosophers, 1975. Ed. Paul A. Schilpp.
'On Blanshard's Ethics', *The Philosophy of Brand Blanshard.* Library of Living Philosophers, 1977. Ed. Paul A. Schilpp.

# INDEX

Action 258
Aesthetics 219
AFOSR xxviii
Anaximenes xxxii
Aristotle xxxiii
Atomic uncertainty 246
Axioms 344

Bacon, R. 211 et seq.
Band, W. 210
Berkeley, G. 278
Bethe, H. xx, xxi
Black hole 389
Born, M. xxiii, 176, 185, 345
Boyle's law 132
Bridgman, P. W. 7, 88, 112
Bucherer, A. M. 198
Bunge, M. 51, 112, 113, 174, 185, 350
Bures, C. E. 132, 142

Carnap, R. 95, 142, 271
Cassirer, E. xxvi, 185, 240, 263
Category 92
Causal description 3, 9, 69
Causality, 21, 39, 107, 175, 279
Color 293
Commandments 347
Commitment 336
Common sense 306
Conscience 373
Consistency principle 22, 45, 46
Constitutive definition 63
Construct 64, 69, 72, 98, 230
Construction, symbolic 59, 61
Convergence, external 201 et seq.
                internal 201 et seq.
Copeland, A. N. 157
Copernicus, N. xxxiii
Correlation 129
Correspondence, rules of 65

Cross, W. xx
Crystal analogy 284, 300
C-field 115, 186, 188

Data 4, 96
Darwin, C. xxxiv
De Broglie, L. 209
Decision 260, 262
Definition 63
Democritus xxxii
Descartes, R. xxxiii
Diffuseness, elementary 77
Dirac, P. 27, 32, 67, 83, 84, 87
Dogmatism 311
Dotterer, H. 88
Ducasse, C. J. 221
Dürer, A. 367

Eddington, A. S. 88
Eidos 318
Eigenfunction 30
Einstein, A. 54
Electrodynamics, 175
Elegance 221
Empiricism 267, 270
Epistemic definition 63
Epistemic correspondence 115
Epistemic feedback 281, 287, 291
Epistemology 320
Ethics 225, 235 et seq. 334
Existence, modes of 71, 276, 376
Existentialism 272
Extensibility 106

Faith 303
Fechner, Th. 23
Feigl, H. 271
Fermi, E. 87
Feynman theory 158 et seq., 173, 178 et seq.

## INDEX

Feynman diagrams 163
Firefly 248
Fizeau, H. 198
Frank, Ph. 42, 271
Freedom, of the will 260, 308, 335
Frequency, relative 23
Frequency theory 143

Galilei, G. xxxiii
Gaussian distribution 201
Gene 259
Gibbs, W. 63
Goethe, J. W. 355
Goldon rule 365
Green's function 159 et seq., 179 et seq.
Grünbaum, A. 198

Hall, D. H. 198
Hallucinations 327
Hamiltonian 76
Heidegger, M. 288
Heisenberg, W. 80, 210, et seq.
Hendel, C. W. 263
Hilbert space 117
Hindu philosophy xxx
Historical Reality 241 et seq., 253, 257
Husserl, E. 317, 328

Imperatives 347
Intuition 379
Invariance 191, 194
Involvement 310

Jordan, P. 176

Kampen, van 185
Kant, I. 40, 47, 54, 95, 216
Kepler, J. xxxiii
Kolmagoroff, A. 132 142

Language 102
Laplace, P. S. 41 et seq.
Laplacian theory of probability 144
Lawrence, E. xix
Lane, M. von 89
Legend of Sais 295, 313
Leucippus xxxii

Lindsay, R. B. xxiii, xxiv, 51, 88, 112, 198
Logic, many-valued 126, 140
Logical positivism, 101, 270
London, F. xxii
Lorentz, H. 14
Lorentz contraction 197
Luther, M. 392

Mach, E. 88
Marcel, G. 387
March, A. 210
Mass point, 175
Materialism 268, 274, 301
Mangham, S. 335
Maxwell, C. xxxiv
McCullagh, J. 14
McKeehan, L. xix
Meaning 353
Measurement 199 et seq.
Mechanical models 307
Mercury 119
Metaethics 339 et seq.
Metaphor 355
Metaphysics 90, 339
Metaphysical principles 111
Metascience 383
Methodology 52
Michelson-Morley experiment 194, 308
Microcosm 35, 243, 249
Mises, R. von 23, 24, 132, 142
Moral principles 361
Moore, B. xviii
Moore, G. E. 341, 350
Mould, R. A. 198
Murphy, G. xxv

Nagel, E. 138, 142, 157
Nature 96, 97
Neumann, von, J. 118
Neutrino 252
Newton, I. xxxiii
Noah 336
Non-communtability 140
Normative values 362
Northrop, F. S. C. xxxiii, xxxvi, 88, 104, 112, 141, 240, 328.

NSF xxviii

Objectivity 61
Obscure movement 266, 302
Observable 27, 32
Obvious movement 266, 303
Ohm's law 60
Ontology 91, 329
Operationalism 247
ONR xxviii

Page, L. 17
Pair annihilation 166
Pair production 166
Panofski, E. 372
Parmenides xxxii
Pauli, W. 67
Personal dedication 366
P-experience 122, 168
Peyre, H. 273, 282
Phenomenology 317
Physicalism 388
Plato xxxiii
Poincaré, H. 69, 89
P-plane 168, 188
Prediction 56
Prescott, C. H. 130
Prima philosophia 212, 217
Primary value 363
Probability 21, 125 et seq., 149, 252, 253
Probability amplitude 173
Probability of theories 134
Projection matrix 204
Projection postulate 205
Proper time 169
Protocol behavior 348
Pythagoras xxxii
Pythagorean theorem 285

Quantity 64
Quantum electrodynamics 172
Quantum mechanics xxxv, 47, 50, 74, 78, 108, 110

Rabbit hunt 244
Ranke, L. 297

Rank-size law 288
Reality 20, 71, 99, 249
Reducibility 95
Reichenbach, H. 135 et seq., 142
Reification (as rule of correspondence) 324
Relative frequency 150, 161
Relativity xxxv, 186 et seq.
Religion 100, 133, 375, 382, 385
Relevance 356
Rig Veda xxx
Rayce, J. 265
Rules of correspondence 103, 148, 346

Sacred cow 298
Sais 295, 313
Sartre, J. P. 387
Scattering processes 158, 165
Schilling, H. 390
Schleiermacher, F. 384
Schrödinger, E. xii, 17, 67, 84, 89 et seq.
Scott, W. xxii
Sensation 258
Sense data 229 et seq.
Significance 351
Simplicity 68, 69, 109
Sommerfeld, A. xx, xxi
Spiritual concerns 301
Standen, A. 298
State function 79, 108
State preparation 206
Steiner, R. xv, 169
Stückelberg, R. von 173
Symmetry 192
System 64, 78, 99

Tendency of passage 102
Thales xxxi
Thermodynamics 270
Time dilation 197
Toynbee, A. 297
Transcendental norms 349
True value 251

Uncertainty principle 282
Uniqueness 120
Universe 71

Unsöld, A. xx

Validation 348
Value 339 et seq.
Verification 237
Vienna Circle 138

Werkmeister, W. H. 329 et seq.
Wesensschau 318, 319

Western culture 225
Weyl, H. 28, 126
Wheeler, J. xxv, 173
Whitehead, A. N. 92, 97, 112
Whittaker, E. 122, 274, 275, 282
Wigner, E. xxv, 122
Will 258
Williams, J. 143 et seq.
World line 164 et seq., 180 et seq.

# EPISTEME

A SERIES IN THE FOUNDATIONAL, METHODOLOGICAL,
PHILOSOPHICAL, PSYCHOLOGICAL, SOCIOLOGICAL,
AND POLITICAL ASPECTS OF THE SCIENCES, PURE AND APPLIED

*Editor:* MARIO BUNGE
*Foundations and Philosophy of Science Unit, McGill University*

---

1. William E. Hartnett (ed.), *Foundations of Coding Theory*. 1974, xiii + 216 pp. ISBN 90-277-0536-4.
2. J. Michael Dunn and George Epstein (eds.), *Modern Uses of Multiple-Valued Logic*. 1977, v + 332 pp. + Index. ISBN 90-277-0747-2.
3. William E. Hartnett (ed.), *Systems: Approaches, Theories, Applications.*
Including the Proceedings of the Eighth George Hudson Symposium, held at Plattsburgh, New York, April 11-12, 1975. 1977, xiv + 197 pp. + Index. ISBN 90-277-0822-3.
4. Władysław Krajewski, *Correspondence Principle and Growth of Science*. 1977, xiv + 138 pp. ISBN 90-277-0770-7.
5. José Leite Lopes and Michel Paty (eds.), *Quantum Mechanics, A Half Century Later.* Papers of a Colloquium on Fifty Years of Quantum Mechanics, held at the University Louis Pasteur, Strasbourg, May 2-4, 1974, x + 303 pp. + Index. ISBN 90-277-0784-7.
6. Henry Margenau, *Physics and Philosophy: Selected Essays*. 1978, xxxviii + 399 pp. + Index. ISBN 90-277-0901-7.
7. Roberto Torretti, *Philosophy of Geometry from Riemann to Poincaré*. Forthcoming. ISBN 90-277-0920-3.
8. Michael Ruse, *Sociobiology: Sense or Nonsense?* Forthcoming. ISBN 90-277-0940-8.